T0296113

LONDON MATHEMATICAL SOCIETY STUDENT TEXTS

Managing Editor: Ian J. Leary,
Mathematical Sciences, University of Southampton, UK

60 Linear operators and linear systems, JONATHAN R. PARTINGTON
61 An introduction to noncommutative Noetherian rings (2nd Edition), K. R. GOODEARL & R. B. WARFIELD, JR
62 Topics from one-dimensional dynamics, KAREN M. BRUCKS & HENK BRUIN
63 Singular points of plane curves, C. T. C. WALL
64 A short course on Banach space theory, N. L. CAROTHERS
65 Elements of the representation theory of associative algebras I, IBRAHIM ASSEM, DANIEL SIMSON & ANDRZEJ SKOWROŃSKI
66 An introduction to sieve methods and their applications, ALINA CARMEN COJOCARU & M. RAM MURTY
67 Elliptic functions, J. V. ARMITAGE & W. F. EBERLEIN
68 Hyperbolic geometry from a local viewpoint, LINDA KEEN & NIKOLA LAKIC
69 Lectures on Kähler geometry, ANDREI MOROIANU
70 Dependence logic, JOUKU VÄÄNÄNEN
71 Elements of the representation theory of associative algebras II, DANIEL SIMSON & ANDRZEJ SKOWROŃSKI
72 Elements of the representation theory of associative algebras III, DANIEL SIMSON & ANDRZEJ SKOWROŃSKI
73 Groups, graphs and trees, JOHN MEIER
74 Representation theorems in Hardy spaces, JAVAD MASHREGHI
75 An introduction to the theory of graph spectra, DRAGOŠ CVETKOVIĆ, PETER ROWLINSON & SLOBODAN SIMIĆ
76 Number theory in the spirit of Liouville, KENNETH S. WILLIAMS
77 Lectures on profinite topics in group theory, BENJAMIN KLOPSCH, NIKOLAY NIKOLOV & CHRISTOPHER VOLL
78 Clifford algebras: An introduction, D. J. H. GARLING
79 Introduction to compact Riemann surfaces and dessins d'enfants, ERNESTO GIRONDO & GABINO GONZÁLEZ-DIEZ
80 The Riemann hypothesis for function fields, MACHIEL VAN FRANKENHUIJSEN
81 Number theory, Fourier analysis and geometric discrepancy, GIANCARLO TRAVAGLINI
82 Finite geometry and combinatorial applications, SIMEON BALL
83 The geometry of celestial mechanics, HANSJÖRG GEIGES
84 Random graphs, geometry and asymptotic structure, MICHAEL KRIVELEVICH *et al.*
85 Fourier analysis: Part I – Theory, ADRIAN CONSTANTIN
86 Dispersive partial differential equations, M. BURAK ERDOĞAN & NIKOLAOS TZIRAKIS
87 Riemann surfaces and algebraic curves, R. CAVALIERI & E. MILES
88 Groups, languages and automata, DEREK F. HOLT, SARAH REES & CLAAS E. RÖVER
89 Analysis on Polish spaces and an introduction to optimal transportation, D. J. H. GARLING
90 The homotopy theory of $(\infty, 1)$-categories, JULIA E. BERGNER
91 The block theory of finite group algebras I, M. LINCKELMANN
92 The block theory of finite group algebras II, M. LINCKELMANN
93 Semigroups of linear operators, D. APPLEBAUM
94 Introduction to approximate groups, M. C. H. TOINTON
95 Representations of finite groups of Lie type (2nd Edition), F. DIGNE & J. MICHEL
96 Tensor products of C*-algebras and operator spaces, G. PISIER
97 Topics in cyclic theory, D. G. QUILLEN & G. BLOWER
98 Fast track to forcing, M. DŽAMONJA
99 A gentle introduction to homological mirror symmetry, R. BOCKLANDT

London Mathematical Society Student Texts 100

The Calculus of Braids

An Introduction, and Beyond

PATRICK DEHORNOY
Université Caen Normandie

Translated by

DANIÈLE GIBBONS
GREG GIBBONS

CAMBRIDGE
UNIVERSITY PRESS

University Printing House, Cambridge CB2 8BS, United Kingdom

One Liberty Plaza, 20th Floor, New York, NY 10006, USA

477 Williamstown Road, Port Melbourne, VIC 3207, Australia

314–321, 3rd Floor, Plot 3, Splendor Forum, Jasola District Centre, New Delhi – 110025, India

103 Penang Road, #05–06/07, Visioncrest Commercial, Singapore 238467

Cambridge University Press is part of the University of Cambridge.

It furthers the University's mission by disseminating knowledge in the pursuit of education, learning, and research at the highest international levels of excellence.

www.cambridge.org
Information on this title: www.cambridge.org/9781108843942
DOI: 10.1017/9781108921121

Originally published in French as *Le calcul des tresses* by Calvage et Mounet, 2019

First published in English by Cambridge University Press 2021

A catalogue record for this publication is available from the British Library

ISBN 978-1-108-84394-2 Hardback
ISBN 978-1-108-92586-0 Paperback

Contents

Preface

Presentation and Motivation

Everyone knows what braids are, whether they be made of hair, knitting wool, or electrical cables. However, it is not at all evident that we can construct a *theory* about them, that is, elaborate a coherent and mathematically interesting corpus of results concerning them. Our goal here is to convince the reader that there is a resoundingly positive response to this question: braids are indeed fascinating objects, with a variety of rich mathematical properties.

For this, we will concentrate on carefully and completely establishing only a few selected results. What they have in common is the sophistication of the proofs they require, in spite of their very simple statements. At the heart of the approach, a natural multiplication operation of braids leads to group structures, the *braid groups*. Combining both algebraic and topological aspects, these groups enjoy multiple interesting properties and are at the same time simple and complex.

A Roadmap for the Reader

As a mere introduction, this short book in no way pretends to be exhaustive, and addresses only a few of the many properties known about braid groups. A special emphasis is made on the algorithmic aspects and on what can be called the 'calculus of braids', in particular the *problem of isotopy*, that is, the question of recognizing if two braid diagrams a priori distinct can be continuously deformed into one another. The interest of this question is that it can be tackled and resolved along multiple paths, none of them evident, but all leading to concrete solutions that can be tested by hand or with a computer.

The text contains eight chapters, certain (Chapters 1, 5, 7) of a more topological nature, and others more algebraic. While Chapters 1 and 2 form a common basis for the rest of the work, Chapters 3 to 7, presenting a diversity of approaches, are quite independent of each other.

More precisely, Chapter 1 is devoted to the modelization of the notion of braids. For each n, we introduce a *space of n-strand braids* whose elements are the equivalence classes of geometric braids with respect to one (in fact several) natural notion(s) of deformation. Next, we project the geometric braids onto planar diagrams, coded by certain words, the 'braid words'. This leads to the *isotopy problem of braids*, that is, the problem of recognizing if two diagrams, or the words coding them, represent the same braid. We remark on the failure of naive attempts to resolve the isotopy problem, and the need to develop more elaborate approaches.

The study of the braid spaces B_n, and in particular the obtention of simple solutions to the isotopy problem, are based on the existence on B_n of a group structure, addressed in Chapter 2. Such an algebraic structure, specific to braids, explains in particular why the associated isotopy problem, while not trivial, is nonetheless simpler than the analogous problem for other topological objects such as knots. As a starting point, we establish a characterization of the group structure in terms of representations by generators and relations. Two appendices complete the chapter, one devoted to representations of monoids, and the other to representations of groups. Each offers a quite complete treatment of the subject, accessible without prior knowledge.

This is not quite enough to answer the questions we might have about B_n, foremost among them being the word problem, or the braid isotopy problem, and we must continue our investigation. In Chapter 3, we describe a first solution, following the approach developed in the 1960s by F.A. Garside in his thesis supervised by G. Higman, and based on an analysis of an auxiliary structure, the monoid B_n^+, later revealed a posteriori to be a submonoid of the group B_n.

As a direct follow-on to the preceding chapter, Chapter 4 continues exploiting the properties of the monoids B_n^+ to study the groups B_n. The new ingredient here is the relation of left divisibility of the monoid B_n^+, which exhibits a lattice structure. This enables us to construct unique normal form, first for positive braids, and then for arbitrary braids, that is, to define, for each n-strand braid, a distinguished decomposition in elementary fragments, essentially permutations of $\{1, ..., n\}$, leading to a new solution, simple and efficient, to the isotopy problem.

In Chapter 5 we return to a topological approach. Developed by E. Artin in the 1920s, it is based on the action of braids on the fundamental group of a punctured disk and leads to a faithful representation of the group B_n in the automorphism group of the free group of rank n. We can thus deduce at the same time a new solution to the isotopy problem, and a proof of the equivalence of diverse notions of geometrical isotopy of braids, left pending in Chapter 1. However, certain results are only established here modulo an additional property of braids (the 'comparison property') left hanging for the moment.

The first goal of Chapter 6 is to provide a proof of this comparison property, thanks to a method known as *handle reduction*. The interest of this method goes even further: it provides a new solution to the isotopy problem, which experimentally seems more efficient than the reduction to a normal form seen in Chapter 4.

In Chapter 5, we used the topological aspects of braids to define an action of the group B_n on the marked disk $\mathbb{D}_n$ (disk marked with n points). In Chapter 7, we action the braids over families of curves traced in an extension of $\mathbb{D}_n$ and, by counting the intersections, we obtain an action over sequences of integers known as the Dynnikov coordinates. This approach leads to surprising formulas bringing into play the operations max and +, and to a new, extremely efficient, solution for the braid isotopy problem.

Finally, the goal of Chapter 8 is to illustrate a few more paths of investigation. We first look at results concerning the groups B_n themselves, with a few words on their dual monoids, on their linear representations, and on potential applications to cryptography. We then consider the numerous extensions and generalizations of braid groups, with a few examples of the braid groups on surfaces and an introduction to the Artin–Tits groups. We conclude with an open conjecture associated with an enticing reward (seriously!)...

Numerous complementary subjects and a few proofs for which there was no room here are proposed as exercises throughout the text. The solutions (in French) are available at the address:
`https://dehornoy.users.lmno.cnrs.fr/Books/Tresses/Exos.pdf`

Prerequisites

The choice made here was to assume a minimum of preliminary knowledge, never above the level of the first couple of years of university mathematics. Most results are established 'from scratch'. In the few cases where more advanced notions are required (monoids and presented groups, Ore's theorem, fundamental group of a punctured disk), a detailed introduction is given in an appendix at the end of the chapter.

Acknowledgements

This text is based on the lecture notes of courses given to a variety of audiences. It benefited from numerous discussions with students in Caen, Paris, and Canton, and with the colleagues of the 'Groupement de Recherche Tresses' of the CNRS (GDR 2105). I send to them all a collective 'Thank you!' My friends Dale ROLFSEN and Seiichi KAMADA deserve a special mention. Finally, for their aid in a situation made difficult by a health problem, I thank Pierre DEHORNOY, John GUASCHI, and especially Jean FROMENTIN.

Arnières-sur-Iton, June 2019

1
Geometric Braids

The goal of this chapter is to define and begin to study braids in a mathematical sense. For this, in Section 1.1, we model a material braid by a family of arcs in $\mathbb{R}^3$, and introduce for every n a *space of n-strand braids* whose elements are the equivalence classes of geometric braids with respect to one of several natural notions of deformation. In Section 1.2, we project the geometric braids onto planar diagrams encoded by certain words (the 'braid words'), and are led to the *braid isotopy problem*, that is, recognizing if two diagrams, or the words that encode them, represent the same braid. In Section 1.3, we observe the failure of a few naive attempts to resolve the isotopy problem, calling for the development of more elaborate approaches.

Important note. The option retained throughout this text is to fully prove the results stated. Those of this chapter are in general natural and comprehensible, but their proofs often require a somewhat heavy geometric formalism. May the readers keep from becoming discouraged, and continue to resolutely plough forward, even if they skip over a few details, especially between Definition 1.1.11 and Section 1.2.2. Once the bases are mastered, the rest of the study, and in particular the algebraic arguments, will no longer pose this type of difficulty.

1.1 The Geometric Braid Space

1.1.1 From Material Braids to Geometric Braids

We propose here to elaborate a theory of braids. At the start, braids are material objects, such as braids of hair, or the sculpture of Figure 1.1: strands that cross each other while conserving a certain constant general orientation, but without forming knots.

Figure 1.1 An ornamental braid in salt dough.

Building a theory for such objects begins with a modelization: we need to decide which aspects to analyse and how to formalize material objects as mathematical objects susceptible to being studied. For the present case, we will neglect the metrical aspects, hence anything concerning the thickness of the strands, their length, their spacing, their curvature... and only retain the topological aspects, that is, the way the strands cross each other, which strand passes over or under another, etc. The theory we will develop is thus first and foremost a theory of crossings.

Since the metrical parameters of the strands are ignored, it is natural to modelize them by curves in the space $\mathbb{R}^3$. In a material braid, the strands have a beginning and an end, and are not broken along the way. We thus consider fragments of continuous curves, referred to as *arcs* in what follows. Consequently, we modelize an n-strand braid by the union of n arcs. And since the strands cross each other, but without cutting each other, we require the arcs to be *pairwise disjoint*.

Next, as we are thinking of finite material braids,[1] it is natural to suppose that the n arcs emanate from n fixed points in the plane $\{0\}\times\mathbb{R}^2$, for example $(0, 1, 0), ..., (0, n, 0)$ and terminate on n fixed points in the plane $\{1\}\times\mathbb{R}^2$, for example $(1, 1, 0), ..., (1, n, 0)$.

Finally, we wish to exclude knots. We thus prohibit the arcs to turn back on themselves, hence conserving a constant general direction. For an arc γ joining the plane $\{0\}\times\mathbb{R}^2$ to the plane $\{1\}\times\mathbb{R}^2$, this condition reduces to requiring γ to be traced in $[0, 1]\times\mathbb{R}^2$ and its intersection with every plane $\{x\}\times\mathbb{R}^2$ for $0 \leqslant x \leqslant 1$ to be a single point: if γ turned back on itself, there would exist a plane $\{x\}\times\mathbb{R}^2$ cutting γ in at least three points.

With this in mind, we are led to the following notion:

[1] At least at first; we could after all imagine infinite braids...

Definition 1.1.1 (Geometric braid) An *n-strand geometric braid* is a union β of n arcs in $\mathbb{R}^3$, pairwise disjoint, linking the n points $(0, i, 0)$, $i = 1, ..., n$, to the n points $(1, i, 0)$, $i = 1, ..., n$, within the band $[0, 1]\times\mathbb{R}^2$, and whose intersection with every plane $\{x\}\times\mathbb{R}^2$ contains exactly n points. The family of n-strand geometric braids is denoted $\mathcal{GB}_n$.

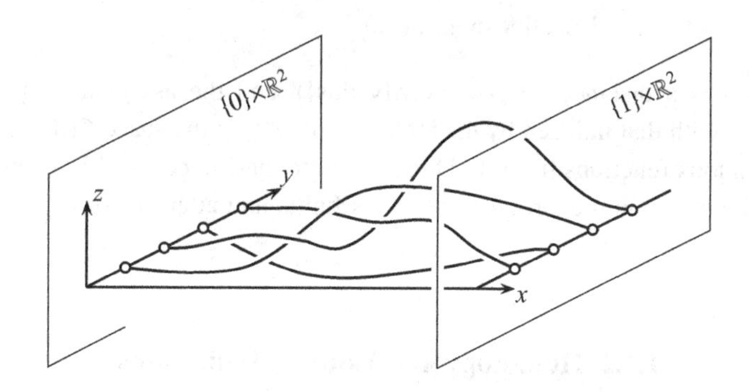

Figure 1.2 A four-strand geometric braid: the union of four disjoint arcs joining the points $(0, i, 0)$, $i = 1, ..., 4$, to the points $(1, i, 0)$, $i = 1, ..., 4$, not cutting each other, and keeping a positive orientation with respect to the x axis.

Before we go any further, note that an n-strand geometric braid can be naturally parametrized by a sequence of $2n$ functions from $[0, 1]$ into $\mathbb{R}$ or, if you prefer, a sequence of n functions from $[0, 1]$ into $\mathbb{R}^2$. Indeed, any arc in $[0, 1]\times\mathbb{R}^2$ can be parametrized by the functions specifying its three coordinates. If this arc cuts each plane $\{x\}\times\mathbb{R}^2$ in one unique point, we can choose the abscissa as the parameter, and hence adapt a parametrization of the form $t \mapsto (t, f(t), g(t))$.

Definition 1.1.2 (Parametrization of $\beta^{[i]}$) If β is an n-strand geometric braid, denote $\beta^{[i]}$ the continuous function from $[0, 1]$ into $\mathbb{R}^2$ such that the arc emanating from $(0, i, 0)$ ('the *ith strand*' of β) is parametrized by the function $t \mapsto (t, \beta^{[i]}(t))$.

By definition, $\beta^{[i]}(t)$ is the tuple formed by the ordinate and the point of intersection of the ith strand of β with $\{t\}\times\mathbb{R}^2$, and is thus determined by β. Conversely, β is the set of points $\beta^{[i]}(t)$ where (i, t) runs across $\{1, ..., n\} \times [0, 1]$. Hence, speaking of β is purely and simply equivalent to speaking of the sequence $(\beta^{[1]}, ..., \beta^{[n]})$.

To conclude these preliminaries, note that the geometric braids can be equipped with a natural notion of distance. For this, we state that two braids

β, β' are close if the strands of β' remain close to the corresponding strands of β.

Definition 1.1.3 (Distance) For β, β' in $\mathcal{GB}_n$, define

$$d(\beta, \beta') := \sup\{\|\beta^{[i]}(t) - \beta'^{[i]}(t)\| \mid i = 1, ..., n,\ t \in [0, 1]\}, \tag{1.1}$$

where $\|\text{-}\|$ is the usual Euclidean norm[2] on $\mathbb{R}^2$.

Then d is a distance on $\mathcal{GB}_n$ (verify this!), and the associated topology coincides with that induced by the inclusion of $\mathcal{GB}_n$ in the space $C([0, 1], \mathbb{R}^{2n})$ of continuous functions from $[0, 1]$ to $\mathbb{R}^{2n}$. Note that, since $[0, 1]$ is a compact space, the supremum expressed in (1.1) is finite, and attained for at least one point.

1.1.2 Homotopy and Isotopy: Definitions

We have not yet finished with the modelization phase, and now tackle its most delicate aspect. We would like to elaborate a theory of braids retaining only their topological aspects. It is thus natural to consider as equivalent geometric braids that are, in some suitable manner, topologically indiscernible; this is what we would like to now formalize.

We might declare geometric braids equivalent when, as subspaces of $\mathbb{R}^3$, they are homeomorphic[3]. This idea quickly fizzles out: as a topological space, every n-strand geometric braid is homeomorphic to the union of n disjoint copies of the interval $[0, 1]$ and hence, two geometric braids are *always* homeomorphic. The problem is clear: it is not a question of considering a geometric braid β as an abstract space, but as an embedding in $[0, 1]\times\mathbb{R}^2$, as specified for example by the functions $\beta^{[i]}$.

We thus seek to translate the idea of a *deformation* transforming one braid embedded in $\mathbb{R}^3$ into another. Several formalizations will be introduced; we then prove their equivalence. The first exploits the notion of a path in the space $\mathcal{GB}_n$. If $\mathcal{X}$ is a topological space, and a and b points of $\mathcal{X}$, a *path from a to b in $\mathcal{X}$* is a continuous mapping ϕ from $[0, 1]$ into $\mathcal{X}$ satisfying $\phi(0) = a$ and $\phi(1) = b$. We saw with Definition 1.1.3 that the geometric braid space $\mathcal{GB}_n$ can be equipped with a distance, hence with a notion of continuity, and thus we can speak of a path in $\mathcal{GB}_n$.

[2] i.e. the norm defined by $\|(y, z)\| := \sqrt{y^2 + z^2}$.

[3] i.e. there exists a continuous bijection with continuous inverse (a 'homeomorphism') mapping one to the other.

Definition 1.1.4 (Homotopic braids) Two geometric braids β, β' of $\mathcal{GB}_n$ are said to be *homotopic*, denoted $\beta \approx^{\mathsf{h}} \beta'$, if there exists a path from β to β' in $\mathcal{GB}_n$.

As paths can be reversed and concatenated, homotopy is an equivalence relation on $\mathcal{GB}_n$.

Example 1.1.5 (One-strand braids) Let β, β' be two geometric braids each with a single strand. For $0 \leqslant s \leqslant 1$, let $\phi(s)$ be the geometric braid whose intersection with the plane $\{t\}\times\mathbb{R}^2$ is the barycentre of $\beta(t)$ and $\beta'(t)$, weighted with coefficients $1 - s$ and s (Figure 1.3). In other words, by identifying here β with $\beta^{[1]}$, we have $\phi(s) := (1 - s)\beta + s\beta'$. Hence ϕ is a path from β to β' in $\mathcal{GB}_1$, and β and β' are homotopic: there is only one single homotopy class in $\mathcal{GB}_1$.

Note that the result does not extend to $\mathcal{GB}_n$ for $n \geqslant 2$: mimicking the interpolation above with several strands at the same time could give arcs that cut each other, hence outside of $\mathcal{GB}_n$.

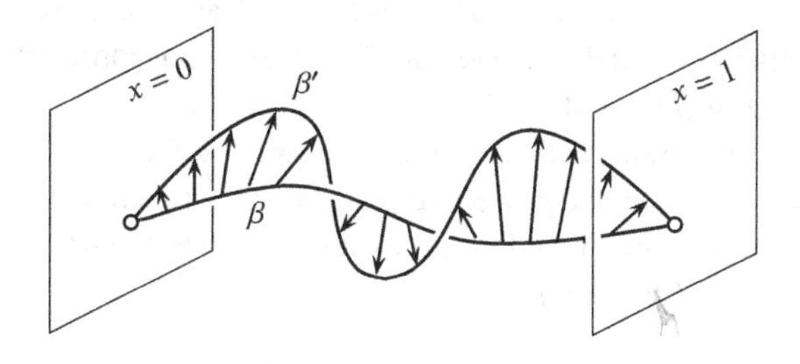

Figure 1.3 Two geometric braids on one strand are always homotopic: a path joining β to β' can be obtained by pushing the intersection of β with $\{t\}\times\mathbb{R}^2$ towards the intersection of β' with this same plane.

The notion of homotopy of Example 1.1.5 is local in the sense that we in fact specify a deformation of the embedded braid, but without a deformation of the whole of the ambient space $[0, 1]\times\mathbb{R}^2$. For a global approach, we can use the homeomorphisms of $[0, 1]\times\mathbb{R}^2$. However, as above, demanding the existence of a homeomorphism of $[0, 1]\times\mathbb{R}^2$ sending β onto β', where β and β' are geometric braids, is not a good idea, as we lose the idea of a continuous deformation[4] of β to β'. Nonetheless, we can amend the idea thanks to the notion of isotopy between homeomorphisms. For any space $\mathcal{X}$, denote $\mathsf{Homeo}(\mathcal{X})$ the set[5] of homeomorphisms of $\mathcal{X}$.

[4] Think about the symmetry with respect to the plane $\{1/2\}\times\mathbb{R}^2$: it is indisputably a homeomorphism of $[0, 1]\times\mathbb{R}^2$, but there is no reason at all for us to be able to continuously deform a figure onto its image by symmetry...

[5] It is in fact a group when equipped with composition.

Definition 1.1.6 (Isotopy) Two homeomorphisms ϕ, ϕ' of a topological space $\mathcal{X}$ are said to be *isotopic* if there exists a path joining ϕ to ϕ' in $\mathsf{Homeo}(\mathcal{X})$.

The idea is then to consider two geometric braids β, β' equivalent if there exists a homeomorphism ϕ of $[0,1]\times\mathbb{R}^2$ sending β onto β' (considered as subsets of $[0,1]\times\mathbb{R}^2$) such that we can pass continuously from the identity to ϕ, that is, that ϕ is isotopic to id. There is an essential condition: we only consider the homeomorphisms that are trivial on the two vertical boundary planes, $\{0\}\times\mathbb{R}^2$ and $\{1\}\times\mathbb{R}^2$. However, if Φ is an isotopy linking id to ϕ, the image of a braid by the homeomorphisms $\Phi(s)$ for $0<s<1$ has no reason to be a braid, since the images of the vertical planes $\{t\}\times\mathbb{R}^2$ have no reason to be vertical planes.[6] Two notions, a priori distinct, emerge.

Definition 1.1.7 (Isotopic braids)
(i) A homeomorphism of $[0,1]\times\mathbb{R}^2$ is said to be *trivial on the boundary* if it leaves $\{0\}\times\mathbb{R}^2$ and $\{1\}\times\mathbb{R}^2$ invariant point by point. It is said to be *stratified* if it is trivial on the boundary and, in addition, leaves each vertical plane $\{t\}\times\mathbb{R}^2$ globally invariant; denote $\mathsf{Homeo}^{\partial}([0,1]\times\mathbb{R}^2)$ and $\mathsf{Homeo}^{\mathsf{st}}([0,1]\times\mathbb{R}^2)$ the two families thus formed.[7]
(ii) Two geometric braids β, β' are said to be *isotopic* (*resp.*, *isotopic in the unrestrained sense*), denoted $\beta \approx \beta'$ (*resp.*, $\beta \approx^{\mathsf{nr}} \beta'$), if there exists a homeomorphism sending β onto β' and linked to id by a path in $\mathsf{Homeo}^{\mathsf{st}}([0,1]\times\mathbb{R}^2)$ (*resp.*, in $\mathsf{Homeo}^{\partial}([0,1]\times\mathbb{R}^2)$).

The relation $\beta \approx \beta'$ implies $\beta \approx^{\mathsf{nr}} \beta'$: a path in the subspace $\mathsf{Homeo}^{\mathsf{st}}([0,1]\times\mathbb{R}^2)$ of $\mathsf{Homeo}^{\partial}([0,1]\times\mathbb{R}^2)$ is, in particular, a path in $\mathsf{Homeo}^{\partial}([0,1]\times\mathbb{R}^2)$. Moreover, since the isotopies between homeomorphisms can be composed or reversed, the relations $\approx$ and $\approx^{\mathsf{nr}}$ are equivalence relations.

Let us recapitulate. Three equivalence relations have been introduced to modelize the idea of the deformation of a geometric braid, all expressed in terms of the existence of a continuous path linking two braids. With the homotopy $\approx^{\mathsf{h}}$, we only consider the strands of the braids, independently from the ambient space; in contrast, with the isotopies, we stipulate the existence of a deformation of the entire ambient space (in fact, we often speak of an 'ambient isotopy'). The default version will here be $\approx$: restricting ourselves to stratified homeomorphisms might seem artificial, but this is the condition to remain within the braid space, which is our reference framework. It is nevertheless legitimate to ask whether removing this restriction would change things, and this is why the unrestrained version was introduced. We will have to wait for

[6] Nor even regular surfaces, as no differentiability condition has ever been mentioned.
[7] The symbol ∂ is commonly used to represent the boundary of a space.

Chapter 5 for all the links to be clarified, but the response will be optimal: the relations $\approx^{\mathsf{h}}$, $\approx$, and $\approx^{\mathsf{nr}}$ are equivalent, thus leading to a unique, natural, and robust notion.

1.1.3 Homotopy and Isotopy: Equivalence

Indeed, the three notions above are equivalent, providing both legitimacy to the unique notion thus exposed, and a technical convenience, as we can use one version or another as convenient. In this section, we establish the equivalence between the first two notions.

Proposition 1.1.8 (Equivalence) *Two geometric braids are isotopic if and only if they are homotopic.*

The case of unrestrained isotopy is left aside for the moment. It will be treated in Chapter 5.

One direction of the equivalence is intuitive: in an isotopy, ignoring the exterior of the braids leads to a homotopy.

Lemma 1.1.9 *Two isotopic geometric braids are homotopic.*

Proof Let β, β' be braids in $\mathcal{GB}_n$ and suppose that Φ is an isotopy attesting $\beta \approx \beta'$, hence satisfying[8] $\Phi(0) = \mathrm{id}$ and $\beta' = \Phi(1) \circ \beta$. For s in $[0,1]$, set $\phi(s) := \Phi(s) \circ \beta$. By definition, $\Phi(s)$ sends any geometric braid to another geometric braid and hence, in particular, $\phi(s)$ belongs to $\mathcal{GB}_n$. The continuity of Φ implies that of ϕ. Finally, we have

$$\phi(0) = \Phi(0) \circ \beta = \beta \quad \text{and} \quad \phi(1) = \Phi(1) \circ \beta = \beta'.$$

Hence ϕ is a path linking β to β' in $\mathcal{GB}_n$, showing $\beta \approx^{\mathsf{h}} \beta'$. □

The other direction requires more care, but is not terribly difficult. We begin with single-stranded braids, which we have seen in Example 1.1.5 are always homotopic.

Lemma 1.1.10 *Two one-stranded geometric braids are isotopic.*

Proof Let β, β' be two geometric braids on one strand.[9] We start with the path ϕ linking β to β' in $\mathcal{GB}_1$ constructed in Example 1.1.5. We seek a path Φ in $\mathsf{Homeo}^{\mathsf{st}}([0,1]{\times}\mathbb{R}^2)$ extending $\phi(s)$ in the sense where, for every t, it sends

[8] Writing $\beta' = \Phi(1)(\beta)$ is tempting, but not formally correct: as β is identified with a family of n mappings from $[0,1]$ into $[0,1]{\times}\mathbb{R}^2$, this says that these mappings composed with $\Phi(1)$ are equal to those constituting β'.

[9] As in Definition 1.1.4, we identify β with the function $\beta^{[1]}$ of $[0,1]$ into $\mathbb{R}^2$.

$\beta(t)$ to $\phi(s)(t)$, that is, to $(1-s)\beta(t)+s\beta'(t)$. For this it suffices to choose the translation that does the work, namely

$$\Phi(s)(x,y,z) := (x,y,z) + s\,(0,\overrightarrow{\beta(x)\beta'(x)}). \tag{1.2}$$

By construction, $\Phi(s)$ is a stratified homeomorphism of $[0,1]\times\mathbb{R}^2$ for every s. Moreover, $\Phi(0)$ is the identity, and $\Phi(1)$ sends β onto β' since, for every x, we find $\Phi(1)(\beta(x)) = \beta(x) + \overrightarrow{\beta(x)\beta'(x)} = \beta'(x)$. Consequently, Φ attests that β and β' are isotopic. □

For braids with at least two strands, it is in general impossible to find a translation sending simultaneously each of the n points of the intersection of β with a plane $\{x\}\times\mathbb{R}^2$ to where we wish them to go, and we must find a more precise construction. For this, we consider the homeomorphisms trivial outside of a neighbourhood of the strands of the braids under consideration.

Definition 1.1.11 (Tubular neighbourhood)
(i) For a in $\mathbb{R}^2$ and $\rho > 0$, denote $D(a,\rho)$ the open disk with centre a and radius ρ in $\mathbb{R}^2$.
(ii) For β in $\mathcal{GB}_n$ and $\rho > 0$, the *ρ-tubular neighbourhood* of β, denoted $V(\beta,\rho)$, is the open subset of $[0,1]\times\mathbb{R}^2$ whose intersection with the plane $\{x\}\times\mathbb{R}^2$ is the union of the n disks $D(\beta^{[i]}(x),\rho)$.

Thus, $V(\beta,\rho)$ is a union of open tubes, each surrounding one of the strands of β with a radius ρ. Note that, if β and β' are n-strand geometric braids, then $d(\beta,\beta') < \rho$ if and only if, as a set of points, β' is contained in $V(\beta,\rho)$.

We can thus amend the result of Lemma 1.1.10 to find a trivial isotopy[10] outside of a neighbourhood of the initial braid.

Lemma 1.1.12 *If β,β' are two one-strand geometric braids satisfying $d(\beta,\beta')<\rho$, then β and β' are isotopic via an isotopy Φ such that, for any s, the homeomorphism $\Phi(s)$ is trivial outside of $V(\beta,\rho)$.*

Proof As in Lemma 1.1.10, we retain the idea of using the translation in the plane $\{x\}\times\mathbb{R}^2$ that sends $\beta(x)$ onto $\beta'(x)$, but modulus a coefficient tending to 0 on the boundary of the disk $D(\beta(x),\rho)$. For this, if D is a disk in $\mathbb{R}^2$ with centre (y_0,z_0) and radius ρ, denote Λ_D the function from $\mathbb{R}^2$ into $\mathbb{R}$ defined by $\Lambda_D(y,z) := 0$ for $(y,z)\notin D$ and

$$\Lambda_D(y,z) := 1 - \frac{1}{\rho}\|(y-y_0, z-z_0)\| \quad \text{for } (y,z)\in D.$$

Thus, Λ_D takes on the value 1 at the centre of D and decreases linearly with

[10] i.e. coincident with the identity mapping.

the distance from the centre to be zero on the boundary and outside: its graph is a Chinese hat on top of (y_0, z_0). Note that Λ_D depends continuously on the parameters y_0, z_0, and ρ, and for any (y, z) and (y_1, z_1) inside D, the translated point $(y, z) + \Lambda_D(y, z)(y_1 - y_0, z_1 - z_0)$ is inside D.

Returning to the braids β and β', we define Φ by

$$\Phi(s)(x, y, z) := (x, y, z) + s\, \Lambda_{D(\beta^{[1]}(x),\rho)}(y, z)\, (0, \overrightarrow{\beta^{[1]}(x)\beta'^{[1]}(x)}). \tag{1.3}$$

Then, Φ is a continuous function of $[0, 1]$ into $\mathsf{Homeo}^{\mathrm{st}}([0, 1]\times\mathbb{R}^2)$. Moreover, $\Phi(0)$ is the identity, and $\Phi(1)$ sends β onto β' since, by construction, $\Lambda_{D(\beta^{[1]}(x),\rho)}(\beta^{[1]}(x))$ is equal to 1 for every x. Furthermore, $\Lambda_{D(\beta^{[1]}(x),\rho)}$ is zero outside of $D(\beta^{[1]}(x), \rho)$, hence $\Phi(s)$ is trivial outside of $V(\beta, \rho)$. Finally, the convexity of $D(\beta^{[1]}(x), \rho)$ guarantees that, for every s, the point $(1 - s)\beta^{[1]}(x) + s\beta'^{[1]}(x)$ is in $D(\beta^{[1]}(x), \rho)$ since, by hypothesis, so is $\beta'^{[1]}(x)$. □

Extending the construction to the case of several strands is simple as long as the distance between the two braids is sufficiently small: it suffices to enclose each strand in an appropriate tubular neighbourhood and to take the sum of the associated tiny translations. The point is that for a geometric braid, two strands can never be arbitrarily close.

Definition 1.1.13 (Minimal strand-spacing) The *minimal strand-spacing* of an n-strand geometric braid β is defined by

$$e(\beta) := \frac{1}{2} \inf\{\|\beta^{[i]}(t) - \beta^{[j]}(t)\| \mid 0 \leqslant t \leqslant 1,\ 1 \leqslant i < j \leqslant n\}.$$

For any i, the domain of definition of $\beta^{[i]}$ is the compact interval $[0, 1]$, hence every function $t \mapsto \|\beta^{[i]}(t) - \beta^{[j]}(t)\|$ attains its minimum. This must be strictly positive, as $\beta^{[i]}(t) = \beta^{[j]}(t)$ would mean that the ith and jth strands of β cross each other in the plane $\{t\}\times\mathbb{R}^2$. The minimal strand-spacing is thus always strictly positive.

Applying the attenuated translations of Lemma 1.1.12, we deduce that two braids sufficiently close are always isotopic.

Lemma 1.1.14 *Two geometric braids β, β' satisfying $d(\beta, \beta') < e(\beta)$ are isotopic.*

Proof Suppose that β and β' have n strands, and set

$$\Phi(s)(x, y, z) := (x, y, z) + s \sum_{i=1}^{i=n} \Lambda_{D(\beta^{[i]}(z),e(\beta))}(y, z)\, (0, \overrightarrow{\beta^{[i]}(x)\beta'^{[i]}(x)}).$$

This barbaric formula is in fact quite simple: $\Phi(s)$ does nothing far from the strands of β, whereas in the neighbourhood of the ith strand of β, it performs a small 'attenuated' translation in the direction of $\beta^{[i]}(x)$ to $\beta'^{[i]}(x)$. The hypothesis that the strands of β are never at a distance less than $2e(\beta)$ guarantees that the translations take place in the interior of tubes surrounding the strands of β; these tubes are disjoint by the definition of $e(\beta)$. In particular, $\Phi(s)$ sends every geometric braid onto a geometric braid. Then $\Phi(0)$ is the identity for $s = 0$, and $\Phi(1)$ sends every point $(t, \beta^{[i]}(t))$ to the corresponding point $(t, \beta'^{[i]}(t))$: Hence Φ attests that β and β' are isotopic. □

Note that, as in Lemma 1.1.12, the above isotopy linking β to β' is trivial outside the tubular neighbourhood $V(\beta, e(\beta))$.

It is now easy to establish the converse of the implication of Lemma 1.1.9 and obtain the equivalence of Proposition 1.1.8.

Proof of Proposition 1.1.8 The implication 'isotopic ⇒ homotopic' was seen in Lemma 1.1.9. Conversely, suppose that ϕ is a path linking β to β' in $\mathcal{GB}_n$. Define β_s to be $\phi(s)(\beta)$ (hence $\beta_0 = \beta$ and $\beta_1 = \beta'$).

If ever we have $d(\beta, \beta') < e(\beta)$, then β and β' are isotopic by Lemma 1.1.14. This condition has no reason in general to be satisfied, but we will reduce to it by sectioning the path ϕ. First, the function $s \mapsto e(\beta_s)$ of $[0, 1]$ into $\mathbb{R}_{>0}$ is continuous and defined on a compact set, hence it attains its minimum ε, strictly positive. By definition, we have $e(\beta_s) \geqslant \varepsilon > 0$ for all s. Next, by hypothesis, ϕ is a continuous function of the compact space $[0, 1]$ in the metric space $(\mathcal{GB}_n, d)$. It is thus uniformly continuous and, consequently, there exists δ such that

$$|s' - s| \leqslant \delta \quad \text{implies} \quad d(\beta_{s'}, \beta_s) < \varepsilon. \tag{1.4}$$

Let $s_0, ..., s_p$ be real numbers satisfying $s_0 = 0$, $s_p = 1$, and $s_j < s_{j+1} < s_j + \delta$ for all j. Consider the braids $\beta_{s_0}, ..., \beta_{s_p}$. For every j, Equation (1.4) implies $d(\beta_{s_j}, \beta_{s_{j+1}}) < \varepsilon$, hence $d(\beta_{s_j}, \beta_{s_{j+1}}) < e(\beta_{s_j})$. Then Lemma 1.1.14 implies that, for every j, the braids β_{s_j} and $\beta_{s_{j+1}}$ are isotopic. By the transitivity of isotopies, $\beta_{s_0} = \beta$ is isotopic to $\beta_{s_p} = \beta'$. □

1.1.4 The Braid Space

Since homotopy and isotopy coincide, there is no reason to differentiate the two equivalence relations on the spaces of geometric braids.[11] In what follows, we will only speak of isotopy of geometric braids, and hence only use the symbol $\approx$. Nevertheless, homotopy remains important, as in practice, to show

[11] Recall that a third relation, unrestrained isotopy, was left aside for the moment. We will come back to it in Chapter 5.

that two geometric braids are isotopic, it suffices to construct a homotopy linking them, without worrying about extending this to an isotopy of the ambient space.

As the notion of deformation is satisfactorily modelized by the isotopy relation $\approx$, it is natural to consider the associated equivalence classes.

Definition 1.1.15 (Braids) An *n-strand braid* is an equivalence class of n-strand geometric braids with respect to the isotopy (or homotopy) relation. We denote B_n the set of n-strand braids.

For every $n \geqslant 1$, we thus have, by definition, the equality:

$$B_n = \mathcal{GB}_n/\approx. \tag{1.5}$$

At this point, we are still far from being able to explicitly describe B_n for every n; however, given Lemmas 1.1.10 and 1.1.12, we can (triumphally!) announce a first result.

Proposition 1.1.16 (Single strand braids) *The space B_1 reduces to a single point.*

Note the importance in the hypothesis that the unique strand conserves a positive orientation with respect to the x-axis: even if, for the moment, the means to prove it are lacking, no one who has ever tried to untangle a string whose endpoints are fixed can doubt that an arc such as the one shown in Figure 1.4 is not isotopic to a one-strand geometric braid.

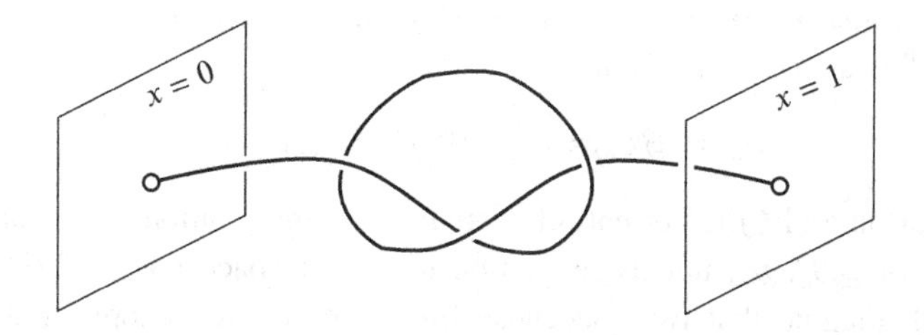

Figure 1.4 An arc that is not a one-strand braid: the arc turns back on itself, and certain planes $\{x\}\times\mathbb{R}^2$ cut it at more than one point. Such a geometric object is known as a 'string link'.

There exist highly regular continuous arcs, for example of class C^1, or C^∞, but we also know the existence of absolute horrors.[12] Another application of the result of Lemma 1.1.14 is that, for the study of classes of isotopy, we can limit ourselves to geometric braids whose strands are arcs of a special type. An

[12] Take for example the Peano curve passing through every single point of a square...

arc is said to be *piecewise linear* (or *polygonal*) if it is a finite union of line segments.

Proposition 1.1.17 (Density) *The piecewise linear geometric braids (resp., of class C^∞) form a dense subspace of $\mathcal{GB}_n$.*

Proof Let β be an arbitrary geometric braid and let ε be a strictly positive real number. We must show that the tubular neighbourhood $V(\beta, \varepsilon)$ includes a piecewise linear geometric braid. However, the coordinate functions of each strand of β are continuous, hence uniformly continuous, on $[0, 1]$: thus there exists a real $\delta > 0$ such that $|t' - t| \leqslant \delta$ implies $d(\beta^{[i]}(t'), \beta'^{[i]}(t)) < \varepsilon$ for $i = 1, ..., n$. Let $t_0, ..., t_q$ be real numbers satisfying $t_0 = 0$, $t_q = 1$ and $t_k < t_{k+1} < t_k + \delta$ for every k. Then, for $1 \leqslant i \leqslant n$ and $0 \leqslant k < q$, the segment between the points $(t_k, \beta^{[i]}(t_k))$ and $(t_{k+1}, \beta^{[i]}(t_{k+1}))$ is included in $V(\beta, \varepsilon)$. Linking these segments provides a geometric braid, piecewise linear, contained in $V(\beta, \varepsilon)$.

The argument is similar for the C^∞ strands. □

From the result of Lemma 1.1.14, two geometric braids sufficiently close are isotopic. The density property of Proposition 1.1.17 immediately implies the following result.

Corollary 1.1.18 (Restriction) *Every geometric braid is isotopic to a piecewise linear geometric braid, and to one of class C^∞.*

Let $\mathcal{GB}_n^{\mathsf{aff}}$ (*resp.*, $\mathcal{GB}_n^\infty$) be the family of piecewise linear (*resp.*, C^∞) n-strand geometric braids. Since any isotopy class of $\mathcal{GB}_n$ contains elements in both $\mathcal{GB}_n^{\mathsf{aff}}$ and $\mathcal{GB}_n^\infty$, we obtain:

$$B_n = \mathcal{GB}_n/{\approx} = \mathcal{GB}_n^{\mathsf{aff}}/{\approx} = \mathcal{GB}_n^\infty/{\approx}. \tag{1.6}$$

The Formula (1.6) is not entirely satisfactory: the relation $\approx$ appearing in it is that of $\mathcal{GB}_n$, but not its restriction to the subspace $\mathcal{GB}_n^{\mathsf{aff}}$ of $\mathcal{GB}_n$. So, we could imagine that two piecewise linear braids are isotopic, but every attesting isotopy involves intermediate braids that are not piecewise linear. The following result shows the contrary.

Lemma 1.1.19 *If two geometric braids β, β' of $\mathcal{GB}_n^{\mathsf{aff}}$ are homotopic, there exists a path linking β to β' in $\mathcal{GB}_n^{\mathsf{aff}}$. This is also the case for $\mathcal{GB}_n^\infty$.*

Proof Let ϕ be a path linking β to β' in $\mathcal{GB}_n$. We write β_s for $\phi(s)$. As in the proof of Proposition 1.1.8, there exists $\varepsilon > 0$ such that we have $e(\beta_s) \geqslant \varepsilon$ for every s, and we can subdivide the domain $[0, 1]$ of ϕ in intervals $[s_j, s_{j+1}]$, $j = 0, ..., p-1$, with $d(\beta_{s_j}, \beta_{s_{j+1}}) < \varepsilon/3$ for every j.

Next, as in the proof of Proposition 1.1.17, there exists a real number $\delta > 0$ such that $|t' - t| < \delta$ implies $d(\beta^{[i]}_{s_j}(t'), \beta^{[i]}_{s_j}(t)) < \varepsilon/3$ for every i and j, and we can subdivide $[0, 1]$ into intervals $[t_k, t_{k+1}]$, $k = 0, ..., q-1$, so that, for every i, j, and k, we have $d(\beta^{[i]}_{s_j}(t_k), \beta^{[i]}_{s_j}(t_{k+1})) < \varepsilon/3$. In addition, we can suppose that the abscissas of the endpoints of the segments making up β and β' are among $t_1, ..., t_{q-1}$.

Set $m^i_{j,k} := (t_k, \beta^{[i]}_{s_j}(t_k))$ and, for $j = 0, ..., p$, let $\widehat{\beta}_j$ be the union of the piecewise linear arcs obtained by concatenating the segments $[m^i_{j,k}, m^i_{j,k+1}]$ for $k = 0, ..., q-1$ and $i = 1, ..., n$. By hypothesis, we have $\widehat{\beta}_0 = \beta$, and $\widehat{\beta}_q = \beta'$. Moreover, let C be the portion of the horizontal cylinder with radius $2\varepsilon/3$ centred at $m^i_{j,k}$ contained between the planes $\{t_k\}\times\mathbb{R}^2$ and $\{t_{k+1}\}\times\mathbb{R}^2$. The choice of parameters guarantees that C is contained in $V(\beta_{s_j}, \varepsilon)$, and that C contains the points $m^i_{j,k}, m^i_{j,k+1}, m^i_{j+1,k}, m^i_{j+1,k+1}$ since we have

$$d(\beta^{[i]}_{s_j}(t_k), \beta^{[i]}_{s_j}(t_{k+1})) < \varepsilon/3, \quad d(\beta^{[i]}_{s_j}(t_k), \beta^{[i]}_{s_{j+1}}(t_k)) < \varepsilon/3,$$
$$\text{and } d(\beta^{[i]}_{s_j}(t_k), \beta^{[i]}_{s_{j+1}}(t_{k+1})) < 2\varepsilon/3.$$

From this, for every i, j, k, the convex hull of the four points is contained in C, hence in $V(\beta_{s_j}, \varepsilon)$ – see Figure 1.5. It thus follows that the n strands of $\widehat{\beta}_j$ are pairwise disjoint, hence $\widehat{\beta}_j$ is a piecewise linear geometric braid. Next, we obtain a path $(\dot{\beta}_s)_{s\in[s_j,s_{j+1}]}$ linking $\widehat{\beta}_j$ to $\widehat{\beta}_{j+1}$ in $\mathcal{GB}^{\mathrm{aff}}_n$ by setting, for s in $[s_j, s_{j+1}]$ and for every t

$$\dot{\beta}^{[i]}_s(t) := \widehat{\beta}^{[i]}_j(t) + \frac{s - s_j}{s_{j+1} - s_j}\overrightarrow{\widehat{\beta}^{[i]}_j(t)\widehat{\beta}^{[i]}_{j+1}(t)},$$

that is, by moving at a uniform speed from $m^i_{j,k}$ to $m^i_{j+1,k}$ and from $m^i_{j,k+1}$ to $m^i_{j+1,k+1}$ and then joining the corresponding points. As everything takes place in the ε-neighbourhood of β_{s_j}, we are certain that the arcs corresponding to distinct values of i are disjoint. By construction, the path $(\dot{\beta}_s)_{s\in[0,1]}$ links $\widehat{\beta}_0$ to $\widehat{\beta}_q$, hence β to β', in $\mathcal{GB}^{\mathrm{aff}}_n$.

The argument is similar for the class C^∞. □

We can thus amend Formula (1.6) to read

$$B_n = \mathcal{GB}_n/\!\approx\; = \mathcal{GB}^{\mathrm{aff}}_n/\!\approx^{\mathrm{aff}}\; = \mathcal{GB}^{\infty}_n/\!\approx^{\infty}, \tag{1.7}$$

where $\approx^{\mathrm{aff}}$ refers to the existence of a path within the space of piecewise linear geometric braids, and $\approx^\infty$ to a path in the braid space C^∞. In conclusion, to study the space B_n, it is permissible to limit ourselves to piecewise linear geometric braids, or to C^∞ geometric braids, as is convenient.

We conclude this (boring?) topological discussion with an embedding of the space B_n in the space B_{n+1}.

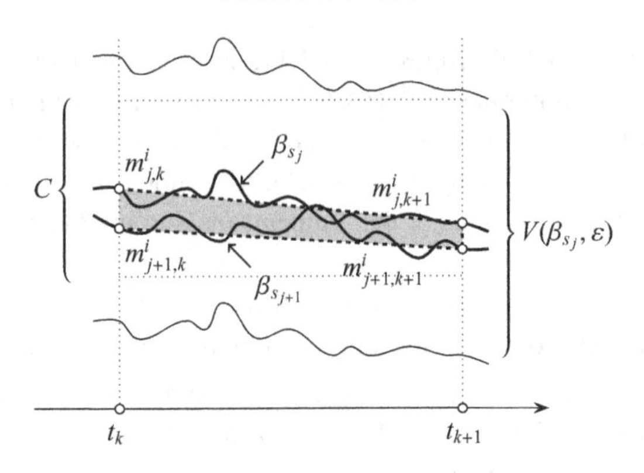

Figure 1.5 Construction of a path in $\mathcal{GB}_n^{\mathrm{aff}}$: once assured that everything takes place in the small cylinder C, itself included in the 'safe neighbourhood' $V(\beta_{s_j}, \varepsilon)$ which avoids all the other strands, we can replace the original arcs by unions of interpolating segments.

Proposition 1.1.20 (Embedding) *For every n, the addition of a trivial $(n{+}1)$th strand induces an embedding of B_n in B_{n+1}.*

Proof Denote $\mathcal{GB}^n_{n+1}$ the subspace of $\mathcal{GB}_{n+1}$ formed by the geometric braids for which the $(n{+}1)$th strand finishes at the position $n + 1$, and $\mathcal{GB}_n^{\mathrm{sp}}$ the subspace of $\mathcal{GB}_n$ formed by the geometric braids contained in the prism $[0,1] \times (-\infty, n{+}0.5) \times \mathbb{R}$. The mapping π_n from $\mathcal{GB}^n_{n+1}$ into $\mathcal{GB}_n$ corresponding to 'forgetting' the $(n{+}1)$th strand is compatible with the isotopy: $\beta \approx \beta'$ implies $\pi_n(\beta) \approx \pi_n(\beta')$. Let C be a strictly increasing bijection of $\mathbb{R}$ to $(-\infty, n{+}0.5)$ which is the identity on $(-\infty, n]$. For any geometric braid β in $\mathcal{GB}_n$, the deformation β^C of β obtained by applying C to the second coordinate belongs to $\mathcal{GB}_n^{\mathrm{sp}}$ and is isotopic to β. Moreover, the mapping e_n from $\mathcal{GB}_n^{\mathrm{sp}}$ into $\mathcal{GB}^n_{n+1}$ corresponding to the addition of an $(n + 1)$th rectilinear strand is compatible with the isotopy: for β, β' in $\mathcal{GB}_n^{\mathrm{sp}}$, if $(\beta_s)_{s\in[0,1]}$ links β with β' in $[0,1]{\times}\mathbb{R}^2$, then $(\beta_s^C)_{s\in[0,1]}$ links β^C to β'^C in $[0,1] \times (-\infty, n{+}0.5] \times \mathbb{R}$. Passing to equivalence classes, we deduce that π_n induces a continuous mapping of B^n_{n+1} onto B_n, and that e_n induces a continuous mapping of B_n into B^n_{n+1}, the right inverse of the former: $\pi_n \circ e_n$ is the identity on B_n. Hence, e_n is an embedding of B_n into B^n_{n+1}, hence into B_{n+1}. □

There is thus no danger in considering the braid spaces as forming an increasing sequence of inclusions:

$$\{1\} = B_1 \subseteq B_2 \subseteq B_3 \subseteq \cdots . \tag{1.8}$$

Hence, we can now introduce the space of all the braids.

Definition 1.1.21 (Space B_∞) The space B_∞ is defined as $B_\infty := \bigcup_{n\geqslant 1} B_n$.

At times, we will use $\mathcal{GB}_\infty$ for the union of the spaces $\mathcal{GB}_n$. Note that the elements of B_∞ are braids with an arbitrary finite number of strands, and not braids with an infinity of strands.

Exercise 1.1.22 (Shifts)
(i) Following the same schema as in Proposition 1.1.20, show that the addition of a first non-braided strand defines, for every n, an embedding shift_n (a 'shift') of the space B_n into B_{n+1}. Is this embedding surjective? Anticipating Definition 1.2.8, what is the image of σ_i by shift_n?
(ii) Show that the embeddings shift_n are compatible between themselves, and hence deduce that they induce an embedding shift of B_∞ into itself.

1.2 Diagrams and Braid Words

1.2.1 Planar Projections of Geometric Braids

As they live in $\mathbb{R}^3$, geometric braids are 3-dimensional objects. We show here how we can, without loss of information, pass to 2-dimensional objects, the *braid diagrams*, and the *braid words* that encode them. We can thus obtain more convenient representations of the braid space B_n.

We start with the idea of *projecting* the geometric braids onto a suitably chosen plane. A natural projection to use is the orthogonal projection onto the plane $\mathbb{R}^2\times\{0\}$ (the plane '$z = 0$'), as it does not squish the figure and hence a priori preserves the maximum of information. Starting from an n-strand geometric braid, we obtain n arcs in the plane, joining the points $(0, 1), ..., (0, n)$ to the points $(1, 1), ..., (1, n)$, as in the left image of Figure 1.6, that we can imagine is the projection of the braid of Figure 1.2.

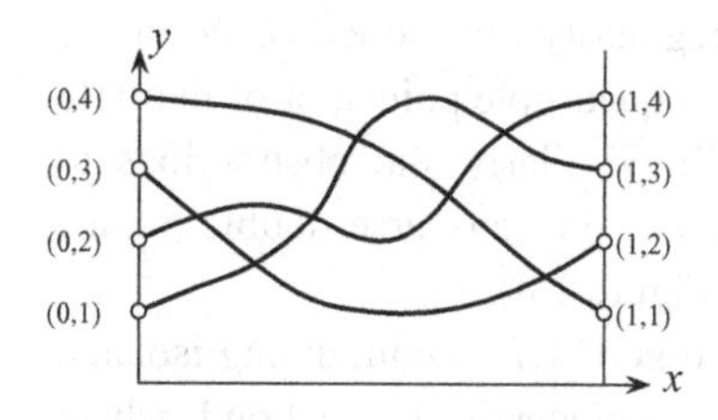

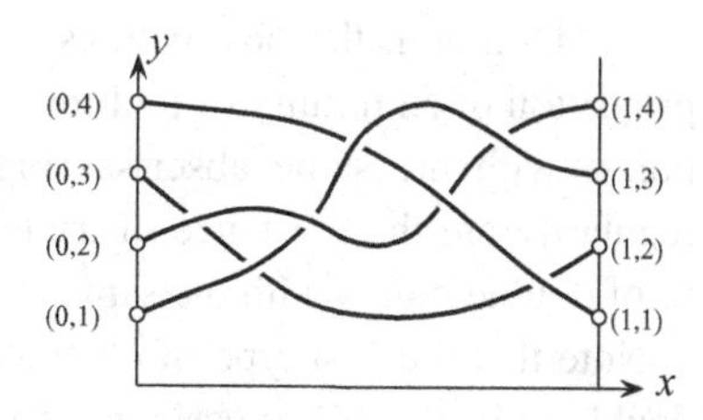

Figure 1.6 Projection of a geometric braid into the plane '$z=0$': on the left, the projection with its double points, and on the right the same, but with supplementary information about the upper strand at each double point.

As in the left image of Figure 1.6, the projection of a geometric braid β can contain double points: the hypothesis of strands of β not cutting each other forbids the existence of t satisfying $\beta^{[i]}(t) = \beta^{[j]}(t)$ with $i \neq j$, but not $\beta^{[i]}(t)$ and $\beta^{[j]}(t)$ having the same first component. In any case, their second coordinates are distinct, hence one is strictly greater than the other: one of the strands passes *above* the other.

Definition 1.2.1 (Braid diagram) An *n-strand braid diagram* is the orthogonal projection onto $\mathbb{R}^2 \times \{0\}$ of an n-strand geometric braid, completed according to the convention that at each double point the strand at the back is interrupted.

The right image of Figure 1.6 is a four-strand braid diagram: when the projections of strands 2 and 3 cross each other for the first time, strand 2 is on top, and hence strand 3 is interrupted.

There remains a difficulty: nothing forbids the projection of a geometric braid to not only have double points, but also triple points – at least three integers i, j, k pairwise distinct, and t such that $\beta^{[i]}(t), \beta^{[j]}(t)$, and $\beta^{[k]}(t)$ have the same first component – or even non-isolated double points, in which case the convention of representation is insufficient. We could amend this, but instead we choose to use the possibility of excluding the geometric braids judged to be undesirable.

Definition 1.2.2 (Regular, semi-regular) A geometric braid is *semi-regular* if its projection on $\mathbb{R}^2 \times \{0\}$ has only a finite number of multiple points, which are at worst triple points and have distinct abscissas, except in the case of pairs of transverse double points; it is *regular* if it is semi-regular and, in addition, the multiple points of its projection are transverse double points with distinct abscissas.

By definition, the obstructions to semi-regularity are the existence in the projection of an infinity of multiple points, of quadruple points, or of multiple points with the same abscissa (Figure 1.7). Similarly, the obstructions to regularity are the existence of triple points, of non-transverse double points, or of double points with the same abscissa (Figure 1.8).

Note that the first type of obstruction in Figure 1.7 (accumulating isolated double points[13]) is impossible in the case of a piecewise linear braid, where the multiple points of the projection can only be isolated or form segments.

[13] The resulting diagram is then said to be 'wild'.

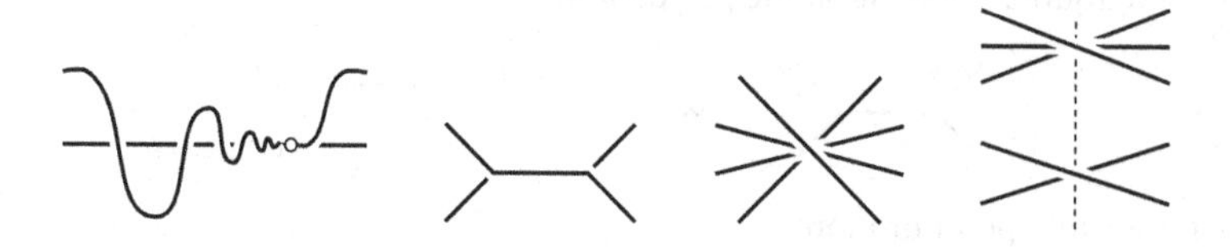

Figure 1.7 Obstructions to semi-regularity: the projection contains an infinity of double points (accumulating, forming a segment, etc), or (at least) one quadruple point, or a triple point and a double point with the same abscissa.

Figure 1.8 Obstructions to regularity of a semi-regular braid: the projection contains (at least) one triple point, a non-transverse double point (the strands touch each other without crossing), or two double points with the same abscissa.

Denote $\mathcal{GB}_n^{\text{aff,semireg}}$ (*resp.*, $\mathcal{GB}_n^{\text{aff,reg}}$)[14] the family of semi-regular (*resp.*, regular) piecewise linear n-strand geometric braids. The following result is exactly what we need in order to ignore the non-semi-regular geometric braids.

Lemma 1.2.3
(i) *Every geometric braid of $\mathcal{GB}_n$ is isotopic to a geometric braid in $\mathcal{GB}_n^{\text{aff,reg}}$.*
(ii) *If two geometric braids β, β' in $\mathcal{GB}_n^{\text{aff,reg}}$ are isotopic, there exists a path linking β to β' in $\mathcal{GB}_n^{\text{aff,semireg}}$, and even included in $\mathcal{GB}_n^{\text{aff,reg}}$ with the exception of a finite number of points.*

Proof
(i) As in Proposition 1.1.17 and Corollary 1.1.18, it suffices to show that $\mathcal{GB}_n^{\text{aff,reg}}$ is dense in $\mathcal{GB}_n$. For this, given an arbitrary geometric braid β and $\varepsilon > 0$, we must show that the neighbourhood $V(\beta, \varepsilon)$ contains a regular piecewise linear braid. We saw in Proposition 1.1.17 that $V(\beta, \varepsilon)$ contains a piecewise linear braid, say $\widehat{\beta}$, and it follows from the construction that any piecewise linear braid obtained by a sufficiently small movement of the segments of $\widehat{\beta}$ is again in $B(\beta, \varepsilon)$. If $\widehat{\beta}$ is regular, there is nothing more to say. If not, $\widehat{\beta}$ contains a finite number of obstructions of the types listed in Definition 1.2.2. However, each of them can be eliminated by a small translation of the strands involved, without introducing new obstructions, as shown below:

[14] A note from the translators: we have kept the notation of the original French text. The 'aff' in these definitions corresponds to the French 'affine par morceaux' for 'piecewise linear'.

- for a double segment on the projection:

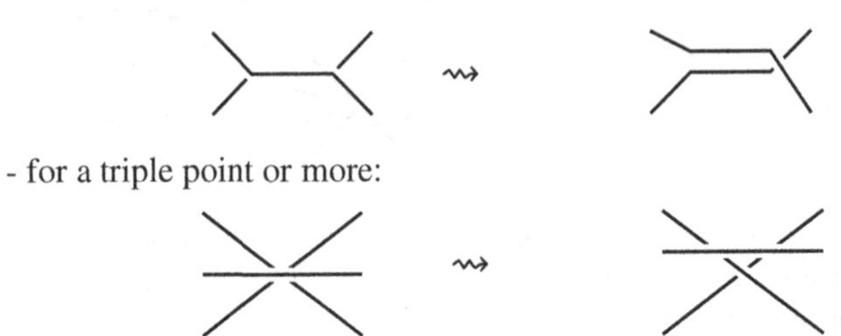

- for a triple point or more:

- for two double points with the same abscissa:

(ii) We begin this time with the construction of Lemma 1.1.19. Suppose that β and β' are two regular piecewise linear braids, and that ϕ is the path linking β to β' in $\mathcal{GB}_n^{\mathrm{aff}}$ constructed in Lemma 1.1.19. Denote β_s for $\phi(s)$.

The first observation is that it suffices to consider the singularities occurring for isolated values of the parameter s. For example, if a double point emerges from the crossing of the projections of two segments between the planes $\{t_k\}\times\mathbb{R}^2$ and $\{t_{k+1}\}\times\mathbb{R}^2$ for a non-isolated value s, then by hypothesis, it persists for a whole interval $[s_j, s_{j+1}]$. Performing on s_j a tiny translation in the plane $\{t_k\}\times\mathbb{R}^2$ and propagating linearly the modification does not suppress the double point, but makes it become one for an isolated value of s.

By (i), for each value of s, one (or several) small translation(s) of a strand (with coordinated movements of the adjacent strands) can remove the obstructions to the regularity of β_s, but this is not the solution if doing so introduces a new obstruction in a neighbouring braid $\beta_{s'}$: we must take into account here the dynamics of the process. However, by construction, the endpoints of the segments are found in a finite number of planes $\{t_k\}\times\mathbb{R}^2$, and by default the segments slide at a constant speed between these planes. We must see if by accelerating or delaying the slippages we can avoid these undesirable singularities.

First consider the case of an isolated quadruple (or worse) point. Suppose that two segments, coming from strands i and j, cross each other in the braid β_s. A triple point will appear in β_s when a third segment, coming from a third strand k, crosses the strands i and j before their crossing for $s' < s$, and traverses them after the crossing for $s' > s$, according to a schema of the type shown below:

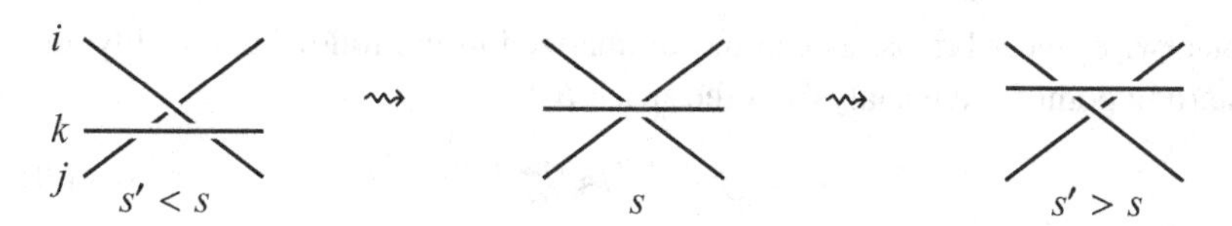

Translating one of the three segments involved in the triple point of β_s will simply cause it to appear either a bit sooner, or a bit later. However, a quadruple point will appear in β_s only if a fourth strand comes and crosses the three others exactly at the point where these three meet: by modifying the translation speed of one of the segments, we replace the quadruple point by three successive triple points, without creating a new quadruple point in the neighbourhood.

The case of a triple point with the same abscissa as a double point is similar. Two double points with the same abscissa appear in β_s when two pairs of segments, originating in strands i, j for one and i', j' for the other, cross each other, and crossing of i and j precedes the other for $s'<s$ and follows it for $s'>s$, or vice versa. Translating one of the segments only delays or advances the singularity. However, a triple point and a double point with the same abscissa appear only if two events coincide: if a third strand comes and crosses the crossing point of i and j, and that this changes position with respect to that of i' and j'. As above, a translation avoids the coincidence.

There remains the case of double segments. Such a double segment appears in β_s when, in the uniform default movement of their endpoints, two pairs of points, one situated in a plane $\{t_k\}\times\mathbb{R}^2$, the other in $\{t_{k+1}\}\times\mathbb{R}^2$, have the same projection. Regardless of the past ($s'<s$) and the future ($s'>s$), delaying one of the two movements avoids the coincidence of the projections without re-creating another one further on.

In this way, we finish with a path linking β to β' all of whose intermediate braids are semi-regular, and even regular for all but eventually a finite number.[15] □

From the preceding results we can conclude that it is sufficient to limit ourselves to regular piecewise linear braids, linked by homotopies involving only regular piecewise linear braids, apart from a finite number of semi-regular

[15] If $t_0 < \cdots < t_q$ is a fixed subdivision of $[0, 1]$, a piecewise linear braid, the endpoints of whose segments are in the vertical planes $\{t_k\}\times\mathbb{R}^2$, is specified by the sequence of $2n(q-1)$ coordinates of these points, and we can thus consider a path passing through these particular braids as an arc in $\mathbb{R}^{2n(q-1)}$ of points. Among these braids, those with triple points form a submanifold of codimension 1, whereas those with quadruple points form a submanifold of codimension 2: the argument sketched out exploits the fact that, while nothing can prevent a path from meeting a submanifold of codimension 1, a small deformation suffices to keep it from meeting a submanifold of codimension 2.

piecewise linear braids, as can be summarized in the following equality, our starting point for the analysis of the space B_n:

$$B_n = \mathcal{GB}_n^{\text{aff,reg}} / \approx^{\text{aff,semireg}}. \tag{1.9}$$

1.2.2 Isotopy of Braid Diagrams

With the projection of a geometric braid onto the plane $\mathbb{R}^2 \times \{0\}$, a portion of the information is lost: an infinite number of geometric braids admit the same projection, in particular braids whose mutual distance is unbounded. What is essential here is that nothing is lost concerning the isotopy class.

Lemma 1.2.4 *Two regular geometric braids admitting as projections the same braid diagram are isotopic.*

Proof Let β, β' be two regular geometric braids of $\mathcal{GB}_n$ having the same projection diagram. For i in $\{1, ..., n\}$ and s in $[0, 1]$, consider the arc $\gamma_{s,i}$ of $\mathbb{R}^3$ defined (on the model of Example 1.1.5) by

$$t \mapsto (t, (1 - s)\beta^{[i]}(t) + s\beta'^{[i]}(t)). \tag{1.10}$$

By construction, $(\gamma_{s,i})_{0 \leqslant s \leqslant 1}$ is a path linking the ith strand of β to the corresponding strand of β'. Thus, the union β_s of the n arcs $\gamma_{s,1}, ..., \gamma_{s,n}$ link β to β', and the question is to ensure that β_s is, for every s, a geometric braid, that is, if its strands are pairwise disjoint. This is where the hypothesis that β and β' are regular and have the same diagram comes into play. Indeed, first this implies that β_s has the same projection as β and β'. Next, if $(t, y, 0)$ is a simple point of this projection, the unique integer i such that $(t, y, 0)$ is the projection of a point of the ith strand of β, and of a point of the ith strand of β', is also, for every s, the unique integer i such that $(t, y, 0)$ is the projection of a point of the ith strand of β_s. If now $(t, y, 0)$ is a double point of the projection, there exist i and j such that $(t, y, 0)$ is the projection of (t, y, z_i) and (t, y, z_j) of the ith and jth strands of β, and, also, the projection of (t, y, z'_i) and (t, y, z'_j) of the ith and jth strands of β'. However, the hypothesis that the projection diagrams (and not only the bare projections) of β and β' coincide implies that we have either $z_i < z_j$ and $z'_i < z'_j$, or $z_i > z_j$ and $z'_i > z'_j$, hence $(1-s)z_i + sz_j < (1-s)z'_i + sz'_j$ or $(1-s)z_i + sz_j > (1-s)z'_i + sz'_j$ and, in any case, $(1-s)z_i + sz_j \neq (1-s)z'_i + sz'_j$, indicating that the ith and jth strands of β_s do not cut each other. Hence, β_s is a geometric braid, thus β and β' are isotopic. □

Given the preceding result, there is no ambiguity in introducing a notion of isotopy for the braid diagrams.

Definition 1.2.5 (Isotopic diagrams) Two braid diagrams D, D' are declared *isotopic*, denoted $D \simeq D'$, if there exist two isotopic regular geometric braids β, β' admitting D and D' as projections.

By Lemma 1.2.4, the choice of the regular braids whose projections produce the diagrams is not important: if two diagrams D, D' are isotopic, every geometric braid projecting onto D is isotopic to every geometric braid projecting onto D'.

Introducing then $\mathcal{BD}_n$ as the set of n-strand braid diagrams, and $\mathcal{BD}_n^{\mathsf{aff}}$ as the subset of $\mathcal{BD}_n$ formed by the piecewise linear diagrams, we obtain a new representation (in fact two) of the space B_n:

$$B_n = \mathcal{BD}_n/\simeq = \mathcal{BD}_n^{\mathsf{aff}}/\simeq . \tag{1.11}$$

Note that at this point the definition of the relation $\simeq$ remains indirect, relying on the isotopy of geometric braids and not expressed in terms of the diagrams themselves.

1.2.3 Coding by Braid Words

After the passage of 3-dimensional geometric braids to 2-dimensional diagrams, we are now going to perform an additional reduction by coding the braid diagrams by *words*, and again showing that no information on the isotopy classes is lost in the passage.

An n-strand braid diagram is comprised of n arcs in the plane linking the points $(0, 1), ..., (0, n)$ with the points $(1, 1), ..., (1, n)$ and cutting each line $\{x\}\times\mathbb{R}$ in exactly n points. As in 1.1.2, we can use the abscissa as the parameter of these arcs.

Notation 1.2.6 (Function $D^{[i]}$) For an n-strand braid diagram D, denote $D^{[i]}$ the continuous function of $[0, 1]$ into $\mathbb{R}$ such that the strand emanating from $(0, i)$ ('ith strand') is parametrized by $t \mapsto (t, D^{[i]}(t))$.

By definition, the strands of a braid diagram cross each other at a finite number of points with pairwise distinct abscissas. We introduce a coding of these diagrams by taking into account the position of these double points as a function of the number of strands with smaller ordinate.

Definition 1.2.7 (Code of a braid diagram)
(i) A double point (x, y) of a braid diagram is said to be *positive* (*resp., negative*) if the strand in front is the one with the larger (*resp.*, smaller) ordinate, in a left neighbourhood of x; its *code* is then the letter σ_i (*resp.*, $\overline{\sigma_i}$), with $i := \#\{j \mid D^{[j]}(x) < y\} + 1$.

(ii) The *code* of a braid diagram is the ordered sequence of its double points, ordered by increasing abscissa.

Example 1.2.8 (Code) We return to the braid diagram of Figure 1.6, which has seven double points:

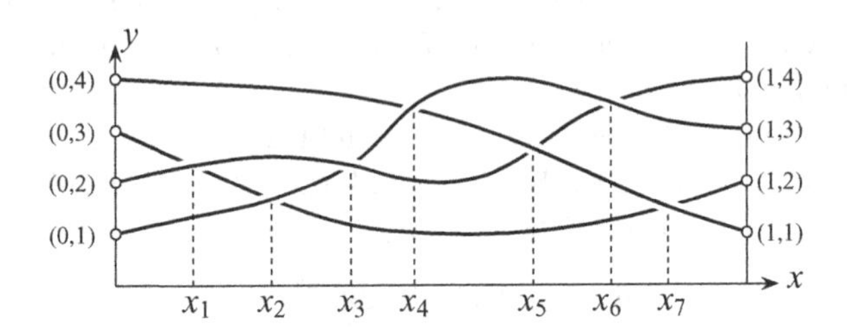

At the first double point, (x_1, y_1), there exists a strand situated beneath the double point, that is, cutting the dotted segment joining (x_1, y_1) and $(x_1, 0)$, and the strand passing on top at (x_1, y_1) is the one with the smaller ordinate to the left of x_1: the code of (x_1, y_1) is thus the letter $\overline{\sigma}_2$. Similarly, the code of the second double point, (x_2, y_2), is $\overline{\sigma}_1$ (no strand cuts the dotted line), whereas that of the third double point, (x_3, y_3), is σ_2: a strand cuts the dotted line, and this time, the strand passing on top at (x_3, y_3) is the one with the larger ordinate to the left of x_3. Continuing in the same manner, we find the codes $\overline{\sigma}_3$, σ_2, σ_3, and σ_1, hence the code of the diagram is the sequence $(\overline{\sigma}_2, \overline{\sigma}_1, \sigma_2, \overline{\sigma}_3, \sigma_2, \sigma_3, \sigma_1)$, usually simply written $\overline{\sigma}_2\overline{\sigma}_1\sigma_2\overline{\sigma}_3\sigma_2\sigma_3\sigma_1$.

If D is an n-strand braid diagram, then the codes of the double points of D are the letters σ_i and $\overline{\sigma}_i$ with $1 \leqslant i \leqslant n-1$, as the maximum number of strands with a smaller ordinate at a double point is between 0 and $n-2$.

Definition 1.2.9 (Braid word) A finite sequence of letters σ_i and $\overline{\sigma}_i$ with $1 \leqslant i \leqslant n-1$ is called an *n-strand braid word*. The set of n-strand braid words is denoted $\mathcal{BW}_n$.

Hence the sequence $\overline{\sigma}_2\overline{\sigma}_1\sigma_2\overline{\sigma}_3\sigma_2\sigma_3\sigma_1$ encountered in Example 1.2.8 is a four-strand braid word. It is also an n-strand braid word for every $n \geqslant 4$: by definition, we have $\mathcal{BW}_1 \subseteq \mathcal{BW}_2 \subseteq \mathcal{BW}_3 \subseteq \cdots$.

The following refinement of Lemma 1.2.4 shows that no information about the associated braid, that is, its isotopy class, is lost in the coding of the diagrams.

Lemma 1.2.10 *Two regular geometric braids admitting as projection two diagrams with the same code are isotopic.*

Proof A braid diagram is said to be *standard* if it is made up of either n horizontal segments (parallel to the x axis), or obtained by juxtaposing ℓ elementary patterns of width $1/\ell$ composed of $n-2$ horizontal segments and two segments with slope $\pm\ell$ that cross each other. For each n-strand braid word w, there exists a unique n-strand standard diagram with code w. For example, for the code $\overline{\sigma}_2\overline{\sigma}_1\sigma_2\overline{\sigma}_3\sigma_2\sigma_3\sigma_1$ of 1.2.8 and for $n = 4$, the standard braid diagram is shown below:

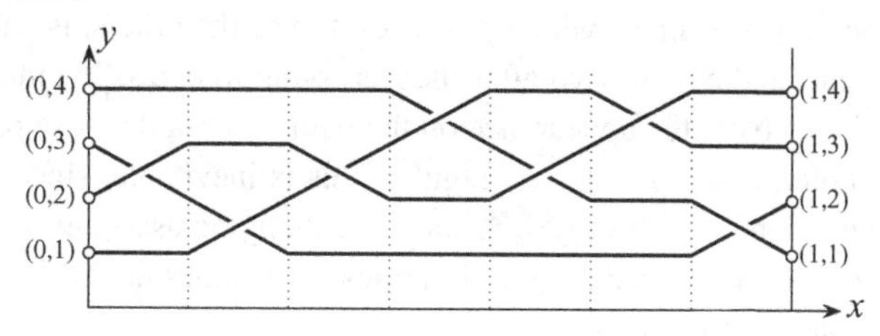

For any n-strand braid diagram D, there exists a continuous path ϕ_D in $\mathcal{BD}_n$ joining D to the standard diagram with the same code: we expand/contract the abscissas to move the double points to the desired abscissas, and then, for every x, we expand/contract the ordinates to move the strands to the desired positions. As the strands of D do not cross each other except at double points, rectifying them does not introduce any new intersections.

Now suppose that β and β' are two regular geometric braids whose projections are two diagrams D, D' admitting the same code w. Let $\widehat{D}$ be the standard diagram with the same code as D and D'. Then $\phi_D \times \mathrm{id}_{\mathbb{R}}$, consisting of applying ϕ_D to the abscissas and ordinates but leaving the sides (front/back) invariant, defines a path in $\mathcal{GB}_n$ linking β to a braid $\widehat{\beta}$ whose projection is $\widehat{D}$. Similarly, $\phi_{D'} \times \mathrm{id}_{\mathbb{R}}$ defines a path in $\mathcal{GB}_n$ linking β' to a braid $\widehat{\beta}'$ whose projection is also $\widehat{D}$. By Lemma 1.2.4, the braids $\widehat{\beta}$ and $\widehat{\beta}'$ are isotopic. By the transitivity of isotopy, so are β and β'. □

As in Definition 1.2.5, we can thus without ambiguity define the notion of isotopy for the braid words.

Definition 1.2.11 (Equivalent words) Two braid words w, w' are said to be *equivalent*, denoted $w \equiv w'$, if there exist two isotopic regular geometric braids admitting w and w' as the codes of their projections.

As every regular braid admits a code, the realization of the braid space B_n in terms of regular braids allows a new realization of B_n in the form:

$$B_n = \mathcal{BW}_n/\equiv. \tag{1.12}$$

Two equal words are of course equivalent. However there exist distinct braid words nonetheless equivalent.

Example 1.2.12 (Equivalent words) Consider for example the words $\sigma_1\sigma_2\sigma_1$ and $\sigma_2\sigma_1\sigma_2$. The sequence of diagrams below suggests a path in $\mathcal{GB}_3^{\text{aff,semireg}}$ linking the braids with codes $\sigma_1\sigma_2\sigma_1$ and $\sigma_2\sigma_1\sigma_2$:

The strand in dotted lines, which passes on top of the others, is backed up, and while it meets the other two after their crossing in $\sigma_1\sigma_2\sigma_1$, it meets them before in $\sigma_2\sigma_1\sigma_2$. Note the appearance on the path of a braid whose projection has a triple point, and hence is non-regular: this is inevitable, since the code cannot change along a path in $\mathcal{GB}_n^{\text{reg}}$, since the mapping associating a regular braid to the code of its projection is continuous with values in $\mathcal{BW}_n$, a discrete topological space (why is this?).

As with the isotopy of braid diagrams, the equivalence of braid words is, for the moment, defined only in terms of reference to the isotopy of geometric braids, and not by an internal characterization of $\mathcal{BW}_n$. One of the first tasks in the following chapters will be to fill this gap.

Remark 1.2.13 The reader could be surprised that the notation $\equiv$, as with $\approx$ or $\simeq$, makes no reference to the number of strands n of the braids under consideration. According to Proposition 1.1.20, this is not a problem for $\approx$: two n-strand geometric braids are isotopic in $\mathcal{GB}_n$ if and only if they are in $\mathcal{GB}_{n+1}$, and hence we can use one unique notation. By projection and coding, this also holds for $\simeq$ and $\equiv$. Similarly, there is no danger in using σ_i and $\overline{\sigma}_i$ to code all the diagrams of at least $i+1$ strands, without distinguishing between a 'σ_i of $\mathcal{BW}_n$' and a 'σ_i of $\mathcal{BW}_{n+1}$'.

1.3 The Braid Isotopy Problem

1.3.1 A Problem Open for the Moment

Definition 1.2.11 is precise and non-ambiguous, but it remains abstract until such time as we can describe a method allowing us to recognize whether or not braid words are equivalent.

Definition 1.3.1 (The isotopy problem) The *braid isotopy problem* is the question of effectively deciding the equivalence of braid words. A *solution* to this problem is an algorithm deciding, for two arbitrary braid words w, w', whether or not the relation $w \equiv w'$ is satisfied, in a finite number of steps.

To make precise Definition 1.3.1, we must make explicit the notion of algorithm, and for this fix a model of computation. We content ourselves here by invoking the somewhat vague notion of 'procedure executable by a computer', and inviting the reader to verify that this is the case for the algorithms described in the rest of this text, and to consult for example (Dehornoy, 1993). Note that a solution to the braid isotopy problem must be a uniform procedure functioning for *all* pairs of input words, not only for certain among them.

The braid isotopy problem is important, as having in hand a solution is a prerequisite to any practical utilization of braid words. We obtained with Equation (1.12) a realization of the braid spaces in terms of classes of braid words with respect to equivalence. This shows progress, as the space $\mathcal{GB}_n$ is enormous, with its elements depending on an infinity of parameters, whereas $\mathcal{BW}_n$ is countable: for each ℓ, the family of n-strand braid words of length ℓ is finite. Nonetheless, the progress is illusory as long as we do not know how to recognize if two words are equivalent, that is, represent the same braid.

The restricted size of the family $\mathcal{BW}_n$ does not imply that the braid isotopy problem will be easy: we know of problems formulated in terms of equivalence classes on families of words that are *undecidable*, that is, there is no algorithm in existence to solve them in every case; see for example the footnotes [17] and [21] of Chapter 2.

As another element inciting us to prudence, we can mention the isotopy problem of links, in every way analogous to the one for braids. We consider this time closed diagrams (the strands close back on themselves)[16] and ask if two (suitably encoded) diagrams such as

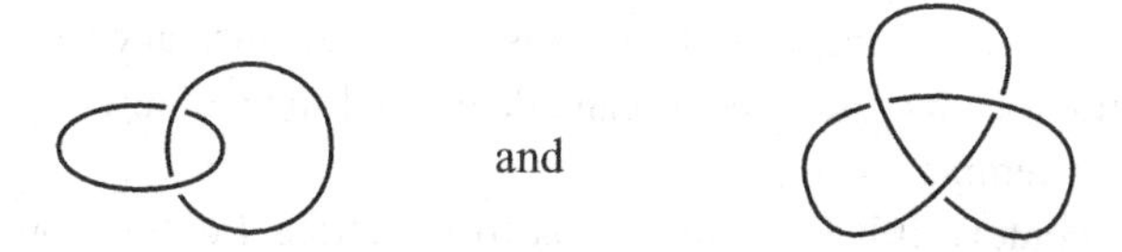

can be seen as projections of isotopic three-dimensional figures.[17] The problem is related to the one for braids, since any link can be realized as the closure of a braid and the closures of isotopic braids are isotopic links, see Figure 1.9. However, it is known that the isotopy problem of links is *very* difficult: solutions exist, but to describe them and prove the results would greatly exceed the level and length of this text.

[16] We only speak of a 'knot' in the case of a single connected component, otherwise we use 'links'.

[17] As it happens, certainly not: isotopy preserves the number of connected components; however, there are two on the left and only one on the right.

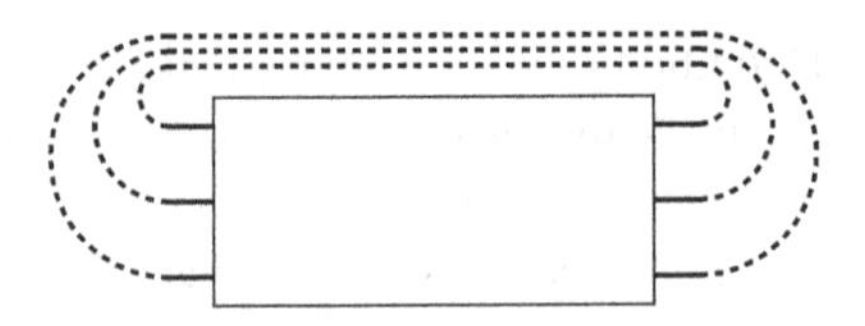

Figure 1.9 The closure of a braid diagram (the white box, with here three strands at the input and the output) is a link diagram (a knot diagram if there is only one connected component).

Exercise 1.3.2 (Closure) Show that the closures of two isotopic braid diagrams are diagrams of isotopic links, but that non-isotopic braid diagrams can have closures corresponding to isotopic link diagrams. [Hint: Compare the closures of the braids of Example 1.3.5.]

Happily, the braid isotopy problem is simpler than the one for knots, and later in this text, we will be able to describe several solutions, some quite simple and easy to implement on a computer.

1.3.2 First Attempts

As for all analogous decision problems, the braid isotopy problem presents two aspects with different flavours, according to whether the response is positive or negative. The first case is a priori simpler: if two words w, w' are equivalent, for proof it suffices to exhibit an isotopy between two braids whose codes are w and w'. If we could systematically test all the candidates of an isotopy,[18] we would finish by finding one if it exists. On the other hand, if such an isotopy does not exist, persevering with the tests is of no use, since at each instant we have only tried a finite number of candidates, and their failure says nothing about future attempts.

In such a context, it is habitual to look for functions I defined on the braid words and such that

$$w \equiv w' \quad \text{implies} \quad I(w) = I(w'), \tag{1.13}$$

called *isotopy invariants*. In this way, if we find $I(w) \neq I(w')$, we can conclude that w and w' are not equivalent. The ideal situation is with an invariant for which the implication of Equation (1.13) is an equivalence: this is known as a *complete* invariant.

Thus we are led to seek isotopy invariants defined on the braid words. Such invariants are easy to find, but to be honest up front, none of the simple

[18] For the time being, the possibility of doing so is not clear, but it will become so after Chapter 2.

invariants presented here are complete: even though less difficult than the problem for knots, the braid isotopy problem is not so simple... A natural idea would be to count the letters of each type. However, we have already seen in Example 1.2.12 that the words $\sigma_1\sigma_2\sigma_1$ and $\sigma_2\sigma_1\sigma_2$ are equivalent. As one of them has two σ_1 and one σ_2, and the other, one σ_1 and two σ_2, so much for that idea!

Similarly, the total number of letters is again not an invariant. Indeed, as in Example 1.2.12, the sequence of diagrams

illustrates an isotopy in which the top strand, with a dotted line, first makes a loop over the bottom strand, and then, sliding upwards, finishes by no longer touching it. In terms of codes, we pass from $\sigma_1\overline{\sigma_1}$ to the empty word ε. Hence we have $\sigma_1\overline{\sigma_1} \equiv \varepsilon$, and a two-letter word is equivalent to one with no letters at all.

A better idea consists of associating each n-strand braid diagram with a permutation of the set $\{1, ..., n\}$.

Definition 1.3.3 (Permutation) If β is an n-strand geometric braid, the *permutation of* β, denoted $\mathsf{perm}(\beta)$, is the permutation f of $\{1, ..., n\}$ such that, for every i, the strand finishing at the position i in β begins[19] at the position $f(i)$.

For example, for the geometric braid of Figures 1.2 and 1.6, we find the permutation $\binom{1\ 2\ 3\ 4}{4\ 3\ 1\ 2}$.

This turns out to be an isotopy invariant.

Lemma 1.3.4 *Isotopic geometric braids have the same permutation.*

Proof By definition, the strands conserve the origin and destination along any path linking the two braids under consideration. □

Consequently, the mapping perm on $\mathcal{GB}_n$ induces a well-defined mapping on the quotient space B_n, hence also on the space of braid diagrams $\mathcal{BD}_n$, and on the space of braid words $\mathcal{BW}_n$. We use perm to denote all of these mappings.

Hence, perm is an isotopy invariant: $w \equiv w'$ implies $\mathsf{perm}(w) = \mathsf{perm}(w')$, allowing us to establish non-equivalence.

Example 1.3.5 (non-equivalence) The braid words $\sigma_1\sigma_2$ and $\sigma_2\sigma_1$ are not equivalent. Indeed, we can read on the diagrams

[19] It could seem more natural to consider the final position of the strand beginning at the position i – which amounts to inverting the permutation – but the choice made here is preferable to assure the homomorphism of Proposition 2.1.13.

the values $\mathsf{perm}(\sigma_1\sigma_2) = \binom{1\,2\,3}{2\,3\,1} \neq \mathsf{perm}(\sigma_2\sigma_1) = \binom{1\,2\,3}{3\,1\,2}$.

But this invariant is not complete: for example, we can verify that the words $\sigma_1\sigma_1$ and ε correspond to the identity permutation, whereas we will see in Example 1.3.10 that they are not equivalent.

Another approach consists in considering how the strands of a geometric braid wrap about each other.

Definition 1.3.6 (Linking number) For β in $\mathcal{GB}_n^\infty$ and i, j distinct in $\{1, ..., n\}$, the *linking number* of the ith and jth strands of β is

$$\lambda_{i,j}(\beta) := \frac{1}{2\pi}\int_0^1 \theta'_{i,j}(t)dt, \tag{1.14}$$

where $\theta_{i,j}(t)$ is the measure of the angle between $\overrightarrow{\beta^{[i]}(t)\beta^{[j]}(t)}$ and $(1, 0)$, determined by continuity starting with $\theta_{i,j}(0) := 0$.

Examples 1.3.7 (Linking number) Let β be a two-strand geometric braid.

$$\beta^{[1]}(t) := (1 + t, 0), \qquad \beta^{[2]}(t) := (1 + t + \cos(\pi t), \sin(\pi t)). \tag{1.15}$$

Then the vector $\overrightarrow{\beta^{[1]}(t)\beta^{[2]}(t)}$ is $(\cos(\pi t), \sin(\pi t))$. We thus have $\theta_{1,2}(t) = \pi$, hence $\lambda_{1,2}(\beta) = 1/2$.

Similarly, for the braid β of Figure 1.2, we can read the values $\lambda_{1,2}(\beta) = 1$, then $\lambda_{2,3}(\beta) = -1/2$, $\lambda_{3,4}(\beta) = 1/2$, $\lambda_{1,3}(\beta) = \lambda_{1,4}(\beta) = -1/2$, $\lambda_{2,4}(\beta) = 1/2$.

Since by hypothesis $\theta_{i,j}(1)$ is a multiple of π, $\lambda_{i,j}(\beta)$ is always an integer or a half-integer.[20]

Exercise 1.3.8 (Symmetry of the linking number) Show that, for any geometric braid β and every i, j, $\lambda_{j,i}(\beta) = \lambda_{i,j}(\beta)$.

In this way we obtain new isotopy invariants.

Lemma 1.3.9 *Isotopic geometric braids have the same linking numbers.*

Proof For each i, j, the mapping $\beta \mapsto \lambda_{i,j}(\beta)$ is a continuous mapping of the metric space $\mathcal{GB}_n^\infty$ into the discrete space $\mathbb{Z}/2$. It is thus constant on each connected component of $\mathcal{GB}_n^\infty$, that is, on each homotopy class. □

[20] Dividing by π in (1.14) would avoid the half-integer values, but this would lose the intuitive notion where $\lambda_{i,j}(\beta) = p$ means that, in a moving frame centred on $(t, \beta^{[i]}(t))$, the point $(t, \beta^{[j]}(t))$ turns p times around the origin when t varies from 0 to 1. Thus, in σ_1, the strand 2 makes a *half*-turn around the strand 1.

We again denote $\lambda_{i,j}$ the isotopy invariant induced by the linking number on the braid diagrams, and on the braid words. We thus obtain a new method to establish non-equivalence.

Example 1.3.10 (Non-equivalence) The projection of the geometric braid β of Example 1.3.7 has a single double point, at $(1/2, 3/2)$, with strand 2 passing over strand 1. Its isotopy class is thus σ_1, and we obtained $\lambda_{1,2}(\sigma_1) = 1/2$. Similarly, we find the values $\lambda_{1,2}(\sigma_1\sigma_1) = 1$, and $\lambda_{1,2}(\varepsilon) = 0$, implying $\sigma_1\sigma_1 \not\equiv \varepsilon$, as stated in Example 1.3.5.

Example 1.3.11 (A challenging example) We have now defined several isotopy invariants for the braid words. They are all non-trivial, in the sense that they can take on several values and can effectively establish the non-equivalence of words. Nonetheless, even considered collectively, they do not constitute a complete invariant. For example, they cannot separate the words $\sigma_1\sigma_1\sigma_2\sigma_2$ and $\sigma_2\sigma_2\sigma_1\sigma_1$, codes of the diagrams

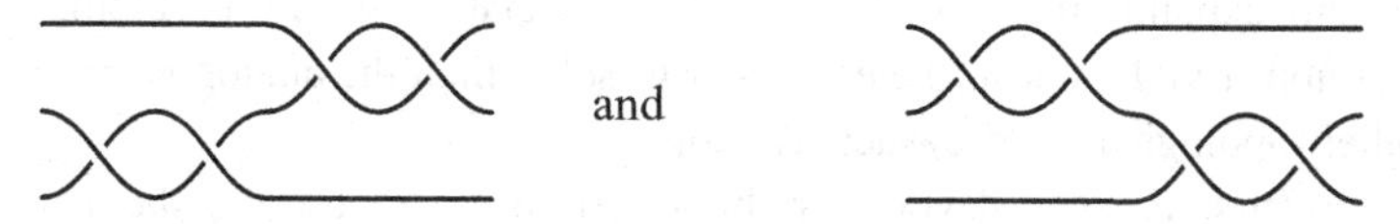

In both cases, their permutation is the identity, whereas the linking numbers $\lambda_{1,2}$ and $\lambda_{2,3}$ are 1, and $\lambda_{1,3}$ is 0. However, it will be shown in Chapter 3 that these words are not equivalent. The braid isotopy problem thus remains open for now, and more work will be necessary to resolve it.

2
Braid Groups

The study of the braid spaces B_n, and in particular finding simple solutions to the isotopy problem, relies on the existence of a group structure on B_n. Such an algebraic structure, specific to braids, underlies most of the results concerning them, and explains in particular why the associated isotopy problem, even though non-trivial, is nevertheless easier to solve than the analogous problem for other topological objects such as knots.

In this first chapter devoted to braid groups, we describe the product operation underlying the group structure, and then establish an intrinsic characterization of this structure in terms of a presentation by generators and relations, the starting point for all the subsequent developments (Section 2.1). As these presentations of algebraic structures are not necessarily familiar to everyone, two appendices complete the chapter, one devoted to monoid presentations (Section 2.2), the other to group presentations (Section 2.3). Both offer a quite complete treatment, accessible without prior knowledge; however, most of the proofs are left as exercises.[1]

2.1 The Group B_n

2.1.1 The Product of Braids

We will now equip the space of n-strand braids with a group structure. For this, we start with the natural operation consisting of concatenating end-to-end two geometric braids, which makes sense since the strands of the braids each have two well-differentiated extremities.

By definition, a braid in $\mathcal{GB}_n$ is included in the band $[0, 1]{\times}\mathbb{R}^2$. To concatenate end-to-end two such braids and obtain a figure traced in $[0, 1]{\times}\mathbb{R}^2$,

[1] Note that their solutions are accessible at the address https://dehornoy.users.lmno.cnrs.fr/Books/Tresses/Exos.pdf.

it is necessary to translate the second braid and to adjust the scale. This is easy.

Definition 2.1.1 (Product of geometric braids) If β_1 and β_2 are two n-strand geometric braids, the *product* $\beta_1 \cdot \beta_2$ of β_1 and β_2 is the image by the map $(x, y, z) \mapsto (x/2, y, z)$ of the union of β_1 and β_2, where the latter is first translated by $(1, 0, 0)$.

Seen as a family of parametrized arcs, the product braid β of β_1 and β_2 is defined by the formulas

$$\beta^{[i]}(t) := \begin{cases} \beta_1^{[i]}(2t) & \text{for } 0 \leqslant t \leqslant 1/2, \\ \beta_2^{[j]}(2t-1) & \text{for } 1/2 \leqslant t \leqslant 1, \end{cases} \tag{2.1}$$

where j is the final position of the ith strand of β_1, hence $\mathsf{perm}(\beta_1)^{-1}(i)$ in the formalism of Definition 1.3.3. When the strands of β_1 and β_2 are joined, the strand of β_2 connected to the ith strand of β_1 is the one facing its final position.

In the projection to the plane $\mathbb{R}^2 \times \{0\}$, the product of geometric braids corresponds to the same operation of concatenating end-to-end the strands. If D_1, D_2 are two n-strand braid diagrams, denote $D_1 \cdot D_2$ the diagram obtained by juxtaposing D_2 with D_1 and re-scaling the figure according to the schema

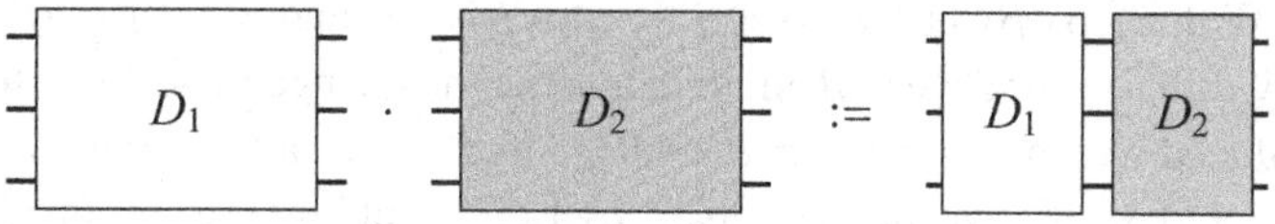

In practice, the scale factor is often ignored, and the product corresponds to a simple end-to-end concatenation.

Example 2.1.2 (Product of braid diagrams) Typically, we could write equalities such as the following.

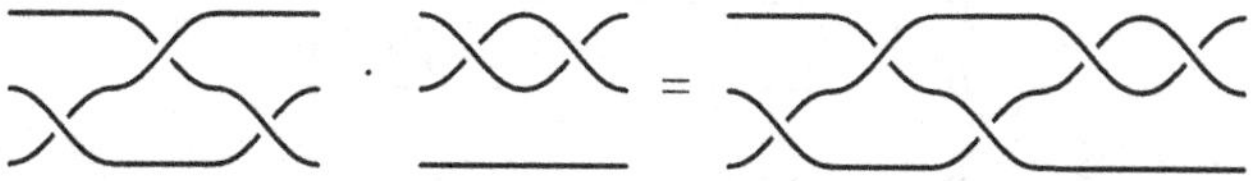

In terms of codes, the code of the product diagram of two diagrams with codes w_1, w_2 is, by construction, the word obtained by writing w_2 after w_1, denoted $w_1 \cdot w_2$ or simply $w_1 w_2$; in scholarly terms this is known as the *concatenation* of w_1 and w_2.

Example 2.1.3 (Concatenation) The codes of the diagrams below are $\sigma_1 \overline{\sigma}_2 \sigma_1$ and $\sigma_2 \sigma_2$, and we obtain as code for the product diagram the concatenated word $\sigma_1 \overline{\sigma}_2 \sigma_1 \cdot \sigma_2 \sigma_2 = \sigma_1 \overline{\sigma}_2 \sigma_1 \sigma_2 \sigma_2$.

We would like to obtain a product on the quotient space B_n. For this, we need a result linking the product with isotopy.

Lemma 2.1.4 *The product of geometric braids is compatible with the isotopy relation of geometric braids.*

Proof If for $i = 1, 2$ ϕ_i is a path connecting β_i with β'_i in $\mathcal{GB}_n$, then ϕ defined by $\phi(s) := \phi_1(s) \cdot \phi_2(s)$ connects $\beta_1 \cdot \beta_2$ with $\beta'_1 \cdot \beta'_2$ in $\mathcal{GB}_n$. □

We of course have an analogous compatibility with the derived relations $\simeq$ on the braid diagrams and $\equiv$ on the braid words.

Consequently, the product of geometric braids induces a well-defined operation on the isotopy classes.

Definition 2.1.5 (Product of braids) The *product* on B_n is the operation induced by the product of geometric braids.

Almost as simple as the result of Lemma 2.1.4, the first properties of the braid product can be read from the diagrams.

Lemma 2.1.6 *The braid product is associative, and admits as neutral element the class of trivial braids.*

Proof With β_1,β_2,β_3 in $\mathcal{GB}_n$, set $\beta := (\beta_1 \cdot \beta_2) \cdot \beta_3$ and $\beta' := \beta_1 \cdot (\beta_2 \cdot \beta_3)$. The only difference between β and β' concerns the change of scale factors: β_1 is reproduced with a factor $1/4$ in β, and $1/2$ in β', whereas β_3 has a factor $1/2$ in β and $1/4$ in β'. We define a path $(\beta_s)_{s\in[0,1]}$ linking β to β' in $\mathcal{GB}_n$ by setting $\beta_s^{[i]}(t) := (1-s)\beta^{[i]}(t) + s\beta'^{[i]}(t)$ for each $i = 1, ..., n$. Along this path, the scale factor increases linearly from $1/4$ to $1/2$ on the portion corresponding to β_1, remains constant on the portion corresponding to β_2, and decreases from $1/2$ to $1/4$ on the portion corresponding to β_3, as suggested in the figure below:

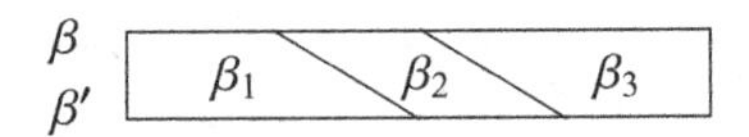

The argument is similar for the isotopy of β, $\beta \cdot \varepsilon_n$, and $\varepsilon_n \cdot \beta$, where ε_n is the trivial n-strand geometric braid. The figures are now:

β / $\beta \cdot \varepsilon_n$ [β | ε_n] and β / $\varepsilon_n \cdot \beta$ [ε_n | β]

□

In what follows, we denote by 1 the trivial braid, that is, the class of the geometric braid ε_n of Lemma 2.1.6 – the index n is omitted (cf. Remark 1.2.13).

Recall that a *monoid* is an algebraic structure composed of a set equipped with an associative operation admitting a neutral element.

Proposition 2.1.7 (Monoid B_n) *For every n, the structure $(B_n, \cdot, 1)$ is a monoid.*

As usual, we will speak of 'the monoid B_n' without specifying the product or the neutral element (necessarily unique).

A subset of a monoid M is said to be *generating* – or X *generates* M – if every element of M, except perhaps the neutral element 1, is a finite product of elements[2] of X. In the case of B_n, it is easy to identify a generating subset.

Lemma 2.1.8 *For every n, the monoid B_n is generated by the family of $2n-2$ braids σ_i and $\overline{\sigma}_i$ with $1 \leqslant i < n$.*

Proof Let σ_i^{+1} designate the letter σ_i and σ_i^{-1} the letter $\overline{\sigma}_i$, and consider an arbitrary braid word $w := \sigma_{i_1}^{e_1} \cdots \sigma_{i_\ell}^{e_\ell}$ with $i_1, ..., i_\ell$ in $\{1, ..., n\}$ and $e_1, ..., e_\ell = \pm 1$. Let g be the product braid $\sigma_{i_1}^{e_1} \cdot \sigma_{i_2}^{e_2} \cdot \cdots \cdot \sigma_{i_\ell}^{e_\ell}$. By construction, the code of the diagram associated with this product is the word w, and hence $g = [w]$. This applies to any braid in a class with at least one non-trivial diagram, hence to any braid other than 1: every non-trivial braid can be expressed as a product of letters of one of these codes (or the braids represented by them). □

Exercise 2.1.9 (Submonoids)
(i) Show that a subset M' of a monoid M is a submonoid of M if and only if M' contains 1 and is closed under the product (the product of two elements of M' belongs to M').
(ii) Deduce that a subset X of M generates a monoid M if and only if every element of $M \setminus \{1\}$ can be written as a finite product of elements of X.

2.1.2 The Group Structure of B_n

As indicated in the title of this chapter, the algebraic structure of the spaces B_n is not only that of a monoid.

Proposition 2.1.10 (Group structure) *For every n, the monoid B_n is a group. As a group, B_n is generated by the family of braids σ_i with $1 \leqslant i < n$.*

Proof Since B_n is a monoid generated by the elements σ_i and $\overline{\sigma}_i$, to show that B_n is a group, it suffices to show that each of the generators σ_i and $\overline{\sigma}_i$ admits an

[2] The definition given above is suitable for here, but it is not the 'true' general definition: a subset X of any structure M (monoid or otherwise) is said to be generating if no substructure of M other than M contains X. Since a subset of a monoid M is a submonoid if and only if it contains 1 and is closed under the product (Exercise 2.1.9), we are led to the form given above.

inverse. However, we observed in Section 1.3.2 the equivalence $\sigma_1\overline{\sigma}_1 \equiv \varepsilon$. An analogous argument gives $\overline{\sigma}_i\sigma_i \equiv \varepsilon$ for any i; in conclusion, in the monoid B_n, the element $\overline{\sigma}_i$ is the inverse of σ_i, and vice versa. Hence, B_n is a group.

Since B_n is generated by the elements σ_i and $\overline{\sigma}_i$ as a monoid, so it is a fortiori as a group. However $\overline{\sigma}_i$, the inverse of σ_i, necessarily belongs to any subgroup containing σ_i, and hence is a redundant generator. Thus, the group B_n is generated[3] by $\sigma_1, \ldots, \sigma_{n-1}$. □

Note that, since for every i, $\overline{\sigma}_i$ is the inverse of σ_i in B_n, from now on we will feel free to use σ_i^{-1} instead of $\overline{\sigma}_i$, at least when we do not wish to insist on $\overline{\sigma}_i$ as a letter.

We deduce that, if a braid g is represented by the word w, then the braid g^{-1} is represented by the word $\overline{w}$ obtained by exchanging everywhere σ_i and $\overline{\sigma}_i$ and then reversing the order of the letters. Thus, if β is an n-strand geometric braid, a representative of the braid $[\beta]^{-1}$ is the geometric braid $\overline{\beta}$ image of β by the orthogonal symmetry with respect to the plane $\{1/2\}\times\mathbb{R}^2$: if β is regular with code w, then, by construction, $\overline{\beta}$ is regular with code $\overline{w}$, and hence represents $[\beta]^{-1}$.

Example 2.1.11 (Inverse) Typically, the inverse of the braid $\sigma_1\overline{\sigma}_2$ is the braid $\sigma_2\overline{\sigma}_1$, corresponding to the diagram below:

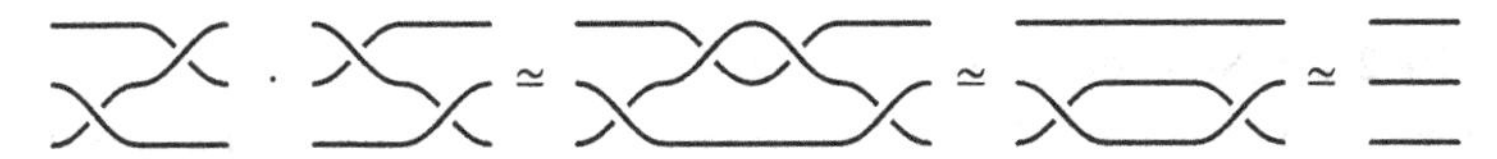

The product of a geometric braid with its mirror image is isotopic to the trivial braid, the crossings 'unwind themselves' one after the other starting from the middle.

Describing the group B_n for any n remains out of reach; however, just as triumphantly as in Proposition 1.1.16 where the case $n = 1$ was resolved, we can now complete the case $n = 2$.

Proposition 2.1.12 (Two-strand braids) *The group B_2 is isomorphic to the group $(\mathbb{Z}, +)$.*

Proof By Proposition 2.1.10, the group B_2 is a monogenic group generated by σ_1. The mapping $1 \mapsto \sigma_1$ thus induces a surjective homomorphism ϕ of the additive group $\mathbb{Z}$ into B_2, and the question is to know whether there exists p with $\sigma_1^p = 1$, that is, whether B_2 has a torsion element. However, we established in Examples 1.3.7 and 1.3.10 the value $\lambda_{1,2}(\sigma_1) = 1/2$, and an analogous calculation gives the value $\lambda_{1,2}(\sigma_1^p) = p/2$ for any $p \geqslant 1$. This implies $\sigma_1^p \neq 1$, and the homomorphism ϕ is thus an isomorphism. □

[3] and also by $\overline{\sigma}_1, \ldots, \overline{\sigma}_{n-1}$, and by any family $\sigma_1^{e_1}, \ldots, \sigma_{n-1}^{e_{n-1}}$ with $e_i = \pm 1$.

As we might suspect, the trivialities stop here, and the groups B_n become more complicated starting with $n=3$ ...

For the moment, we limit ourselves to observing how the group structure on B_n allows us to sharpen several results established in Chapter 1. For example, we have seen in Definition 1.3.3 that every n-strand braid g is associated with a permutation $\mathsf{perm}(g)$ of $\{1, ..., n\}$.

Proposition 2.1.13 (Permutation) *For every n, the mapping* perm *is a surjective homomorphism of the group B_n onto the symmetric group $\mathfrak{S}_n$.*

Proof Let β_1, β_2 be n-strand geometric braids, and f_1, f_2 the associated permutations. By definition, for every i between 1 and n, the strand finishing at position i in β_2 starts at position $f_2(i)$, whereas the strand finishing at position $f_2(i)$ in β_1 starts at position $f_1(f_2(i))$. Since a representative of the braid $[\beta_1]\cdot[\beta_2]$ is the geometric braid $\beta_1 \cdot \beta_2$, we deduce the equality $\mathsf{perm}(\beta) = f_1 \circ f_2$. Thus we have, in the group B_n, the equality

$$\mathsf{perm}([\beta_1] \cdot [\beta_2]) = \mathsf{perm}([\beta_1]) \circ \mathsf{perm}([\beta_2]). \tag{2.2}$$

Hence perm is a homomorphism of $(B_n, \cdot)$ to $(\mathfrak{S}_n, \circ)$.

This homomorphism is surjective. Indeed, for every i between 1 and $n-1$, the permutation of the braid σ_i is the transposition $(i, i+1)$. The image of the homomorphism perm thus includes the subgroup of $\mathfrak{S}_n$ generated by the $n-1$ transpositions $(i, i+1)$, which is the whole of $\mathfrak{S}_n$, as the transpositions are known to generate the group $\mathfrak{S}_n$. □

Note that, given the definition of the composition of mappings, only the definition of $\mathsf{perm}(\beta)(i)$ as the initial position of the strand finishing at position i guarantees Equation (2.2): using the final position of the strand starting at position i would give an *anti*homomorphism.[4]

This naturally leads to introducing a subgroup of B_n, namely the kernel of the homomorphism perm, that is, the family of braids whose permutation is the identity.

Definition 2.1.14 (Pure braids) A braid is said to be *pure* if its permutation is the identity. The distinguished subgroup of B_n formed by the pure n-strand braids is denoted PB_n.

By Proposition 2.1.13, we have the existence for every n of an exact sequence of groups

$$1 \to PB_n \to B_n \to \mathfrak{S}_n \to 1.$$

Since $\mathfrak{S}_n$ has $n!$ elements, the index of PB_n in B_n is $n!$.

[4] i.e. satisfying $\mathsf{perm}(g_1 \cdot g_2) = \mathsf{perm}(g_2) \circ \mathsf{perm}(g_1)$.

Other isotopy invariants were defined in Example 1.3.7 – the linking numbers $\lambda_{i,j}(\beta)$. In contrast with the permutation of a braid, these numbers do not induce a homomorphism for the algebraic structure of B_n: there exists a formula for $\lambda_{i,j}(\beta_1 \cdot \beta_2)$, but it involves only $\lambda_{i,j}(\beta_1)$ and $\lambda_{i,j}(\beta_2)$.

Lemma 2.1.15 *Let $\beta_1, \beta_2 \in \mathcal{GB}_n$ and $1 \leqslant i < j \leqslant n$. Then*

$$\lambda_{i,j}(\beta_1 \cdot \beta_2) = \lambda_{i,j}(\beta_1) + \lambda_{f(i),f(j)}(\beta_2), \tag{2.3}$$

with $f := \mathsf{perm}(\beta_1)^{-1}$.

Proof Set $\beta := \beta_1 \cdot \beta_2$. Let $\theta_{i,j}(t)$ be the measure of the angle between $\overrightarrow{\beta^{[i]}(t)\beta^{[j]}(t)}$ and $(1,0)$ starting from 0, and let $\theta_{i,j,1}(t)$ and $\theta_{i,j,2}(t)$ be the corresponding measures for β_1 and β_2. By definition,

$$\lambda_{i,j}(\beta) = \frac{1}{2\pi} \int_0^1 \theta'_{i,j}(t)dt = a + b,$$

$$\text{with } aa : = \frac{1}{2\pi} \int_0^{1/2} \theta'_{i,j}(t)dt, \text{ and } b := \frac{1}{2\pi} \int_{1/2}^1 \theta'_{i,j}(t)dt.$$

However, by the definition of the product of geometric braids, we have $\theta_{i,j}(t) = \theta_{i,j,1}(2t)$ for $0 \leqslant t \leqslant 1/2$, and, since $\theta_{i,j}(1/2)$ is zero, $\theta_{i,j}(t) = \theta_{i',j',2}(2t-1)$ for $1/2 \leqslant t \leqslant 1$, where i' and j' are the final positions in β_1 of the strands starting at positions i and j, that is, by definition, $f(i)$ and $f(j)$. By a change of variables, we conclude $a = \lambda_{i,j}(\beta_1)$, and $g = \lambda_{f(i),f(j)}(\beta_2)$, hence Formula (2.3). □

Formula (2.3) shows that, for fixed i, j, the mapping $\lambda_{i,j}$ induces a homomorphism of PB_n to $(\mathbb{Z}/2, +)$. If β_1 is not a pure braid, the linking numbers appearing in Formula (2.3) are not the same, and hence $\lambda_{i,j}$ does not induce a homomorphism of B_n to $(\mathbb{Z}/2, +)$. Nevertheless, we obtain a homomorphism by summing all the linking numbers.

Lemma 2.1.16 (Sum of exponents) *For β in $\mathcal{GB}_n$, denote $\mathsf{exp}(\beta)$ the double of the sum for $1 \leqslant i < j \leqslant n$ of the numbers $\lambda_{i,j}(\beta)$. Then for every n exp induces a homomorphism from B_n to $(\mathbb{Z}, +)$, with value 1 on each braid σ_k.*

Proof Formula (2.3) implies

$$\mathsf{exp}(\beta_1 \cdot \beta_2) = \mathsf{exp}(\beta_1) + 2 \sum_{1 \leqslant i < j \leqslant n} \lambda_{f(i),f(j)}(\beta_2),$$

with $f := \mathsf{perm}(\beta_1)^{-1}$. As f is a permutation of $\{1, ..., n\}$, each pair of distinct integers appears once and only once in the above sum. Since $\lambda_{i,j}(\beta)$ and $\lambda_{j,i}(\beta)$ are equal (Exercise 1.3.8), this sum is equal to $2 \sum_{1 \leqslant i < j \leqslant n} \lambda_{i,j}(\beta_2)$, that is, to $\mathsf{exp}(\beta_2)$. Hence, exp induces a homomorphism. Moreover, from Example 1.3.10, $2\lambda_{i,j}(\sigma_k)$ is $+1$ for $\{i, j\}$ equal to $\{k, k+1\}$, and 0 otherwise, hence $\mathsf{exp}(\sigma_k) = 1$. □

By definition, the integer $\mathsf{exp}(g)$ is the algebraic sum of the exponents of the generators σ_i of an arbitrary expression of g.

Example 2.1.17 (Sum of exponents) We have observed that the length of braid words does not induce an invariant, since, for example, the words $\sigma_i\sigma_i^{-1}$ and ε are equivalent; the sum of exponents is an algebraic version of the length, where the contributions of the generators σ_i^{-1} are -1. Thus, we indeed find $\mathsf{exp}(\sigma_i\sigma_i^{-1}) = \mathsf{exp}(\varepsilon) = 0$.

We saw in Proposition 1.1.20 that, for every n, the addition of an $(n+1)$th trivial strand induces a topological embedding of the space B_n into the space B_{n+1}, in particular excusing the (abusive) convention of using the same notation σ_i for 'the braid σ_i of B_n' for every $n > i$. This embedding extends to the algebraic structures.

Proposition 2.1.18 (Embedding) *For every n, the group B_n is isomorphic to the subgroup of B_{n+1} generated by $\sigma_1, ..., \sigma_{n-1}$.*

Proof With the notation of Proposition 1.1.20, and for β in $\mathcal{GB}_n^{sp}$, denote $\widehat{\beta}$ the braid obtained by adding an $(n+1)$th trivial strand. The definition of the product implies $\widehat{\beta_1 \cdot \beta_2} = \widehat{\beta_1} \cdot \widehat{\beta_2}$ for every β_1, β_2. The embedding ι_n of B_n into B_{n+1} induced by the mapping $\beta \mapsto \widehat{\beta}$ is thus a homomorphism for multiplication, which we know is injective. Consequently, the group B_n is isomorphic to the subgroup of B_{n+1}, the image of ι_n. However, we know that B_n is generated by $\sigma_1, ..., \sigma_{n-1}$. The image of ι_n in B_{n+1} is thus generated by $\iota_n(\sigma_1), ..., \iota_n(\sigma_{n-1})$: these braids are, by virtue of our convention of notation, the elements of B_{n+1} denoted $\sigma_1, ..., \sigma_{n-1}$. □

We deduce that the union B_∞ of the groups B_n is itself equipped with a group structure, where the product of two elements g, h of B_∞ is the common value of the product gh in any of the groups B_n containing them.[5]

Exercise 2.1.19 (Shift)
(i) Show that the embedding shift_n of B_n into B_{n+1} defined in Exercise 1.1.22 is a homomorphism.
(ii) Deduce that the embedding shift is a non-surjective endomorphism of B_∞ into itself, characterized by the fact that it sends σ_i to σ_{i+1} for every i.

[5] In more formal terms, the sequence of groups B_n generates with the embeddings ι_n an inductive system, and B_∞ is the limit of this inductive system.

2.1.3 The Artin Presentation of the Group B_n

We saw in Proposition 2.1.10 that, for every n, the family $\{\sigma_1, ..., \sigma_{n-1}\}$ generates the group B_n. The generators σ_i are called the *Artin generators* of B_n. Moreover, we know from Proposition 1.1.16 that the group B_1 is the trivial group, and from Proposition 2.1.12 that B_2 is a free group isomorphic to $(\mathbb{Z}, +)$. However, for $n \geqslant 3$, the group B_n is not free: we will see below that relations other than the relations $\sigma_i\overline{\sigma}_i = 1$ and $\overline{\sigma}_i\sigma_i = 1$ connect the generators σ_i.

Lemma 2.1.20 *The following relations hold in B_n:*

$$\sigma_i\sigma_j = \sigma_j\sigma_i \qquad \text{for } |i-j| \geqslant 2, \tag{2.4}$$

$$\sigma_i\sigma_j\sigma_i = \sigma_j\sigma_i\sigma_j \quad \text{for } |i-j| = 1. \tag{2.5}$$

Proof See Figure 2.1. In the first case, the crossings concern pairs of disjoint strands, and we easily see how to go from the left side to the right side by pushing the left crossing to the right and the right crossing to the left. In the second case, the strand starting at position 3 passes over the two other strands, which themselves also cross: in the diagram on the left, strand 3 traverses the strands 1 and 2 after they cross, while on the right, it traverses before they cross. However, as strand 3 remains above the other two in both cases, we can deform one figure into the other, as already sketched out in Example 1.2.12. □

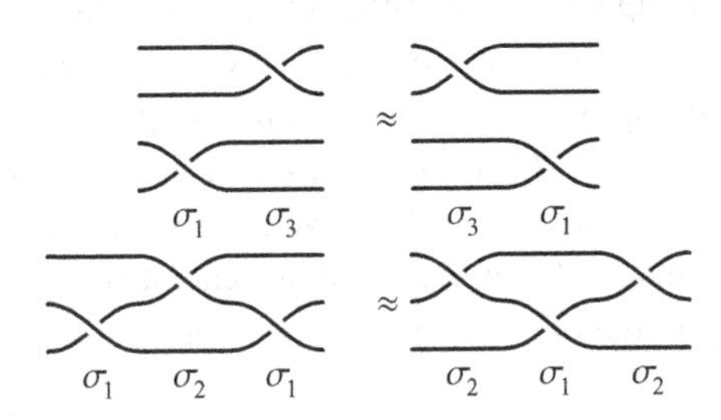

Figure 2.1 Two types of relations between Artin generators, here in the particular cases $i = 1, j = 3$ on the top, $i = 1, j = 2$ on the bottom.

The relations of Lemma 2.1.20 are satisfied in B_n, hence it is a quotient of the group admitting the presentation

$$\left\langle \sigma_1, ..., \sigma_{n-1} \;\middle|\; \begin{array}{ll} \sigma_i\sigma_j = \sigma_j\sigma_i & \text{for } \; |i-j| \geqslant 2, \\ \sigma_i\sigma_j\sigma_i = \sigma_j\sigma_i\sigma_j & \text{for } \; |i-j| = 1 \end{array} \right\rangle \tag{2.6}$$

(see 2.3.16 in Appendix B). We will show here that the quotient is in fact an equality, that is, the relations of Equation (2.6) generate all the possible relations between the generators σ_i.

Proposition 2.1.21 (Artin's theorem, 1925) *For every n, the group B_n admits the presentation* (2.6).

Proof Recall that $\mathcal{BW}_n$ is the set of signed braid words on $\sigma_1, ..., \sigma_{n-1}$. Denote $\equiv'$ the congruence on $\mathcal{BW}_n$ generated by the relations of (2.6) and the free group relations. By Equation (1.12), the group B_n is the quotient $\mathcal{BW}_n/\equiv$, where $\equiv$ is the equivalence of Definition 1.2.11; showing that (2.6) is a presentation of B_n consists of showing that the congruences $\equiv$ and $\equiv'$ coincide. By Lemma 2.1.20, $w \equiv' w'$ implies $w \equiv w'$, and it remains to prove the converse implication.

Suppose then that w and w' belong to $\mathcal{BW}_n$ and satisfy $w \equiv w'$. We want to show $w \equiv' w'$. By definition, there exist isotopic regular geometric braids β, β' with respective codes w and w'. By Lemma 1.2.3, we can suppose β and β' piecewise linear, with a path $(\beta_s)_{0 \leqslant s \leqslant 1}$ linking β to β' in $\mathcal{GB}_n^{\mathsf{aff,semireg}}$, included in $\mathcal{GB}_n^{\mathsf{aff,reg}}$ except for a finite set of values $s_1 < \cdots < s_m$. Since the function $s \mapsto \mathsf{code}(\beta_s)$ has values in the discrete space $\mathcal{BW}_n$ and is continuous everywhere where it defined, s is constant on each of the intervals $[0, s_1[$, $]s_1, s_2[$, ..., $]s_m, 1]$. For $0 \leqslant k \leqslant m$, and defining $s_0 := 0$ and $s_{m+1} := 1$, denote w_k the value of $\mathsf{code}(\beta_s)$ for $s_k \leqslant s \leqslant s_{k+1}$. To establish $w \equiv' w'$, since $\equiv'$ is transitive, it suffices to establish $w_k \equiv' w_{k+1}$ for each k.

By hypothesis, the braid β_{s_k} is piecewise linear and semi-regular, and we can suppose that it is not regular (otherwise we would simply have $w_k = w_{k+1}$), and hence its orthogonal projection onto $\mathbb{R}^2 \times \{0\}$ possesses a singularity of one of three types: 'triple point', 'non-transverse double point', or 'two double points with the same abscissa'.

We first consider the case of a triple point. Without loss of generality, we can suppose that it appears on the projection when two strands, say in positions i and $i+1$, cross each other, and that a third strand, initially in position $i+2$ or $i-1$, crosses the two strands to the left of their crossing for $s < s_k$, and to the right for $s > s_k$. It then suffices to compare the codes corresponding to two situations. In the case where the strand with initial position $i+1$ passes over the strand with initial position i, there are six possibilities, according to their order of intersections and their orientation:

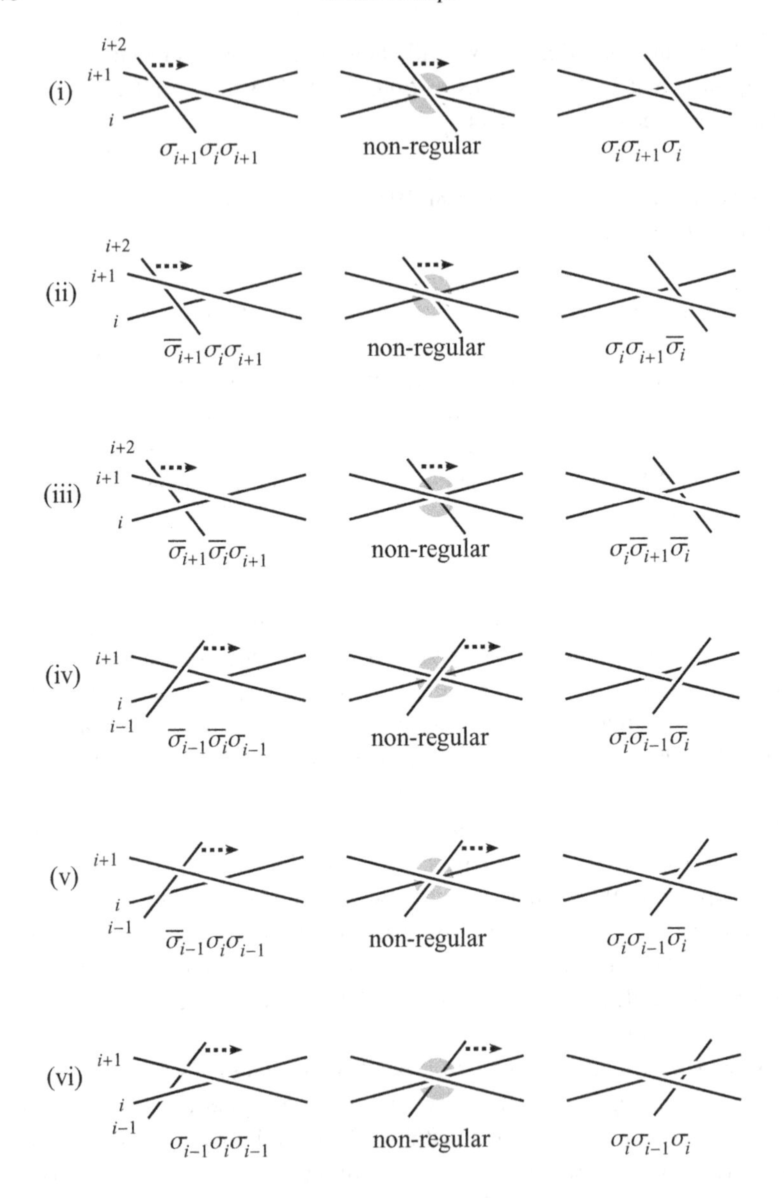

Each time, we observe that the corresponding (fragments of) codes are $\equiv'$-equivalent words: this is evident for (i) and (vi) and, for the case (ii), for example, we find

$$\overline{\sigma}_{i+1}\sigma_i\sigma_{i+1} \equiv' \overline{\sigma}_{i+1}\sigma_i\sigma_{i+1}\sigma_i\overline{\sigma}_i \equiv' \overline{\sigma}_{i+1}\sigma_{i+1}\sigma_i\sigma_{i+1}\overline{\sigma}_i \equiv' \sigma_i\sigma_{i+1}\overline{\sigma}_i.$$

Note that the 'third' strand can be either over the other two, or between them, or under them, but it is impossible for it to be at the same time above the

$(i+1)$th and below the ith, which would generate a problematic relation $\sigma_{i+1}\overline{\sigma}_i\sigma_{i+1} \equiv? \sigma_i\overline{\sigma}_{i+1}\sigma_i$. The other cases are similar, as are the six symmetric cases where the strand with initial position $i+1$ passes under the strand with initial position i.

The case of a non-transverse double point is treated in the same manner. Here it involves studying the change of code induced by the passage through s_k of a non-transverse double point: for $s < s_k$, the $(i+1)$th or the $(i-1)$th strand crosses twice the ith strand; for $s > s_k$, it no longer cuts them (or vice versa). There are this time four possibilities of orientation:

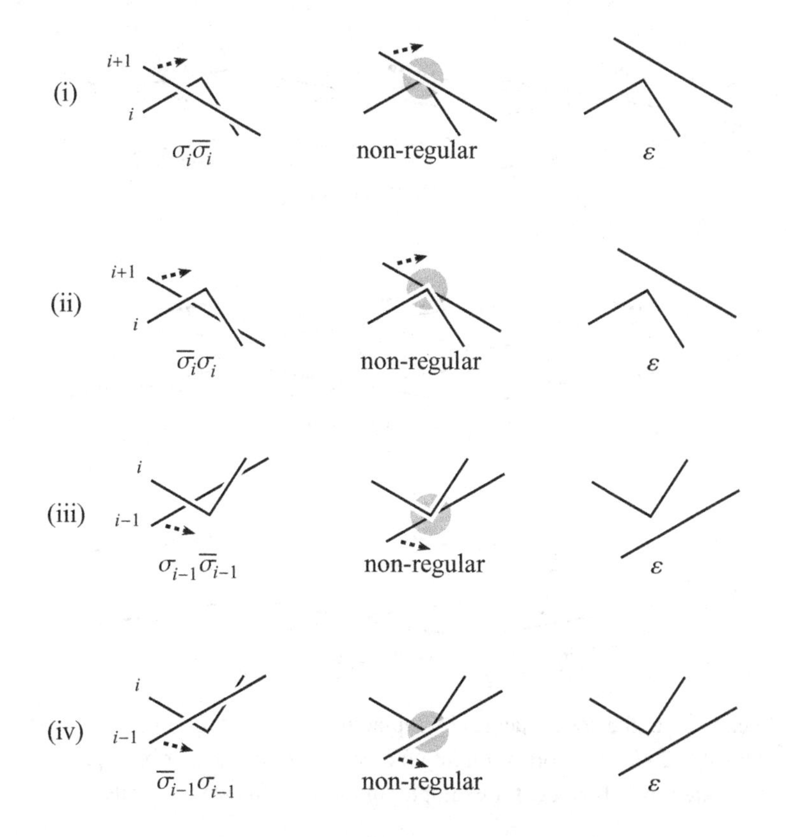

In each case, the code change eliminates a factor of the type $\sigma_i\overline{\sigma}_i$ or $\overline{\sigma}_i\sigma_i$, (or by inserting such a factor in the symmetric cases), and hence, once again, leads to $\equiv'$-equivalent words. We refer the reader to Sections 2.3.1 and 2.3.2 for the general context of presented groups and, in particular, the relations of the free group $\mathsf{Sym}(S)$.

Finally, there remains the case of two double points with the same abscissa. This now involves the change of code induced by the passage over s_k of two double points with the same abscissa: for $s < s_k$, the crossing of the strands at position j and $j{+}1$ precedes that of the strands at position i and $i{+}1$; for $s > s_k$, it follows them. There are again four possibilities up to symmetry:

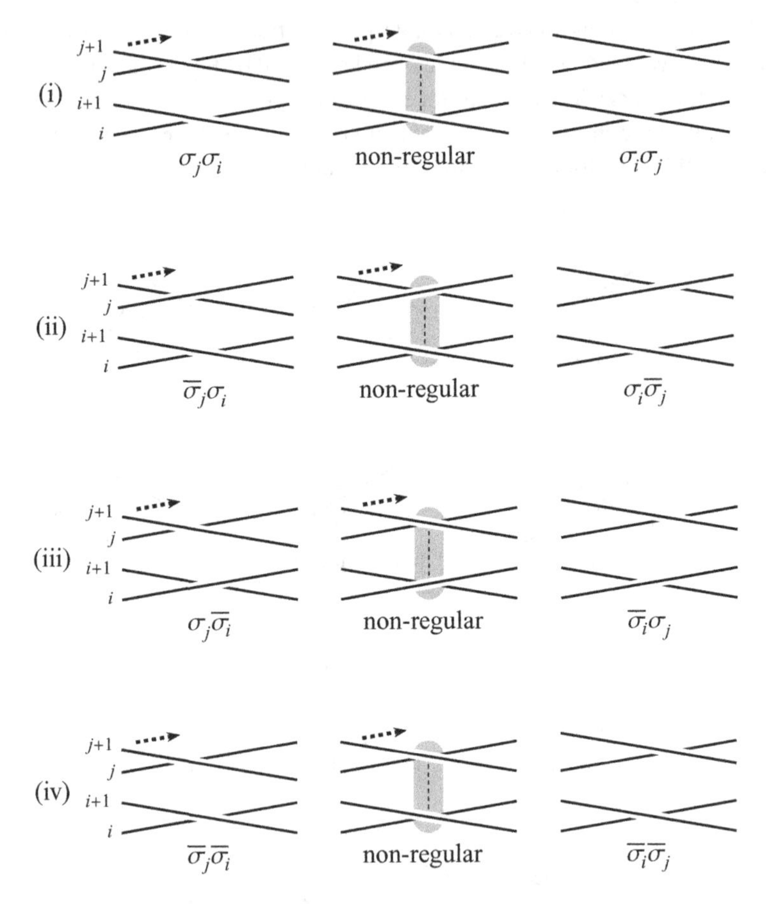

In each case, the code changes by replacing a factor $\sigma_j\sigma_i$, $\overline{\sigma}_j\sigma_i$, $\sigma_j\overline{\sigma}_i$, or $\overline{\sigma}_j\overline{\sigma}_i$ with $|i-j| \geqslant 2$ with the corresponding factor $\sigma_i\sigma_j$, $\sigma_i\overline{\sigma}_j$, $\overline{\sigma}_i\sigma_j$, or $\overline{\sigma}_i\overline{\sigma}_j$, producing $\equiv'$-equivalent words since, for example, in the case (ii) we can write

$$\overline{\sigma}_j\sigma_i \equiv' \overline{\sigma}_j\sigma_i\sigma_j\overline{\sigma}_j \equiv' \overline{\sigma}_j\sigma_j\sigma_i\overline{\sigma}_j \equiv' \sigma_i\overline{\sigma}_j.$$

Consequently, the congruences $\equiv$ and $\equiv'$ coincide, hence the relations of (2.6) furnish a presentation of the group B_n. □

Without any further effort, we can deduce that the Artin presentation with an infinity of generators constitutes a presentation of B_∞.

Corollary 2.1.22 (Group B_∞) *The group B_∞ admits the presentation*

$$\left\langle \sigma_1, \sigma_2, \ldots \;\middle|\; \begin{array}{lll} \sigma_i\sigma_j = \sigma_j\sigma_i & \text{for} & |i-j| \geqslant 2, \\ \sigma_i\sigma_j\sigma_i = \sigma_j\sigma_i\sigma_j & \text{for} & |i-j| = 1. \end{array} \right\rangle \tag{2.7}$$

Proof All the relations of (2.7) are satisfied in B_∞. Conversely, if two braid words w, w' represent the same element of B_∞, there exists n such that w and w' belong to $\mathcal{BW}_n$. Hence, by definition of B_∞, the words w and w' represent the same element of B_n, and thus, by Artin's theorem, are equivalent with respect to the congruence generated by the relations of (2.7). These then furnish a presentation. □

2.1.4 A Few Consequences

A direct consequence of the existence of the group structure on the space B_n is that the isotopy problem of n-strand braids, initially purely topological, is now reduced to the word problem for the presented group B_n: two geometric braids, supposed regular, are isotopic if and only if the braid words coding their diagrams are equivalent with respect to the presentation (2.6).

For the moment, this is not sufficient to resolve the question; however, what is originally a problem of two input variables is already reduced to a problem with a single variable. Indeed, the relation $w \equiv w'$ is equivalent to $w^{-1}w' \equiv \varepsilon$, and the question of recognizing whether two braids are isotopic is reduced to that of recognizing whether or not a single braid is trivial.

Example 2.1.23 (Triviality) We return to the two braids β_1, β_2, whose isotopy was left open in Section 1.3.11, that is, the braids coded by $\sigma_1^2\sigma_2^2$ and $\sigma_2^2\sigma_1^2$. Then β_1 and β_2 are isotopic if and only if the braid $\sigma_2^{-2}\sigma_1^{-2}\sigma_2^2\sigma_1^2$ is trivial, or in other words, if and only if the word $\overline{\sigma}_2^2\overline{\sigma}_1^2\sigma_2^2\sigma_1^2$ represents 1 in the group B_n.

Moreover, the knowledge of not only the existence of a group structure, but indeed its explicit presentation, allows us to demonstrate properties of elements of the group by using the fact that two words represent the same element if and only if they are related by a derivation involving only the relations of the presentation and those of the free group. We present here examples.

Proposition 2.1.24 (Flip) *For every n, the mapping $\phi_n \colon \sigma_i \mapsto \sigma_{n-i}$ induces an involutive automorphism of the group B_n.*

Proof Denote ϕ_n^* the involutive automorphism of the free monoid $\mathcal{BW}_n$ into itself associating with each braid word w the word obtained by substituting everywhere σ_{n-i} for σ_i (cf. Proposition 2.2.8). Observe that for any relation[6]

[6] Recall that, by Convention 2.2.25, $u = v$ is nothing but a notation for the pair of words (u, v).

$u = v$ that is either a braid relation or a free group relation, the pair of words $(\phi_n^*(u), \phi_n^*(v))$ is again a relation of the same type. It thus results from Propositions 2.2.8 and 2.3.5 that the mapping $w \mapsto [\phi_n^*(w)]$ of $\mathcal{BW}_n$ into B_n induces a homomorphism of B_n into itself. By construction, this homomorphism is involutive, and hence bijective. □

We denote by ϕ_n the above automorphism of B_n, commonly called the '*flip*'.

Exercise 2.1.25 (Flip) What transformation of geometric braids corresponds to the automorphism ϕ_n?

Another example exploits the palindromic character of the braid relations, unchanged when read from right to left.

Proposition 2.1.26 (Reversings) *For a braid word w, let $\widetilde{w}$ be the word obtained by reversing the order of the letters of w. For every n, the mapping $w \mapsto \widetilde{w}$ induces an involutive antiautomorphism of B_n.*

Proof As above, observe that, for any relation $u = v$ that is either a braid relation or a free group relation, the pair of words $(\widetilde{u}, \widetilde{v})$ is again a relation of the same type. It follows from Propositions 2.2.8 and 2.3.5 (adapted for *anti*homomorphisms) that the mapping $w \mapsto [\widetilde{w}]$ of $\mathcal{BW}_n$ into B_n induces an antihomomorphism of B_n into itself. By construction, this antihomomorphism is involutive, and hence bijective. □

We will use $\widetilde{g}$ for the image of an element g of B_∞ by reversing. For example, $\widetilde{g} = \sigma_2\sigma_1^{-2}$ for $g = \sigma_1^{-2}\sigma_2$.

Exercise 2.1.27 (Reversing) What transformation of geometric braids corresponds to a reversing?

Exercise 2.1.28 (Coxeter presentation of $\mathfrak{S}_3$) Show that the quotient G of the group B_3 by the congruence generated by the supplementary relations $\sigma_1^2 = \sigma_2^2 = 1$ is (isomorphic to) the symmetric group $\mathfrak{S}_3$. [Hint: Use the relations to show that the cardinality of G is at most 6.]

2.2 Appendix A: Presented Monoids

A reader not very familiar with presentations of structures can find in the next sections the definitions and basic results; otherwise, these sections can be skipped. We begin with monoids, where all the ideas can be found, but where certain details are simpler.

2.2.1 Monoids of Words

A monoid is a set equipped with an associative binary operation and a neutral element. In the same way as for groups, we usually write 'the monoid M' for 'the monoid $(M, \cdot, 1)$', where by default the operation is denoted multiplicatively,[7] and the neutral element denoted 1.

Exercise 2.2.1 (Neutral element) Show that a monoid has only a single neutral element.

A particular family of monoids plays an important role in what follows. The starting point is the notion of a *word.*

Definition 2.2.2 (Word, S^*) Let S be a nonempty set.[8] A *word over S*, or *S-word*, of *length ℓ*, is a sequence of elements of S of length ℓ. Denote $|w|$ the length of a word w, and S^* the set of all the S-words, completed by the empty sequence ε, known as the *empty word*, whose length, by convention, is 0.

Example 2.2.3 (Word) Thus, $(\mathsf{a}, \mathsf{b}, \mathsf{a}, \mathsf{a})$ is an $\{\mathsf{a}, \mathsf{b}\}$-word, of length 4. Formally, we consider a finite sequence of length ℓ as a mapping of the interval $\{1, 2, ..., \ell\}$ into S, and denote $w(i)$ the ith letter of the word w. At least when the elements of S correspond to unique characters, it is customary to represent a word by its sequence of letters, so abaa instead of $(\mathsf{a}, \mathsf{b}, \mathsf{a}, \mathsf{a})$ above.

We now introduce a binary operation on words.

Definition 2.2.4 (Concatenation) For $u, v \in S^*$, the *concatenation* of u and v, denoted $u \cdot v$, or uv, is the word w of length $|u| + |v|$ defined by $w(i) := u(i)$ for $i \leqslant |u|$, and $w(i) := v(i - |u|)$ for $i > |u|$.

The concatenation of u and v is the word obtained by writing v after u. For example, $\mathsf{abaa} \cdot \mathsf{ba} = \mathsf{abaaba}$ (verify this!).

The words and their concatenations form a monoid.

Proposition 2.2.5 (Monoid) *For any nonempty set S, the structure $(S^*, \cdot, \varepsilon)$ is a monoid.*

Exercise 2.2.6 Prove the result. [Hint: To show that $(uv)w$ and $u(vw)$ coincide, it suffices to verify that the words have the same length and that, for every i, their ith letters are the same.]

[7] There exist a few exceptions, for example the monoid $(\mathbb{N}, +, 0)$ of non-negative integers under addition, whose neutral element is 0, or again the monoid $(\mathbb{Z}, +, 0)$ of integers (which is a group).

[8] In this context, S is often called the *alphabet*, and its elements are then known as *letters*.

The concatenation of words is in general not commutative: as soon as S contains two distinct letters s, t, the words st and ts are distinct: they have neither the same first letter, nor the same second.

Every word is the product of the words of length 1 corresponding to its successive letters.[9] Consequently, S generates the monoid S^*. It is in fact the smallest generating set.

Exercise 2.2.7 (Generating subsets of S^*) Show that a subset X of a monoid S^* is a generating set S^* if and only if X includes S. In particular, S generates S^*.

2.2.2 Universal Property of the Monoid S^*

Among the monoids generated by S, the monoid S^* has a special property – any such monoid is its homomorphic image. The key point is the following universal property.[10]

Proposition 2.2.8 *For any mapping ϕ of S into a monoid M, there exists a unique homomorphism $\widehat{\phi}$ of S^* into M extending ϕ, defined by $\widehat{\phi}(\varepsilon) := 1$ and, for $w = s_1 \cdots s_\ell$ with $s_1, ..., s_\ell$ in S, by $\widehat{\phi}(w) := \phi(s_1) \cdots \phi(s_\ell)$.*

Exercise 2.2.9 Prove the result.

Applying Proposition 2.2.8 to id_S, we obtain the stated result.

Proposition 2.2.10 (Homomorphic image) *Any monoid M generated by S is an image of S^* by a surjective homomorphism.*

Proof The homomorphism id_S^* obtained is surjective, since by construction its image contains the element 1 and all the elements expressible as products in M of elements of S, as a result of the hypothesis that S generates M (Exercise 2.1.9). □

As is the case for groups or other algebraic structure, we can introduce the notion of a *quotient monoid*. Passing to a quotient means regrouping the elements into equivalence classes and organizing these into a monoid. However, in order to obtain a well-defined product on the classes, a compatibility is necessary.

Definition 2.2.11 (Congruence) A *congruence* on a monoid M is an equivalence relation $\equiv$ on M compatible with the product: the conjunction of $a' \equiv a$ and $b' \equiv b$ implies $a'b' \equiv ab$.

[9] From this, the notation without parentheses introduced in Definition 2.2.2 comes down to identifying the word (s), of length 1, with the letter s: since the mapping $s \mapsto (s)$ is injective, this is not dangerous.

[10] Also expressed as qualifying S^* as the *free monoid with base S*.

Recall that for an equivalence relation $\equiv$ on E, the *quotient set* $E/{\equiv}$ is the family of equivalence classes.

Lemma 2.2.12 *If M is a monoid and $\equiv$ an equivalence relation on M, then the mapping sending an element to its class for $\equiv$ induces a quotient monoid structure on $M/{\equiv}$ if and only if $\equiv$ is a congruence on M.*

Exercise 2.2.13 Prove the result.

The link between homomorphic images and quotients is simple.

Lemma 2.2.14 *The existence of a surjective homomorphism of a monoid M onto a monoid M' is equivalent to that of a congruence $\sim$ on M such that M' is isomorphic to the quotient $M/{\sim}$.*

Exercise 2.2.15 Prove the result.

We then deduce from Proposition 2.2.10 another statement of the fact that S^* is the most general of the monoids generated by S.

Proposition 2.2.16 (Quotient) *Any monoid generated by a set S is a quotient of the monoid S^*.*

It is easy to specify the congruence in question. The homomorphism id_S^* of Proposition 2.2.10 corresponds to *evaluating*[11] in M the words of S^* using the operation of M. If $(M, *)$ is a monoid containing S, we denote eval_M the mapping of S^* into M defined by $\mathsf{eval}_M(\varepsilon) := 1$ and, for $w = s_1 \cdots s_\ell$ with $s_1, \ldots, s_\ell$ in S,

$$\mathsf{eval}_M(w) = s_1 * \cdots * s_\ell. \tag{2.8}$$

Proposition 2.2.17 (Quotient') *Any monoid generated by a set S is isomorphic to $S^*/{\sim}$, where $\sim$ is the congruence defined*[12] *by*

$$w \sim w' \quad \Leftrightarrow \quad \mathsf{eval}_M(w) = \mathsf{eval}_M(w'). \tag{2.9}$$

2.2.3 Presentation of a Monoid

The monoids of words and the congruences on these monoids are infinite, hence impossible to enumerate exhaustively. We thus seek more economical

[11] This can be done as soon as S is included in M, whether or not it generates M: simply, S generates M if and only if every element of M is the evaluation of a word of S^*.

[12] Note that Equation (2.9) provides no effective description of $\sim$: it is simply a formal definition.

means, in particular finite, to specify a congruence. This brings us to the notions of a congruence *generated* by relations, and then of *presentations*.

The starting point is the notion of *derivation*.

Definition 2.2.18 (Derivation) Let M be a monoid and R a subset of $M \times M$. For a, a' in M, an *R-derivation of a to a'* is defined as a finite sequence $(a_0, ..., a_m)$ in M with $a_0 = a$ and $a_m = a'$ such that, for $1 \leqslant i < m$, there exists (g, g') in R and x, y in M satisfying $a_{i-1} = xgy$ and $a_i = xg'y$, or $a_{i-1} = xg'y$ and $a_i = xgy$.

The definition is simpler than it looks. If, for example, M is a monoid of words S^*, then R is a family of pairs of words (u_p, v_p), and an R-derivation is a sequence of words where we pass from one word to the next by replacing a factor[13] u_p by the corresponding word v_p , or vice versa.

Example 2.2.19 (Derivation) Suppose $S = \{\mathtt{a}, \mathtt{b}\}$ and $R = \{(\mathtt{ab}, \mathtt{ba})\}$ (a single element, the pair (ab, ba)). Then the sequence

$$(\mathtt{ab\underline{ba}aa}, \mathtt{\underline{ab}abaa}, \mathtt{ba\underline{ab}aa}, \mathtt{babaaa})$$

is a typical R-derivation: each time, a factor (underlined above) ab is replaced by ba, or vice versa.

The link between derivations and congruences is simple.

Lemma 2.2.20 *If M is a monoid, then for any subset R of $M \times M$, there exists a smallest congruence including R, namely the relation 'there exists an R-derivation from a to a''.*

Exercise 2.2.21 Prove the result.

The smallest congruence including R is said to be *generated*[14] by R.

Exercise 2.2.22 (Generated congruence) Show that, in the case of Example 2.2.19, the congruence generated by the pair (ab, ba) is the relation 'having the same numbers of the letters a and b'.

Since, by Proposition 2.2.16, every monoid is a quotient of a monoid of words by a suitable congruence, we obtain a compact specification by presenting the elements generating the monoid, and the pairs of words generating the associated congruence.

[13] A word u is a *factor* of a word w if there exists w_1, w_2 satisfying $w = w_1uw_2$, i.e. if u appears as a fragment of the word w.

[14] Note that this definition conforms with the usage of the expression 'generated' for 'the smallest such that...'.

Definition 2.2.23 (Presentation, generators, relations) A pair (S, R) is said to be a *presentation* of a monoid M, denoted[15]

$$M = \langle S \mid R \rangle^{+}, \tag{2.10}$$

if S generates M and R is a subset of $S^* \times S^*$ generating the congruence $\sim$ such that M is isomorphic to $S^*/\sim$. The elements of S are called the *generators*, those of R the *relations*.

Example 2.2.24 (Presentation) For example, the congruence on S^* generated by the empty set is the smallest of all congruences, namely equality. Hence the monoid $\langle S \mid \emptyset \rangle^{+}$ is $S^*/=$, that is, S^* itself.

Convention 2.2.25 (Relations) To simplify, we suppress the set brackets in a (finite) presentation. Moreover, if (u, v) is a relation of a presentation (S, R), then the words u and v represent the same element of $\langle S \mid R \rangle^{+}$, hence have the same evaluation therein. For this, common usage in this context is to write $u = v$ for (u, v).[16] With this, $\langle \{\mathsf{a}, \mathsf{b}\} \mid \{(\mathsf{ab}, \mathsf{ba})\} \rangle^{+}$ simplifies to $\langle \mathsf{a}, \mathsf{b} \mid \mathsf{ab} = \mathsf{ba} \rangle^{+}$, much more readable.

The following exercises propose a few more examples.

Exercise 2.2.26 (Presented monoids)
(i) Let S be arbitrary, and R the set of relations $s = 1$ for s in S. Show that $\langle S \mid R \rangle^{+}$ is the trivial monoid of one element.
(ii) Let S be a nonempty set, and R the set of relations $s = t$ for s, t in S. Show that $\langle S \mid R \rangle^{+}$ is isomorphic to $(\mathbb{N}, +)$. [Hint: Show that the congruence on S^* generated by R is the relation 'have the same length'.]
(iii) Show that $\langle \mathsf{a}, \mathsf{b} \mid \mathsf{ab} = \mathsf{ba} \rangle^{+}$ is the direct product of two copies of the monoid $(\mathbb{N}, +)$. [Hint: Use Exercise 2.2.22 and show that the mapping $w \mapsto (\|w\|_{\mathsf{a}}, \|w\|_{\mathsf{b}})$, where $\|w\|_s$ denotes the number of s in w, induces a bijection of $\{\mathsf{a}, \mathsf{b}\}^*/\equiv$ onto $\mathbb{N}^2$, and then that this bijection is a homomorphism when $\mathbb{N}^2$ is equipped with addition coordinate by coordinate.]
(iv) Show that the monoid $\langle \mathsf{a} \mid \mathsf{a}^p = 1 \rangle^{+}$ is isomorphic to the cyclic group $\mathbb{Z}/p\mathbb{Z}$. [Hint: Show that the congruence $\equiv$ on $\{\mathsf{a}\}^*$ generated by $(\mathsf{a}^p, \varepsilon)$ is the relation $|w| = |w'| \pmod{p}$.] Describe in the same manner the monoid $\langle \mathsf{a} \mid \mathsf{a}^p = \mathsf{a}^q \rangle^{+}$ for fixed $p, q \geqslant 0$.

[15] While usual, the use of the equal sign in (2.2.23) is abusive, as the monoid $\langle S \mid R \rangle^{+}$ is only defined up to an isomorphism.

[16] A strange usage, since u and v are words and, as such, are not in general equal: the symbol = is thus not used in its usual sense. This is regrettable, but not terribly dangerous in that this alternative sense is only used in the context of relations of a presentation.

It is often useful to recognize the monoids that are quotients of a presented monoid. If M is a monoid containing S and if u, v are words of S^*, the equality $\mathsf{eval}_M(u) = \mathsf{eval}_M(v)$ is often expressed as '*the relation $u = v$ is satisfied in M*'.

Proposition 2.2.27 (Quotient) *A monoid M generated by a set S is a quotient of the monoid $\langle S \mid R\rangle^+$ if and only if all the relations of R are satisfied in M.*

Exercise 2.2.28 Prove the result.

Similarly, it is easy to characterize the homomorphisms of a presented monoid towards another monoid.

Lemma 2.2.29 *A homomorphism ϕ defined on a monoid S^* to a monoid M induces a homomorphism of $\langle S \mid R\rangle^+$ into M if and only if $\phi(u) = \phi(v)$ for each relation $u = v$ of R.*

Exercise 2.2.30 Prove the result.

2.2.4 The Word Problem of a Presented Monoid

If a monoid M is specified by a presentation (S, R), every element of M is the evaluation of one or several words of S^*. However, the knowledge of S and R does not a priori provide a practical method to determine for two words of S^* whether or not they represent the same element of M, that is, if they are equivalent with respect to the congruence on S^* generated by R.

Definition 2.2.31 (Word problem) If M is a monoid generated by S, the *solution to the word problem* for M with respect to S is an algorithm that, for any pair of S-words (u, v), can decide in finite time whether u and v represent the same element of M.

If (S, R) is a presentation, we speak of the 'word problem of $\langle S \mid R\rangle^+$' for the 'word problem of $\langle S \mid R\rangle^+$ with respect to S'.

As for the isotopy problem in Section 1.3.1, in order to make precise Definition 2.2.31, we must fix a model of computation. We hope that the reader will be satisfied with intuitive ideas, and agree for example to say that the pseudo-code below specifies a (trivial) solution for $\langle \mathsf{a}, \mathsf{b} \mid \mathsf{ab} = \mathsf{ba}\rangle^+$.

Algorithm 2.2.32 (Word problem for $\langle \mathsf{a}, \mathsf{b} \mid \mathsf{ab} = \mathsf{ba}\rangle^{+}$)

Input: two words w, w' of $\{\mathsf{a}, \mathsf{b}\}^*$

Output: **yes** if w and w' represent the same element of $\langle \mathsf{a}, \mathsf{b} \mid \mathsf{ab} = \mathsf{ba}\rangle^{+}$,
no otherwise

1: $p \leftarrow$ number of letters a in w
2: $p' \leftarrow$ number of letters a in w'
3: $q \leftarrow$ number of letters b in w
4: $q' \leftarrow$ number of letters b in w'
5: **if** $p = p'$ **and** $q = q'$ **then**
6: **return yes**
7: **else**
8: **return no**
9: **end if**

In contrast with what might be suggested by Algorithm 2.2.32, the word problem of a monoid presentation, even finite (i.e. with a finite set of generators and a finite list of relations), is often very difficult.[17] Following the general schema alluded to in Section 1.3.2, the fundamental problem is, given a finite presentation (S, R), whether we can recognize if a finite sequence of words is an R-derivation, and from there, exhaustively enumerate the R-derivations of given length. If two words u, v are equivalent and we enumerate all the derivations starting from u of lengths 1, then 2, then 3, etc.: a derivation of u to v will end up appearing, and we could conclude. However, if u and v are not equivalent, not having found a derivation at a given instant says nothing about the later attempts, and never allows us to conclude.

A situation where we can resolve a word problem is one – to be revisited in Chapter 4 – where a distinguished element can be identified in each equivalence class. We then often speak of a 'normal' element, or of a 'normal form'.

Proposition 2.2.33 (Normal form) *Suppose M is a monoid generated by a set S, and L is a subset of S^* containing exactly one element per class of the congruence $\sim$ where M is $S^*/\sim$. For w in S^*, let $\mathrm{NF}(w)$ be the unique element $\sim$-equivalent to w in L. Then M is isomorphic to $(L, *, \mathrm{NF}(\varepsilon))$, with $*$ defined by $u * v := \mathrm{NF}(uv)$. Moreover, if NF is computable, the word problem of M with respect to S is decidable.*

Exercise 2.2.34 Prove the result.

[17] In 1944 Markov and Post proved the existence of finite presentations for which the word problem is *undecidable*, i.e. for which *no* algorithmic solution can exist.

It is often quite simple to identify a set L containing at least one element per class: for this, it suffices to show how to transform an arbitrary word into a word of L passing through the relations of R. However, this is not sufficient to obtain a normal form, as it remains to show that two distinct words of L are not equivalent – which, as we have seen above, is in general more difficult. If ever we could find a $\sim$-invariant[18] injective on L, then we could conclude L provides a normal form for M.

Example 2.2.35 (Normal form) In the case $M := \langle \mathsf{a}, \mathsf{b} \mid \mathsf{ab} = \mathsf{ba} \rangle^{+}$ of Exercise 2.2.26(iii), set $L := \{\mathsf{a}^p \mathsf{b}^q \mid p, q \geqslant 0\}$. We can easily verify that every word is equivalent to a word of L, and then, defining I by $I(w) := (|w|_{\mathsf{a}}, |w|_{\mathsf{b}})$, observe that I is injective on L, hence conclude that M is isomorphic to $(\mathbb{N}, +)^2$ and the mapping (effectively computable) defined by $\mathrm{NF}(w) := \mathsf{a}^{|w|_{\mathsf{a}}} \mathsf{b}^{|w|_{\mathsf{b}}}$ is a normal form.

Exercise 2.2.36 (Normal form) Describe the monoid $\langle \mathsf{a}, \mathsf{b} \mid \mathsf{ab}^2 = \mathsf{ba} \rangle^{+}$ following the model given above.

2.3 Appendix B: Presented Groups

What has been said for the presentations of monoids applies to groups seen as a special subclass of monoids. However, the existence of inverses influences the construction and the statements. Here again, we do not propose to present the complete theory, but merely to arrive rapidly to the notion of a presented group and its associated word problem, since these are the notions at the heart of our study of braid groups.

2.3.1 Free Groups

As with monoids, we begin with the existence of a special family of groups that can be qualified as universal.

Definition 2.3.1 (Free group) For any set S, define

$$F_S := \langle S \cup \overline{S} \mid \mathsf{Sym}(S) \rangle^{+}, \tag{2.11}$$

where $\overline{S}$ is a disjoint copy of S containing, for each letter s of S, a copy denoted $\overline{s}$, and where $\mathsf{Sym}(S)$ is the family of all the relations[19] $s\overline{s} = 1$ and $\overline{s}s = 1$ for s in S. For every word w over $S \cup \overline{S}$, denote $[w]$ the image of w in F_S.

[18] i.e. a mapping $I : S^* \to M$ such that $w \sim w$ implies $I(w) = I(w')$.

[19] i.e. following Convention 2.2.25, the pairs of words $(s\overline{s}, \varepsilon)$ and $(\overline{s}s, \varepsilon)$.

In the context of Definition 2.3.1, the letters of S are said to be *positive*, those of $\overline{S}$ *negative*, and a word over $S \cup \overline{S}$ is also called a *signed S-word.*[20] A word containing only letters of S (*resp.*, of $\overline{S}$) is said to be *positive* (*resp.*, *negative*). It is convenient to extend the notation 'bar' to all the signed words.

Notation 2.3.2 (Word $\overline{w}$) For $\overline{s}$ in $\overline{S}$, define $\overline{\overline{s}}$ as s and, for any word w over $S \cup \overline{S}$, set

$$\overline{w} := \overline{x_\ell} \cdots \overline{x_1} \text{ for } w = x_1 \cdots x_\ell \text{ with } x_1, ..., x_\ell \in S \cup \overline{S}. \tag{2.12}$$

Thus, 'bar' becomes an involutive antiautomorphism of the monoid $(S \cup \overline{S})^*$: for every w, w', we have $\overline{ww'} = \overline{w'} \cdot \overline{w}$, and $\overline{\overline{w}} = w$.

By the relations of Equation (2.11), in the monoid F_S the class $[\overline{s}]$ of $\overline{s}$ is an inverse for the class $[s]$ of s: we force the existence of inverses. The following result should not be terribly surprising.

Proposition 2.3.3 (Group) *For any set S, the monoid F_S is a group, and as a group, it is generated by $[S]$. Moreover, the mapping $x \mapsto [x]$ is an injection of $S \cup \overline{S}$ into F_S.*

Exercise 2.3.4 Prove the result.

Based on the preceding result of injectivity, there is no danger in considering S and $\overline{S}$ as included in F_S, hence in identifying a letter s of S with its image $[s]$ in F_S and, similarly, $\overline{s}$ with $[s]^{-1}$, thus with s^{-1}. This is what we will adopt in what follows. However, there is no question of identifying an arbitrary signed word with its class in F_S.

The following fundamental result is, for groups, the exact counterpart of the universal property of Proposition 2.2.8.

Proposition 2.3.5 (Universal property) *For any mapping ϕ of S to a group G, there exists a unique homomorphism $\widehat{\phi}$ of F_S into G extending ϕ. It is defined by $\widehat{\phi}(1) := 1$, $\widehat{\phi}(\overline{s}) := \phi(s)^{-1}$, and $\widehat{\phi}(w) := \phi(x_1) \cdots \phi(x_\ell)$ for $w = x_1 \cdots x_\ell$ with $x_1, ..., x_\ell \in S \cup \overline{S}$.*

Exercise 2.3.6 Prove the result.

As with the monoids, F_S is said to be a *free group based on S*: the notion of 'free' referred to here is the property that, starting from F_S, we can freely send the elements of S wherever we want by a homomorphism, without encountering an obstruction.

[20] Note that a signed S-word is not an S-word if it contains at least one negative letter.

Exercise 2.3.7 (Cyclic groups)
(i) Show that $\mathbb{Z}$ is a free group with a single generator.
(ii) Show that $\mathbb{Z}/n\mathbb{Z}$ is not free.

Adapting the results of Section 2.2.2 for groups is simple. Applying Proposition 2.3.5 to the identity on S provides the following:

Proposition 2.3.8 (Image) *Any group generated by a set S is the image of the group F_S by a surjective homomorphism.*

We now pass to quotients of groups. A *congruence* on a group G is an equivalence relation on G compatible with the operations product and inverse, and this is what is necessary to obtain a well-defined operation on the quotient set. There are two remarkable facts here. First, a monoid congruence on a group is always a group congruence.

Exercise 2.3.9 (Congruence) Show that, if G is a group and $\equiv$ is a monoid congruence on G, then $\equiv$ is a group congruence, that is, it is necessarily compatible with the inverse operation.

The other remarkable point, specific to groups, is the correspondence between congruences and distinguished subgroups.

Proposition 2.3.10 (Group congruences) *Let G be a group.*
(i) *If $\equiv$ is a congruence on G, the equivalence class of* 1 *is a distinguished subgroup H of G, and for every g, g' in G, the relation $g \equiv g'$ is equivalent to $g^{-1}g' \in H$.*
(ii) *Conversely, if H is a distinguished subgroup of G, the relation $\equiv_H$ defined by $g^{-1}g' \in H$ is a congruence on G, hence H is the class of* 1.

Exercise 2.3.11 Prove the result.

This result explains the usage of omitting the congruences in the case of groups, and calling the quotient of G by a distinguished subgroup H the quotient of G by the congruence $\equiv_H$ associated with H.

The link between homomorphic images and quotients is analogous for groups and for monoids.

Exercise 2.3.12 (Images and quotients) Show that, if G and G' are groups, then there exists a surjective homomorphism of G onto G' if and only if there exists a congruence $\equiv$ on G such that G' is isomorphic to $G/{\equiv}$.

We can thus express Proposition 2.3.8 in terms of quotients.

Proposition 2.3.13 (Quotient) *Any group generated by a set S is a quotient of the group F_S.*

The congruence (or the distinguished subgroup) in play can once again be expressed in terms of evaluation, this time with values in a group. For S a subset of a group $(G, *)$, define the mapping eval_G of $(S \cup \overline{S})^*$ into G as in Equation (2.8), extended by $\mathsf{eval}_G(\overline{s}) := s^{-1}$. By Proposition 2.2.16, any group G generated by S is isomorphic to $(S \cup \overline{S})^*/{\sim}$, where $\sim$ is defined by

$$w \sim w' \quad \Leftrightarrow \quad \mathsf{eval}_G(w) = \mathsf{eval}_G(w'). \tag{2.13}$$

This is not exactly the desired result, as we want to express G as a quotient of the group F_S and not of the monoid $(S \cup \overline{S})^*$. This takes nothing more than an additional passage to a quotient .

Lemma 2.3.14 *Two signed S-words representing the same element of F_S have the same evaluation in every group including S.*

Exercise 2.3.15 Prove the result.

We continue to use eval_G for the evaluation homomorphism induced on F_S: this notation is not so abusive if we consider F_S as included in $(S \cup \overline{S})^*$ via the realization described in Corollary 2.3.38.

We can thus restate Proposition 2.3.8.

Proposition 2.3.16 (Quotient) *Every group G generated by a subset S is isomorphic to $F_S/{\simeq}$, where $\simeq$ is the congruence on F_S defined by*

$$g \simeq g' \quad \Leftrightarrow \quad \mathsf{eval}_G(g) = \mathsf{eval}_G(g'). \tag{2.14}$$

To conclude these generalities, it is easy to characterize the homomorphisms of a presented group towards another group.

Lemma 2.3.17 *A homomorphism ϕ defined on a monoid $(S \cup \overline{S})^*$ towards a group G induces a homomorphism of $\langle S \mid R \rangle$ into G if and only if $\phi(u) = \phi(v)$ for every relation $u = v$ of R and every free group relation.*

Exercise 2.3.18 Prove the lemma.

2.3.2 Group Presentations

Since every group generated by a set S can be expressed as a quotient of both the free monoid $(S \cup \overline{S})^*$ and the free group F_S, there are two ways to introduce the notion of a presented group, that in fact turn out to be equivalent. We start here with signed words; the link with free groups will be verified later.

Definition 2.3.19 (Presentations) A pair (S, R) is said to be a *presentation* of a group G, denoted

$$G = \langle S \mid R \rangle, \tag{2.15}$$

if $(S \cup \overline{S}, R \cup \mathsf{Sym}(S))$ presents G as a monoid.

In other words, we have the equality (or more precisely the isomorphism)

$$\langle S \mid R \rangle = \langle S \cup \overline{S} \mid R \cup \mathsf{Sym}(S) \rangle^{+}. \tag{2.16}$$

If (S, R) is a group presentation, then, as in the case of a monoid presentation, the elements of S are called the *generators* and those of R the *relations*.

Example 2.3.20 (Presented group) If R is empty, we find for $\langle S \mid R \rangle$ the monoid $\langle S \cup \overline{S} \mid R \cup \mathsf{Sym}(S) \rangle^{+}$, hence $\langle S \mid \ \rangle = F_S$.

Exercise 2.3.21 (Presented groups) Revisit the presentations of Exercise 2.2.26 and, in each case, identify the corresponding group $\langle S \mid R \rangle$.

It is not difficult to construct the group $\langle S \mid R \rangle$ starting from the free group F_S rather than from the monoid $(S \cup \overline{S})^*$ as in Definition 2.3.19. We must simply take into account the fact that the elements of the free group F_S are themselves equivalence classes. Recall that the class of a signed S-word w in F_S is denoted $[w]$.

Proposition 2.3.22 (Presentation) *For any presentation* (S, R), *the group* $\langle S \mid R \rangle$ *is isomorphic to* $F_S / {\approx_R}$, *where* $\approx_R$ *is the congruence on* F_S *generated by the relations* $[u] = [v]$ *for* $u = v$ *in* R.

Exercise 2.3.23 Prove the result. [Hint: In a $(\mathsf{Sym}(S) \cup R)$-derivation between two words, distinguish the steps using $\mathsf{Sym}(S)$ and those using R, and pass to the quotient by the free group congruence.]

For a monoid, only relations of the type $u = v$ make sense. For a group, $u = v$ is equivalent to $\overline{u}v = 1$, and we can limit ourselves to presentations with relations of the type $w = 1$, sometimes called the 'w relation'. Note that, if R is a family of relations $w = 1$ on S, the congruence generated corresponds to the distinguished subgroup of F_S generated by the elements $[w]$.

The group $\langle S \mid R \rangle$ behaves, in the world of groups, like the monoid $\langle S \mid R \rangle^{+}$ in the world of monoids. In particular, it is often useful to invoke the following result, similar to Proposition 2.2.27. Naturally, a relation $u = v$ is said to be *satisfied* in a group G if the equality $\mathsf{eval}_G(u) = \mathsf{eval}_G(v)$ holds.

Proposition 2.3.24 (Quotient) *A group G generated by a set S is a quotient of the group* $\langle S \mid R \rangle$ *if and only if all the relations of R are satisfied in G.*

Exercise 2.3.25 Prove the result.

However, it is often delicate to show that a pair (S, R) constitutes a presentation of a group – as for example for the group B_n in Section 2.1.3. The difficulty is not to guess the relations, but to show that these generate *all* the possible relations. The following general schema can be useful.

Proposition 2.3.26 (Presentation) *Suppose G is a group generated by a set S, R is a list of relations satisfied in G, and E is a set of signed S-words such that*
(i) *every signed S-word is R-equivalent to a word of E,*
(ii) *the mapping eval_G is injective on E.*
Then G admits the presentation $\langle S \mid R \rangle$.

Exercise 2.3.27 Prove the result.

2.3.3 The Word Problem of a Presented Group

If a group G is generated by a subset S, every element of G is an evaluation of one or more signed S-words, and the question is how to recognize which words represent which elements. The problem is a special case of word problem for monoids as described in Section 2.2.4. Nevertheless, the existence of inverses implies that, in a group, $[w] = [w']$ is equivalent to $[\overline{w}\, w'] = 1$: consequently, a problem with two parameters can be reduced to a problem with only a single parameter.

Definition 2.3.28 (Word problem) If G is a group generated by a set S, a *solution to the word problem* for G with respect to S is an algorithm determining, for any signed S-word w, if w represents 1 in G (in finite time).

It is common to say 'the word problem of a presentation (S, R) (or $\langle S \mid R \rangle$)' for the word problem of the group $\langle S \mid R \rangle$ with respect to S, that is, the question of deciding whether a signed S-word is $\equiv_R$-equivalent to the empty word, where $\equiv_R$ is the congruence on $(S \cup \overline{S})^*$ generated by $R \cup \mathsf{Sym}(S)$.

Exercise 2.3.29 Following the model of Algorithm 2.2.32, give a solution to the word problem for the presented group $\langle \mathsf{a}, \mathsf{b} \mid \mathsf{ab} = \mathsf{ba} \rangle$.

The word problem of a group presentation, even finite (i.e. with a finite set of generators and a finite list of relations), is often very difficult.[21] As always, the difficulty for a presentation (S, R) is to show that a signed S-word does not

[21] Pavel Novikov proved in 1952 – hence several years after the analogous result for monoids – the existence of a finite presentation whose word problem is undecidable.

represent 1, which no exhaustive enumeration of derivations can guarantee in finite time.

As with the monoids, we can often obtain a solution to the word problem of a presented group by identifying a distinguished (or 'normal') element in each class.

Proposition 2.3.30 (Normal form) *Suppose that G is a group generated by a set S, and L is a set of signed S-words containing exactly one element per class of the congruence $\sim$ such that G is $(S \cup \overline{S})^*/{\sim}$. Then G is isomorphic to $(L, *, \mathrm{NF}(\varepsilon))$, where $\mathrm{NF}(w)$ is the unique word $\sim$-equivalent to w in L and where $*$ is defined by $u * v := \mathrm{NF}(uv)$. If NF is calculable, the word problem of G with respect to S is decidable.*

The proof is, mutatis mutandis, the same as for Proposition 2.2.33.

2.3.4 Free Groups and Reduced Words

In Chapter 5, we will need a precise description of the elements of the free group F_S. Since F_S is a quotient of the free monoid $(S \cup \overline{S})^*$, its elements are equivalence classes of words, and, following the approach of Proposition 2.3.30, we define a distinguished representative for each class.

Definition 2.3.31 (Reduced word) A signed S-word is said to be *reduced* if it does not contain any factor of the form $s\overline{s}$ or $\overline{s}s$.

For example, $\mathtt{ab\overline{a}}$ is reduced, but not $\mathtt{ab\overline{b}}$, with a factor $\mathtt{b\overline{b}}$.

To show that the reduced words constitute a normal form in F_S, we introduce two binary relations on $(S \cup \overline{S})^*$.

Definition 2.3.32 (Reduction) If w, w' are signed S-words, w is said to be *reducible* to w' *in one step*, denoted $w \rightarrow w'$, if there exist words u, v and a letter s in S satisfying

$$w = us\overline{s}v \quad \text{or} \quad w = u\overline{s}sv \quad \text{and} \quad w' = uv.$$

A word w is said to be *reducible* to w', denoted $w \rightarrow^* w'$, if there exist $w_0, \ldots, w_m$ satisfying $w_0 = w$, $w_m = w'$ and $w_{i-1} \rightarrow w_i$ for every $0 < i < m$.

Reducing a word means eliminating factors $s\overline{s}$ or $\overline{s}s$. For example, we have $\mathtt{\overline{a}b\overline{b}a} \rightarrow \mathtt{\overline{a}a} \rightarrow \varepsilon$, hence $\mathtt{\overline{a}b\overline{b}a} \rightarrow^* \varepsilon$.

If w can be reduced to w' in one step, the length of w' is less than the length of w by a count of two. Hence, starting from a word of length ℓ, we must arrive to a reduced word after at most $\ell/2$ steps. However, a word can contain several non-reduced factors, and it is not a priori evident that all the reductions lead to the same word. This is nevertheless the case.

Lemma 2.3.33 *Every reduction of a word w in $(S \cup \overline{S})^*$ leads to the same reduced word*[22].

Exercise 2.3.34 Prove the result. [Hint: Establish by induction on $\ell \geqslant 0$ the property 'If $w \rightarrow^* w'$ and $w \rightarrow^* w''$ with w' reduced and $|w| - |w'| = 2\ell$, then $w'' \rightarrow^* w'$'.]

Definition 2.3.35 (Function red) For a signed word w on S, denote $\mathsf{red}(w)$ the unique reduced word to which w reduces.

For example, $\mathsf{red}(\mathsf{ab\bar{a}a\bar{b}a}) = \mathsf{aa}$, or $\mathsf{red}(\mathsf{\bar{a}bb\bar{b}\bar{b}a}) = \varepsilon$.

By Lemma 2.3.33, we can easily deduce a normal form for the elements of the free group F_S.

Proposition 2.3.36 (Normal form) *The reduced words provide a normal form for F_S.*

Exercise 2.3.37 Prove the result. [Hint: Denote $w \sim w'$ for $\mathsf{red}(w) = \mathsf{red}(w')$, and show that $\sim$ is a congruence on $(S \cup \overline{S})^*$ coincident with the congruence $\equiv$ generated by the relations of (2.11).]

Following Proposition 2.3.30, we deduce a realization of free groups as sets of reduced words.

Corollary 2.3.38 (Realization) *Denote Red_S the set of reduced words on the alphabet S, and for u, v in Red_S, set*

$$u * v := \mathsf{red}(uv).$$

Then $(\mathsf{Red}_S, *)$ is a realization of the free group F_S.

Example 2.3.39 (Realization) When the elements of S are lowercase characters, it is convenient to use for their counterparts in $\overline{S}$ the corresponding uppercase characters, so we write abBBA for $\mathsf{ab\bar{b}\bar{b}\bar{a}}$. Thus, the free group[23] $F_{\{\mathsf{a},\mathsf{b}\}}$ can be identified with the set of words on $\{\mathsf{a}, \mathsf{A}, \mathsf{b}, \mathsf{B}\}$ without factors aA, Aa, bB, or Bb, and we find for example

$$\mathsf{abA} * \mathsf{aBBa} = \mathsf{red}(\mathsf{ab\underline{Aa}BBa}) = \mathsf{red}(\mathsf{a\underline{bB}Ba}) = \mathsf{aBa}.$$

Note that the product $*$ is *not* the concatenation of reduced words, which in general has no reason to be a reduced word, but is concatenation followed by a reduction.

[22] This is often called the *convergence* of the reduction.

[23] Also called the *free group of rank* 2, as it has a base of cardinality 2.

Exercise 2.3.40 (Presented group) Show that the group $\langle S \mid R \rangle$ is the quotient of the free group F_S by the distinguished subgroup generated by the elements $\mathsf{red}(u^{-1}v)$ for the relations $u = v$ of R.

3
Braid Monoids

We now know that the n-strand braids form a group B_n, and have moreover obtained for them a presentation by generators and relations. This does not in practice suffice to respond to questions about B_n, least of all the word problem, that is, the braid isotopy problem, and we must continue our efforts. We describe here a first solution, following the approach developed in the 1960s by F.A. Garside (in his thesis prepared at Oxford under the direction of G. Higman), based on the analysis of an auxiliary structure, the monoid B_n^+, a posteriori revealed as a submonoid of the group B_n.

3.1 The Monoid B_n^+

3.1.1 Definition and First Properties

The braid relations such as those stated in (2.6) are of the type $u = v$ where u and v are positive words, with no letters σ_i^{-1} appearing. We are thus naturally led to consider the monoid presented by these relations.

Definition 3.1.1 (Monoid B_n^+) For $n \geqslant 1$, denote B_n^+ the monoid admitting the presentation

$$\left\langle \sigma_1, \ldots, \sigma_{n-1} \;\middle|\; \begin{array}{ll} \sigma_i\sigma_j = \sigma_j\sigma_i & \text{for } |i-j| \geqslant 2, \\ \sigma_i\sigma_j\sigma_i = \sigma_j\sigma_i\sigma_j & \text{for } |i-j| = 1 \end{array} \right\rangle^+, \tag{3.1}$$

and B_∞^+ the monoid admitting the (infinite) presentation

$$\left\langle \sigma_1, \sigma_2, \ldots \;\middle|\; \begin{array}{ll} \sigma_i\sigma_j = \sigma_j\sigma_i & \text{for } |i-j| \geqslant 2, \\ \sigma_i\sigma_j\sigma_i = \sigma_j\sigma_i\sigma_j & \text{for } |i-j| = 1 \end{array} \right\rangle^+. \tag{3.2}$$

Example 3.1.2 (Monoid B_n^+) The monoid B_1^+ is the trivial monoid with a single element. The monoid B_2^+ is the monoid $\langle \sigma_1 \mid \ \rangle^+$, hence the free monoid generated by σ_1: this is a copy of the monoid $(\mathbb{N}, +)$.

The monoid B_3^+ is $\langle \sigma_1, \sigma_2 \mid \sigma_1\sigma_2\sigma_1 = \sigma_2\sigma_1\sigma_2 \rangle^+$. Playing with the relation, we can for example verify equalities such as $\sigma_1\sigma_2\sigma_1\sigma_2\sigma_1 = \sigma_2\sigma_1\sigma_2^2\sigma_1 = \sigma_1^2\sigma_2\sigma_1^2 = \sigma_1\sigma_2^2\sigma_1\sigma_2$ in B_3^+... but this does not say much about the structure of B_3^+. A fortiori, this is the same for B_n^+ where $n \geqslant 4$, and for B_∞^+.

The title of this chapter as well as the resemblance of the presentation (3.1) with the Artin presentation of the braid groups B_n might suggest calling B_n^+ the 'braid monoid'. This would be premature as no clear link relates the braids to elements of B_n^+: for the moment, these are merely elements of a presented monoid, hence equivalence classes of words.

Notation 3.1.3 (The set $\mathcal{BW}_n^+$) The set of words in $\sigma_1, ..., \sigma_{n-1}$ (*resp.*, in $\sigma_1, \sigma_2, ...$) is denoted $\mathcal{BW}_n^+$ (*resp.*, $\mathcal{BW}_\infty^+$).

We thus have the free monoid inclusions:

$$\{\varepsilon\} = \mathcal{BW}_1^+ \subseteq \mathcal{BW}_2^+ \subseteq \cdots \subseteq \mathcal{BW}_\infty^+. \tag{3.3}$$

Notation 3.1.4 (Relation $\equiv^+$, class $[w]^+$) Denote $\equiv^+$ the congruence on $\mathcal{BW}_\infty^+$ generated by the relations (3.2). For w in $\mathcal{BW}_\infty^+$, denote $[w]^+$ the equivalence class of w with respect to $\equiv^+$.

Thus, by definition, $B_\infty^+ = \mathcal{BW}_\infty^+/\equiv^+$. Recall that, for $[w]^+ = a$, the word w is said to *represent* the element a of B_∞^+.

We begin here with some observations based on the particular form of the relations defining B_∞^+.

Lemma 3.1.5 *Two $\equiv^+$-equivalent words of $\mathcal{BW}_\infty^+$ have the same length and are made up of the same letters.*[1]

Proof By Lemma 2.2.20, two words w, w' are $\equiv^+$-equivalent if and only if there exists a derivation linking w to w', hence a finite sequence of words $w_0, ..., w_m$ where the first is w, the last w', and each is obtained from its predecessor by applying a braid relation. However, whether we use $\sigma_i\sigma_j = \sigma_j\sigma_i$ or $\sigma_i\sigma_j\sigma_i = \sigma_j\sigma_i\sigma_j$, the lengths of the two words of a relation and the letters it contains are the same. Hence, step by step, the lengths of w and w' and the letters appearing coincide. □

[1] Note that we say *nothing* about the multiplicity of the letters in the equivalent words: the words $\sigma_1\sigma_2\sigma_1$ and $\sigma_2\sigma_1\sigma_2$ are $\equiv^+$-equivalent, and one contains two letters σ_1 while the other contains only one.

A consequence of Lemma 3.1.5 is that, for every i, the word σ_i (of length 1) is only equivalent to itself: it is hence without danger to use the same notation for the letter σ_i and the element of B_∞^+ represented. Another consequence is the following result.

Lemma 3.1.6
(i) *For every word w in $\mathcal{BW}_n^+$, the class $[w]^+$ is included in $\mathcal{BW}_n^+$.*
(ii) *The congruence on $\mathcal{BW}_n^+$ generated by the relations of Equation (3.1) is the restriction of $\equiv^+$ to $\mathcal{BW}_n^+$.*

Proof
(i) By Lemma 3.1.5, every word $\equiv^+$-equivalent to a word in $\mathcal{BW}_n^+$ is in $\mathcal{BW}_n^+$ since the only letters that can appear are among $\sigma_1, \ldots, \sigma_{n-1}$.
(ii) Let R_n be the family of braid relations on $\mathcal{BW}_n^+$ (including for $n = \infty$), and $\equiv_n^+$ the congruence on $\mathcal{BW}_n^+$ generated by R_n. Every R_n-derivation is an R_∞-derivation, hence two $\equiv_n^+$-equivalent words are $\equiv^+$-equivalent. A priori, the problem is in the other direction: the inclusion of R_n in R_∞ is strict and there are more R_∞-derivations than R_n-derivations. Now, suppose $w \equiv^+ w'$ with w, w' in $\mathcal{BW}_n^+$. Then there exists an R_∞-derivation $w_0, \ldots, w_m$ linking w to w'. By (i), the hypothesis that w_0 is in $\mathcal{BW}_n^+$ implies that each word w_k is in $\mathcal{BW}_n^+$. The derivation is hence an R_n-derivation, and thus $w \equiv_n^+ w'$. □

We can now make clear the link between B_n^+ and B_∞^+.

Proposition 3.1.7 (Inclusion) *For every n, the monoid B_n^+ is the submonoid of B_∞^+ generated by $\sigma_1, \ldots, \sigma_{n-1}$.*

Proof With the notations of Lemma 3.1.6 and by definition, B_n^+ is the quotient $\mathcal{BW}_n^+/\equiv_n^+$. Let a be arbitrary in B_n^+, and w a word of $\mathcal{BW}_n^+$ representing a. Then a is the equivalence class of w for $\equiv_n^+$, hence is the set of words of $\mathcal{BW}_n^+$ $\equiv_n^+$ equivalent to w. By Lemma 3.1.6(ii), this class is also the set of words of $\mathcal{BW}_\infty^+$ $\equiv^+$-equivalent to w, which is the element $[w]^+$ of B_∞^+. Hence a is in B_∞^+, and we have $B_n^+ \subseteq B_\infty^+$. Moreover, 1, which is $[\varepsilon]^+$, is in B_n^+, and the product of B_n^+ is the restriction of the product of B_∞^+. Hence B_n^+ is a submonoid of B_∞^+. Finally, $\sigma_1, \ldots, \sigma_{n-1}$ generate B_n^+, so B_n^+ is the submonoid of B_∞^+ generated by these elements. □

As a result, the monoids B_n^+ form an increasing sequence:

$$\{1\} = B_1^+ \subseteq B_2^+ \subseteq B_3^+ \subseteq \cdots \tag{3.4}$$

with B_∞^+ as their union. We also deduce that, for any $n \leqslant n'$, B_n^+ is the submonoid of $B_{n'}^+$ generated by $\sigma_1, \ldots, \sigma_{n-1}$. Moreover, there is no danger to use $\equiv^+$ for the restriction of $\equiv^+$ to $\mathcal{BW}_n^+$, and hence to write, for any n,

$$B_n^+ = \mathcal{BW}_n^+/\equiv^+. \tag{3.5}$$

Another consequence of Lemma 3.1.5 is that the mapping 'length' on $\mathcal{BW}_\infty^+$ induces a well-defined mapping on B_∞^+.

Notation 3.1.8 (Length $|a|$) For a in B_∞^+, denote $|a|$ the unique value of $|w|$ for any w representing a.

By construction, the mapping $a \mapsto |a|$ is a homomorphism of B_∞^+ into $(\mathbb{N}, +)$ (hence also, by restriction for any n, of B_n^+).

Exercise 3.1.9 (Inverse) Show that for every $n \leqslant \infty$ the unit element 1 is the only invertible element of B_n^+.

Note that the existence of the length mapping allows us to resolve the word problem in a completely 'stupid' manner.[2]

Proposition 3.1.10 (Decidability) *For every $n \leqslant \infty$, the word problem for the monoid B_n^+ with respect to $\{\sigma_1, ..., \sigma_{n-1}\}$ is decidable.*

Proof We describe an effective procedure deciding whether two words w, w' in $\mathcal{BW}_\infty^+$ are $\equiv^+$-equivalent or not. For this, set $\ell := |w|$, select an integer n such that w belongs to $\mathcal{BW}_n^+$, and set $m := (n-1)^\ell$. Define $C_0 := \{w\}$ and then, inductively for $0 \leqslant k < m$, define C_{k+1} as the union of C_k with the set of words obtained by starting with a word of C_k and applying a relation of the presentation. For each k, and in particular for $k = m$, the set C_k can effectively be determined. By construction, for any k, C_k is included in the set of words of length ℓ in $\sigma_1, ..., \sigma_{n-1}$, which is finite, of cardinality at most m. The sequence $(C_k)_{k \geqslant 0}$ thus cannot grow more than m times. Hence, every word $\equiv^+$-equivalent to a word of C_m is in C_m. We thus have $C_m = [w]^+$, and $w' \equiv^+ w$ is satisfied if and only if w' belongs to C_m, which is effectively testable. □

Exercise 3.1.11 (Word problem) Write the above solution in the form of a pseudocode algorithm.

Apart from the fact that the members of each of the braid relations are words with the same length, exploited since Lemma 3.1.5, another syntactic property of the braid relations is their palindromic character, already exploited in Proposition 2.1.26. Recall that for any word w in $\mathcal{BW}_\infty$, hence a fortiori in $\mathcal{BW}_\infty^+$, the word obtained by reversing the order of the letters in w is denoted $\widetilde{w}$.

Lemma 3.1.12 *For every $n \leqslant \infty$, the mapping $w \mapsto \widetilde{w}$ induces an involutive antiautomorphism of the monoid B_n^+.*

[2] 'Stupid' in that the solution is not in any way based on an analysis of the specific relations of the presentation; moreover the efficiency of the procedure described is calamitous, and much better solutions will be presented later.

The proof is exactly the same as for Proposition 2.1.26. As in the case of the group B_∞, we use $\widetilde{a}$ for the image of an element a of B_∞^+ by the above antiautomorphism (we will see later that there is no risk of ambiguity). The benefit of the reversal for the study of the monoids B_n^+ is to allow the swapping of left and right in lateralized properties.

Another property of the braid relations is their invariance by the symmetry $\sigma_i \mapsto \sigma_{n-i}$, exploited in Proposition 2.1.24.

Lemma 3.1.13 *For every n, the mapping $\sigma_i \mapsto \sigma_{n-i}$ induces an involutive automorphism of the monoid B_n^+.*

The proof is the same as for Proposition 2.1.24. By analogy with B_n, we use ϕ_n for this automorphism of B_n^+.

Exercise 3.1.14 (Shift)
(i) Show that the mapping $\sigma_i \mapsto \sigma_{i+1}$ induces an endomorphism dec^+ of B_∞^+ into itself.
(ii) Show that dec^+ is injective, but not surjective.

3.1.2 The Embedding Problem

The possibility of using the monoid B_n^+ to study the group B_n requires an elucidation of the relation between these two structures. By Proposition 2.1.21 for the one and by definition for the other, the group B_n and the monoid B_n^+ admit, one as a group, the other as a monoid, the same presentation by the braid relations with generators $\sigma_1, ..., \sigma_{n-1}$. This results in the existence of a homomorphism of B_n^+ into B_n.

Lemma 3.1.15 *The identity mapping of $\{\sigma_1, ..., \sigma_{n-1}\}$ induces a homomorphism ι of B_n^+ into B_n.*

Proof First, by Proposition 2.2.8, the identity mapping of $\{\sigma_1, ..., \sigma_{n-1}\}$ extends to a homomorphism id^* of $\mathcal{BW}_n^+$ into the group B_n, simply the restriction of $w \mapsto [w]$ to positive words. Next, the braid relations defining B_n^+ are satisfied in B_n. By Proposition 2.2.27, the homomorphism id^* can thus be factored by $\equiv^+$, inducing a well-defined homomorphism of B_n^+ into B_n, that is, the mapping $\iota : [w]^+ \mapsto [w]$. □

Exercise 3.1.16 (Enveloping group)
(i) Show that, for any monoid presentation (S, R), the identity of S induces a homomorphism ι of the monoid $\langle S \mid R \rangle^+$ into the group $\langle S \mid R \rangle$.
(ii) Hence show that any homomorphism of $\langle S \mid R \rangle^+$ towards a group can be factored by ι.

The preceding results are essentially trivial. This is not the case for the following problem.

Question 3.1.17 (Embedding) Is the morphism of Lemma 3.1.15 injective?

An injective homomorphism is also called an *embedding*, hence the name of the problem.

The existence of the morphism ι is based on the implication

$$w \equiv^{+} w' \quad \Longrightarrow \quad w \equiv w', \tag{3.6}$$

which is valid for any positive braid words w, w'. Since ι is the mapping $[w]^{+} \mapsto [w]$, the injectivity of ι is the same as Equation (3.6) being an equivalence. The question is difficult: a priori, two positive words can be $\equiv$-equivalent, so linked by means of the braid relations and the free group relations, without being $\equiv^{+}$-equivalent, meaning linked only by the braid relations.

Exercise 3.1.18 (Non-embedding) Let M be the monoid $\langle \mathsf{a}, \mathsf{b}, \mathsf{c} \mid \mathsf{ab} = \mathsf{ac} \rangle^{+}$ and G the group $\langle \mathsf{a}, \mathsf{b}, \mathsf{c} \mid \mathsf{ab} = \mathsf{ac} \rangle$. Show that M cannot be embedded into G. [Hint: Show that in the group, we have $\mathsf{b} = \mathsf{c}$, whereas in the monoid, $\mathsf{b} \neq \mathsf{c}$.]

In general, the embedding problem is difficult.[3] There nonetheless exist cases where simple (but not necessary) conditions guarantee an embedding; these are known as *Ore's conditions*. In order to state them, two auxiliary notions are necessary.

Definition 3.1.19 (Cancellative) A monoid M is said to be *left-cancellative* (*resp., right-cancellative*) if $ab = ac$ (*resp.,* $ba = ca$) implies $b = c$ in M; it is *cancellative* if it is both left- and right-cancellative.

Exercise 3.1.20 (Cancellative) Show that a free monoid is cancellative.

Next, we will need the notions of divisors and multiples in B_n^{+}. These are direct extensions of those for numbers. However, as B_n^{+} is not commutative, there exist two relations, depending on whether we multiply on the right or on the left.

Definition 3.1.21 (Divisors, multiples) If M is a monoid, with a, b elements of M, a is said to be a *left-divisor of* b, and b a *right-multiple of* a, denoted $a \preccurlyeq b$, if there exists x in M satisfying $ax = b$.

[3] The counterexample of Exercise 3.1.18 is 'stupid' (a monoid which is not cancellative cannot be embedded in a group), but there exist others, much more subtle. A complete solution is provided by a result of Malcev giving an infinite list of necessary and sufficient conditions, but these are often quite difficult to establish in practice.

We will study the divisibility relations of the monoid B_n^+ in detail in Section 4.1.1. For the time being, we only need a result concerning the existence of common multiples: c is a *common* right-multiple of a and b if at the same time $a \preccurlyeq c$ and $b \preccurlyeq c$.

In the case of the monoids B_n^+, the solution of the embedding problem, and even the precise description of the link between the monoid B_n^+ and the group B_n, relies on the following result.

Proposition 3.1.22 (Ore's theorem for B_n^+ and B_n) *Suppose that the following statements have been shown:*
(i) *the monoid B_n^+ is cancellative,*
(ii) *two arbitrary elements of B_n^+ always admit a common right-multiple.*
Then B_n^+ can be embedded in B_n. Moreover, any element of B_n can be written as a fraction ab^{-1} with a and b in (the image of)[4] *B_n^+.*

This result is an instance of a general result, Ore's theorem, whose statement and proof are given in Appendix C at the end of this chapter.

3.1.3 Common Multiples in B_n^+

By Proposition 3.1.22, to apply Ore's theorem to the monoid B_n^+, we must prove that B_n^+ is cancellative, and that two arbitrary elements of B_n^+ admit a common right-multiple. We begin with this second condition.

By definition, the monoid B_n^+ admits the presentation (3.1). Consequently, establishing that two arbitrary elements of B_n^+ admit a common right-multiple is equivalent to showing

$$\forall u, v \in \mathcal{BW}_n^+ \; \exists u', v' \in \mathcal{BW}_n^+ \; (uv' \equiv^+ vu'). \tag{3.7}$$

The proof first treats the case where u and v are letters σ_i, and then uses an induction on the length of the words.

We begin by accumulating the equivalence results for $\equiv^+$. It is permissible (and a useful guide for our intuition) to verify the validity of the analogous relations bringing into play the isotopies of braid diagrams, which must be satisfied according to (3.6), but this does *not* constitute a proof as long as we have not established that $\equiv^+$ is the restriction of $\equiv$ to B_n^+.

Lemma 3.1.23 *Define $\underline{\sigma}_{i,i} := \varepsilon$ for every i and, for $j \neq i$,*

$$\underline{\sigma}_{i,j} := \begin{cases} \sigma_i \sigma_{i+1} \cdots \sigma_{j-1} & \text{for } i < j, \\ \sigma_{i-1} \cdots \sigma_{j+1} \sigma_j & \text{for } i > j. \end{cases} \tag{3.8}$$

[4] Once B_n^+ can be embedded in B_n, there is no danger in identifying B_n^+ with its image in B_n, in particular not distinguishing between 'the σ_i of B_n^+' and 'the σ_i of B_n'.

Then, for $j \geqslant i+2$ and $i < k < j$,

$$\sigma_k \, \underline{\sigma}_{i,j} \equiv^+ \underline{\sigma}_{i,j} \, \sigma_{k-1}, \tag{3.9}$$

$$\sigma_{k-1} \, \underline{\sigma}_{j,i} \equiv^+ \underline{\sigma}_{j,i} \, \sigma_k. \tag{3.10}$$

Proof We establish (3.9) by induction on $|j-i|$, which is at least 2. For $j = i+2$, we necessarily have $k = i+1$, and (3.9) reduces to the known relation $\sigma_{i+1}(\sigma_i\sigma_{i+1}) \equiv^+ (\sigma_i\sigma_{i+1})\sigma_i$.

Suppose $j \geqslant i+3$. First consider $k = i+1$. Then $\sigma_{i+1}\underline{\sigma}_{i,j-1} \equiv^+ \underline{\sigma}_{i,j-1}\sigma_i$ and $|(j-1)-i| < |j-i|$. Applying the induction hypothesis, we find

$$\sigma_k \, \underline{\sigma}_{i,j} = \sigma_{i+1} \, \underline{\sigma}_{i,j-1} \, \sigma_{j-1} \equiv^+ \underline{\sigma}_{i,j-1} \, \sigma_i \, \sigma_{j-1} \equiv^+ \underline{\sigma}_{i,j-1} \, \sigma_{j-1} \, \sigma_i = \underline{\sigma}_{i,j} \, \sigma_{k-1},$$

which is (3.9).

Now consider $k \geqslant i+2$. Then $\sigma_{j+1}\sigma_i \equiv^+ \sigma_i\sigma_{j+1}$ and $|j-(i+1)| < |j-i|$, so, by applying the induction hypothesis,

$$\sigma_k \, \underline{\sigma}_{i,j} = \sigma_k \, \sigma_i \, \underline{\sigma}_{i+1,j} \equiv^+ \sigma_i \, \sigma_k \, \underline{\sigma}_{i+1,j} \equiv^+ \sigma_i \, \underline{\sigma}_{i+1,j} \, \sigma_{k-1} = \underline{\sigma}_{i,j} \, \sigma_{k-1},$$

again the desired result.

The proof of (3.10) is entirely symmetric (do it!). □

We now introduce a particular element Δ_n of B_n^+ destined to play a fundamental role in all that follows.

Definition 3.1.24 (Word $\underline{\Delta}_n$, Element Δ_n) Define $\underline{\Delta}_1 := \varepsilon$ and then

$$\underline{\Delta}_n := \underline{\sigma}_{1,n} \, \underline{\Delta}_{n-1} = \sigma_1\sigma_2 \cdots \sigma_{n-1} \, \underline{\Delta}_{n-1} \quad \text{for } n \geqslant 2. \tag{3.11}$$

Denote Δ_n the element of B_n^+ represented by the word $\underline{\Delta}_n$.

Then $\underline{\Delta}_2 = \sigma_1$, $\underline{\Delta}_3 = \sigma_1\sigma_2\sigma_1$, $\underline{\Delta}_4 = \sigma_1\sigma_2\sigma_3\sigma_1\sigma_2\sigma_1$, etc.

Lemma 3.1.25 *For every $n \geqslant 2$,*

$$\underline{\Delta}_n \equiv^+ \underline{\Delta}_{n-1} \, \underline{\sigma}_{n,1}. \tag{3.12}$$

Proof We reason by induction on $n \geqslant 1$. The result is clear for $n = 1$ and $n = 2$. Suppose $n \geqslant 3$. Using the induction hypothesis and the fact that $\underline{\Delta}_{n-2}$ contains only letters σ_i with $i \leqslant n-3$, hence satisfying $\sigma_i\sigma_{n-1} \equiv^+ \sigma_{n-1}\sigma_i$, we find

$$\begin{aligned}
\underline{\Delta}_n = \underline{\sigma}_{1,n} \, \underline{\Delta}_{n-1} = \underline{\sigma}_{1,n-1}\sigma_{n-1} \, \underline{\Delta}_{n-1} &\equiv^+ \underline{\sigma}_{1,n-1}\sigma_{n-1} \, \underline{\Delta}_{n-2} \, \underline{\sigma}_{n-1,1} \\
&\equiv^+ \underline{\sigma}_{1,n-1} \, \underline{\Delta}_{n-2} \, \sigma_{n-1}\underline{\sigma}_{n-1,1} \\
&= \underline{\Delta}_{n-1} \, \underline{\sigma}_{n,1}.
\end{aligned}$$

□

Lemma 3.1.26 *For* $1 \leqslant i \leqslant n-1$,

$$\sigma_i \underline{\Delta}_n \equiv^+ \underline{\Delta}_n \sigma_{n-i}. \tag{3.13}$$

Proof We again proceed by induction on $n \geqslant 1$. The result is trivial for $n = 1$. For $n = 2$, it reduces to $\sigma_1^2 \equiv^+ \sigma_1^2$. Suppose now $n \geqslant 3$ and, to start, $i = 1$. Applying (3.12), the induction hypothesis, and then (3.10) and again (3.12), we find

$$\begin{aligned}\sigma_1 \underline{\Delta}_n \equiv^+ \sigma_1 \underline{\Delta}_{n-1} \underline{\sigma}_{n,1} &\equiv^+ \underline{\Delta}_{n-1} \sigma_{n-2} \underline{\sigma}_{n,1}\\ &\equiv^+ \underline{\Delta}_{n-1} \underline{\sigma}_{n,1} \sigma_{n-1}\\ &\equiv^+ \underline{\Delta}_n \sigma_{n-1}.\end{aligned}$$

Suppose now $2 \leqslant i \leqslant n-1$. Applying (3.9) for $1 < i < n$ and the induction hypothesis, we find

$$\begin{aligned}\sigma_i \underline{\Delta}_n = \sigma_i \underline{\sigma}_{1,n} \underline{\Delta}_{n-1} &\equiv^+ \underline{\sigma}_{1,n} \sigma_{i-1} \underline{\Delta}_{n-1}\\ &\equiv^+ \underline{\sigma}_{1,n} \underline{\Delta}_{n-1} \sigma_{(n-1)-(i-1)}\\ &= \underline{\Delta}_n \sigma_{n-i}.\end{aligned}$$

□

Exercise 3.1.27 (Automorphism) Show that, for any a in B_n^+, we have the equality $a\Delta_n = \Delta_n \phi_n(a)$.

This brings us to the first important result, that is, that Δ_n is a right-multiple of each of the generators σ_i in B_n^+.

Lemma 3.1.28 *For* $1 \leqslant i \leqslant n-1$, *there exists in* $\mathcal{BW}_n^+$ *a word* $\underline{\partial}_n(\sigma_i)$ *satisfying* $\underline{\Delta}_n \equiv^+ \sigma_i \underline{\partial}_n(\sigma_i)$.

Proof We use an induction on $n \geqslant 1$. The result is empty for $n = 1$ and trivial for $n = 2$ with $\underline{\partial}_2(\sigma_1) := \varepsilon$. Suppose $n \geqslant 3$. For $i \leqslant n-2$, the induction hypothesis gives a word $\underline{\partial}_{n-1}(\sigma_i)$ satisfying

$$\underline{\Delta}_{n-1} \equiv^+ \sigma_i \underline{\partial}_{n-1}(\sigma_i).$$

By (3.11), we have $\underline{\Delta}_n \equiv^+ \underline{\Delta}_{n-1} \underline{\sigma}_{n,1} \equiv^+ \sigma_i \underline{\partial}_{n-1}(\sigma_i) \underline{\sigma}_{n,1}$: the result is thus verified for $\underline{\partial}_n(\sigma_i) := \underline{\partial}_{n-1}(\sigma_i) \underline{\sigma}_{n,1}$.

There remains the case of σ_{n-1}. By construction, $\underline{\Delta}_{n-1}$ contains only the letters σ_i with $1 \leqslant i \leqslant n-2$. For each, Lemma 3.1.23 implies $\sigma_i \underline{\sigma}_{n,1} \equiv^+ \underline{\sigma}_{n,1} \sigma_{i+1}$. Hence

$$\underline{\Delta}_n \equiv^+ \underline{\Delta}_{n-1} \underline{\sigma}_{n,1} \equiv^+ \underline{\sigma}_{n,1} w = \sigma_{n-1} \underline{\sigma}_{n-1,1}\, w,$$

where w is the word obtained from $\underline{\Delta}_{n-1}$ by shifting all the indices by $+1$, that is, by everywhere replacing σ_i by σ_{i+1}. This gives the desired result for $\underline{\partial}_n(\sigma_{n-1}) := \underline{\sigma}_{n-1,1}\, w$. □

Exercise 3.1.29 (Words $\underline{\partial}_4(\sigma_i)$) Determine the words $\underline{\partial}_n(\sigma_i)$ for $n \geqslant 5$ and $i < n$.

The result of Lemma 3.1.28 implies (3.7) when u and v have length 1, that is, are letters σ_i. Indeed, for i and j between 1 and $n-1$, we obtain

$$\sigma_i \underline{\partial}_n(\sigma_i) \equiv^+ \underline{\Delta}_n \equiv^+ \sigma_j \underline{\partial}_n(\sigma_j), \tag{3.14}$$

hence Δ_n is a common right-multiple of σ_i and σ_j.

We now pass to the general case.

Lemma 3.1.30 *For any word u of length at most ℓ in $\mathcal{BW}_n^+$, there exists a word w in $\mathcal{BW}_n^+$ satisfying $uw \equiv^+ \underline{\Delta}_n^\ell$.*

Proof We proceed by induction on $\ell \geqslant 0$. The result is trivial for $\ell = 0$. Suppose $\ell \geqslant 1$. The result is again trivial if u is the empty word. Otherwise, there exists u' of length at most $\ell - 1$ and i satisfying $u = u'\sigma_i$. By the induction hypothesis, there exists w' satisfying $u'w' \equiv^+ \underline{\Delta}_n^{\ell-1}$. Let $\phi_n(w')$ be the word obtained from w' by everywhere exchanging the letters σ_i and σ_{n-i}. By 3.1.26, we have $\underline{\Delta}_n \phi_n(w') \equiv^+ w'\underline{\Delta}_n$. Moreover, with the notation of Lemma 3.1.28, $\sigma_i \underline{\partial}_n(\sigma_i) \equiv^+ \underline{\Delta}_n$. We thus set $w := \underline{\partial}_n(\sigma_i)\phi_n(w')$. Hence

$$\begin{aligned} uw = u'\, \sigma_i \underline{\partial}_n(\sigma_i)\phi_n(w') &\equiv^+ u'\, \underline{\Delta}_n\, \phi_n(w') \\ &\equiv^+ u'w'\underline{\Delta}_n \equiv^+ \underline{\Delta}_n^{\ell-1}\underline{\Delta}_n = \underline{\Delta}_n^\ell. \end{aligned}$$ □

Formula (3.7) has now been shown: if u and v are words in $\mathcal{BW}_n^+$, there exist words u' and v' satisfying

$$uv' \equiv^+ \underline{\Delta}_n^\ell \equiv^+ vu', \tag{3.15}$$

as soon as $\ell \geqslant \max(|u|, |v|)$. This implies the existence of common right-multiples in B_n^+, with the following even stronger result.

Proposition 3.1.31 (Common right-multiples) *The element Δ_n^ℓ is a right-multiple of any element of B_n^+ of length $\leqslant \ell$.*

In other words, not only do there exist common right-multiples in B_n^+, but Δ_n^ℓ is a common right-multiple for *every* short enough element.

The result of Proposition 3.1.31 is what we sought and suffices in order to apply Ore's theorem. Given the minimum amount of effort required, we mention here the symmetric result. In the monoid B_∞^+, as in any monoid, we can certainly say that a is *right-divisor* of b, or equivalently, that b is a *left-multiple* of a if there exists x satisfying $xa = b$.

Proposition 3.1.32 (Common left-multiples) *The element Δ_n^ℓ is a left-multiple of any element of B_n^+ of length $\leqslant \ell$.*

Proof We begin by showing by induction on $n \geqslant 1$ the relation

$$\widetilde{\Delta_n} = \Delta_n, \tag{3.16}$$

where $\widetilde{\ }$ refers to the reversal of Lemma 3.1.12. The result is trivial for $n = 1$. Suppose $n \geqslant 2$. By (3.13), the fact that $\widetilde{\ }$ is an antihomomorphism, and the induction hypothesis, we obtain

$$\widetilde{\Delta_n} = (\Delta_{n-1}\,\sigma_{n,1})\widetilde{\ } = \sigma_{1,n}\,\widetilde{\Delta_{n-1}} \equiv^+ \sigma_{1,n}\,\underline{\Delta}_{n-1} = \Delta_n,$$

where $\sigma_{i,j}$ is the element of B_∞^+ represented by $\underline{\sigma}_{i,j}$. Now let a be an element of B_n^+ of length $\leqslant \ell$. Then $\widetilde{a}$ is also of length $\leqslant \ell$, and hence by Lemma 3.1.30 there exists an element b satisfying $\widetilde{a}\,b = \Delta_n^\ell$. We deduce the equality $\widetilde{b}\,a = \widetilde{\Delta_n^{\ell}}$, hence also, by (3.16), $\widetilde{b}\,a = \Delta_n^\ell$. □

Thus, not only do there exist common left-multiples in B_n^+, but Δ_n^ℓ is a common left-multiple of any element of length at most ℓ.

3.2 Cancellation in B_n^+

The first of the two conditions required to apply Ore's theorem to the monoid B_n^+ is thus fulfilled, and we pass to the second, the cancellativity. There does not exist a simple proof for the case of the monoid B_n^+; we will use a method whose general version is presented in (Dehornoy, 2019) and that can be seen as an elaboration of F.A. Garside's approach in his thesis (Garside, 1965).

3.2.1 Reversing Grids

The central notion is a certain type of planar diagram composed of horizontal and vertical arrows labelled by the letters σ_i or by the symbol ε.

Definition 3.2.1 (Grid) A *grid*[5] is a planar diagram obtained by juxtaposing rectangular cells of the six following types in such a way that the labels of their common arrows coincide:

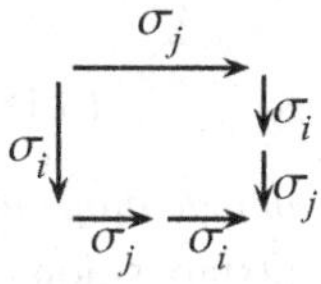

$$\begin{array}{ccc} & \xrightarrow{\sigma_j} & \\ \sigma_i\big\downarrow & & \big\downarrow\sigma_i \\ & & \big\downarrow\sigma_j \\ & \xrightarrow[\sigma_j]{}\xrightarrow[\sigma_i]{} & \end{array} \text{ with } |i-j| = 1, \qquad \begin{array}{ccc} & \xrightarrow{\sigma_i} & \\ \sigma_j\big\downarrow & & \big\downarrow\sigma_j \\ & \xrightarrow[\sigma_i]{} & \end{array} \text{ with } |i-j| \geqslant 2,$$

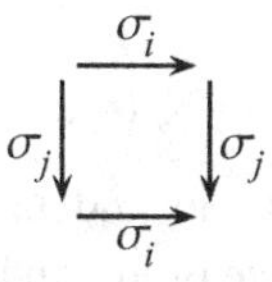

[5] Or, more precisely, a 'reversing grid', see Section 3.3.2.

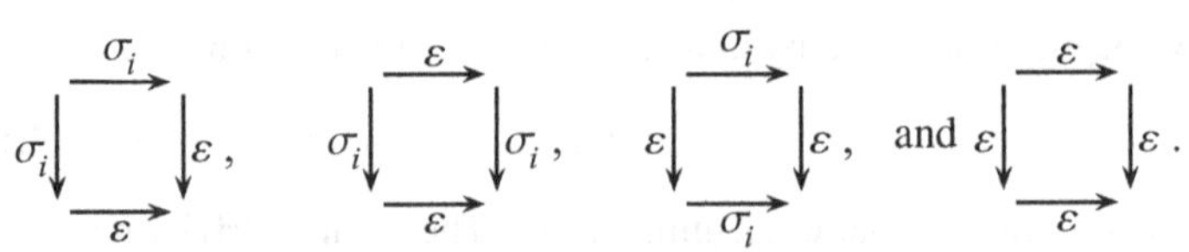

For u, v, u_1, v_1 in $\mathcal{BW}^+_\infty$, a grid Γ is said to admit (u, v) for *source* and (u', v') as *target*, if the products of the labels of the left and top boundaries of Γ form the words u and v, whereas those of the right and bottom form the words u' and v'. If there exists (at least one) a grid with source (u, v) and target (u', v'), we write $(u, v) \searrow (u', v')$, or again, in the form of a diagram, $u \Big\downarrow \overset{v}{\underset{v'}{\searrow}} \Big\downarrow u'$.

Example 3.2.2 (Grid) The grid of (3.17) contains eight cells, of which five correspond to the types of Definition 3.2.1. Its source is the pair $(\sigma_1, \sigma_2\sigma_3\sigma_2)$, and its target the pair $(\sigma_1\sigma_2\sigma_3, \sigma_2\sigma_1\sigma_3\sigma_2\sigma_1)$. It witnesses the relation $(\sigma_1, \sigma_2\sigma_3\sigma_2) \searrow (\sigma_1\sigma_2\sigma_3, \sigma_2\sigma_1\sigma_3\sigma_2\sigma_1)$, alias $\sigma_1 \Big\downarrow \overset{\sigma_2\sigma_3\sigma_2}{\underset{\sigma_2\sigma_1\sigma_3\sigma_2\sigma_1}{\searrow}} \Big\downarrow \sigma_1\sigma_2\sigma_3$.

Here is the grid. Note that the size of the arrows varies: in a cell of the first type, there is only one arrow on the left or on top, whereas there are two shorter ones on the right and on the bottom.

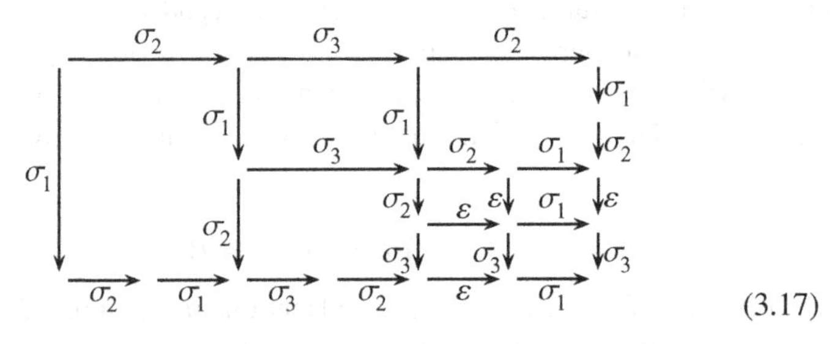

(3.17)

The first observation is easy: in the six types of cells of Definition 3.2.1, the labels of the two possible paths between the top left and bottom right corners form $\equiv^+$-equivalent words. An induction on the number of cells (do it!) gives an analogous result for any grid.

Lemma 3.2.3 *For any u, v, u_1, v_1, the relation $(u, v) \searrow (u_1, v_1)$ implies $uv_1 \equiv^+ vu_1$. In particular,*

$$(u, v) \searrow (\varepsilon, \varepsilon) \quad \textit{implies} \quad u \equiv^+ v. \tag{3.18}$$

A grid with source (u, v) and target (u_1, v_1) is a *van Kampen diagram* witnessing the equivalence of uv_1 and vu_1: such a diagram is a juxtaposition of cells analogous to those of Definition 3.2.1, but without the constraint of being drawn on a rectangular grid and hence having at most two arrows starting at each corner. It is easy to show that for any presented monoid, if two words are equivalent, there exists a van Kampen diagram linking them. However, it is not at all evident that there exists a grid witnessing their equivalence, that is, a rectangular diagram.

The second observation is that any grid containing more than one cell can be decomposed into subgrids

Lemma 3.2.4 *For any u, v', v'', u_1, v_1, the relation $(u, v'v'') \searrow (u_1, v_1)$ is satisfied if and only if there exist words u', v_1', v_1'' satisfying $(u, v') \searrow (u', v_1')$, $(u', v'') \searrow (u_1, v_1'')$, and $v_1 = v_1'v_1''$.*

Proof Let Γ be a grid going from $(u, v'v'')$ to (u_1, v_1). By definition, Γ is a juxtaposition of elementary cells as in Definition 3.2.1. By regrouping cells first under v', and then under v'', we split Γ into two grids Γ' and Γ''. By construction, the source of Γ' is (u, v'). Let (u', v_1') be its target. By construction, the source of Γ' is (u', v''), and its target has the form (u_1, v_1''), with $v_1 = v_1'v_1''$. The condition is hence necessary.

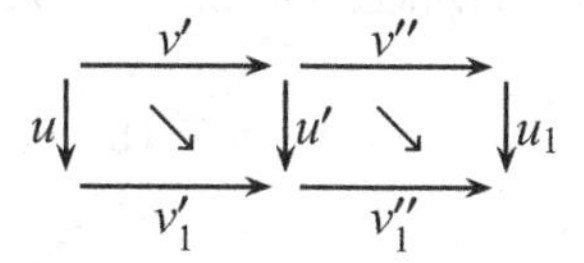

Conversely, by juxtaposing a grid from (u, v') to (u', v_1') with a grid from (u', v'') to (u_1, v_1''), we obtain a grid from $(u, v'v'')$ to $(v_1'v_1'', u_1)$, and the condition is also sufficient. □

The statement has a 'vertical' counterpart (two superposed grids instead of two grids juxtaposed horizontally).

3.2.2 Pairs of Stable Words

We arrive here to crucial notions expressing a compatibility of grids with the relation $\equiv^+$.

Definition 3.2.5 (Equivalence, stability)
(i) Two grids Γ, Γ' are said to be *equivalent* if the labels of the four corners of Γ form $\equiv^+$-equivalent words to their counterparts in Γ'.
(ii) A pair of braid words (u, v) is said to be *stable* if, for any grid Γ with source (u, v) and for any u', v' satisfying $u' \equiv^+ u$ and $v' \equiv^+ v$, there exists a grid with source (u', v') equivalent to Γ.

Therefore, (u, v) is stable if, when replacing u and v by equivalent words u', v', any grid with source (u, v) admits a counterpart with source (u', v') whose target words are equivalent.

We will show in Lemma 3.2.10 that every pair of braid words is stable. For the moment, we observe that the pairs formed by a letter and a member of a braid relation are stable. We begin with the braid relations of length 2.

Lemma 3.2.6 *For any i, j, k with $|j-k| \geqslant 2$, the pairs $(\sigma_i, \sigma_j\sigma_k)$ and $(\sigma_i, \sigma_k\sigma_j)$ are stable.*

Proof What counts is the relative position of the integers i, j, k and their mutual differences: 0, 1, or at least 2. It suffices to examine the various possible cases, which can be grouped into four families. In each case, we will find that there exists one and only one grid with source $(\sigma_i, \sigma_j\sigma_k)$, and one and only one grid with source $(\sigma_i, \sigma_k\sigma_j)$, and that these grids are equivalent, thus establishing the stability since $\{\sigma_i\}$ is the equivalence class of σ_i for $\equiv^+$, whereas $\{\sigma_j\sigma_k, \sigma_k\sigma_j\}$ is that of $\sigma_j\sigma_k$ and $\sigma_k\sigma_j$.

To improve the readability, we suppose $j = 1$, $k = 3$, and choose typical values for i: it is easy to be convinced that all possible configurations are thus handled.

Case 1: $|i - j| \geqslant 2$ and $|i - k| \geqslant 2$, say $i = 5$. The associated grids are

$$\begin{array}{ccccc} \cdot & \xrightarrow{\sigma_1} & \cdot & \xrightarrow{\sigma_3} & \cdot \\ \sigma_5\downarrow & & \downarrow\sigma_5 & & \downarrow\sigma_5 \\ \cdot & \xrightarrow[\sigma_1]{} & \cdot & \xrightarrow[\sigma_3]{} & \cdot \end{array} \quad \text{and} \quad \begin{array}{ccccc} \cdot & \xrightarrow{\sigma_3} & \cdot & \xrightarrow{\sigma_1} & \cdot \\ \sigma_5\downarrow & & \downarrow\sigma_5 & & \downarrow\sigma_5 \\ \cdot & \xrightarrow[\sigma_3]{} & \cdot & \xrightarrow[\sigma_1]{} & \cdot \end{array},$$

equivalent since $\sigma_1\sigma_3 \equiv^+ \sigma_3\sigma_1$.

Case 2: $|i - j| \geqslant 2$ and $|i - k| = 1$, say $i = 4$. The grids are

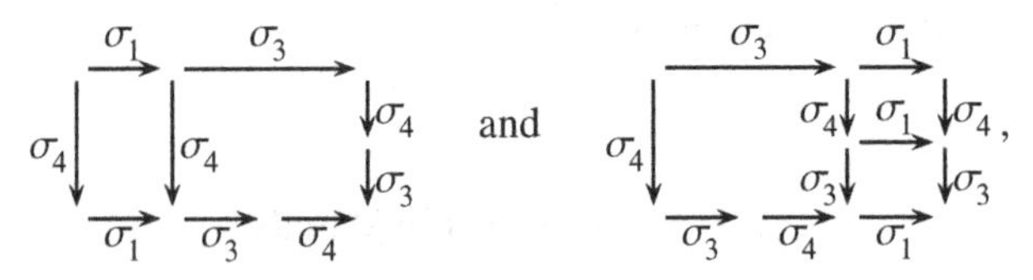

equivalent since $\sigma_1\sigma_3\sigma_4 \equiv^+ \sigma_3\sigma_4\sigma_1$.

Case 3: $|i - j| \geqslant 2$ and $i = k$, so $i = 3$. The grids are

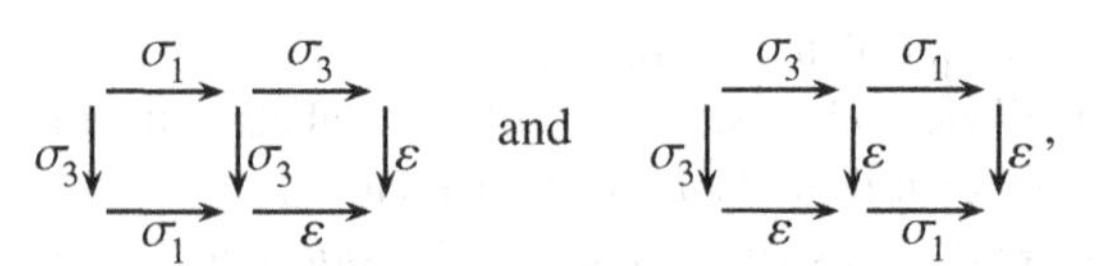

equivalent as $\sigma_1\varepsilon = \sigma_1 = \varepsilon\sigma_1$.

Case 4: $|i - j| = 1$ and $|i - k| = 1$, so $i = 2$. The grids are

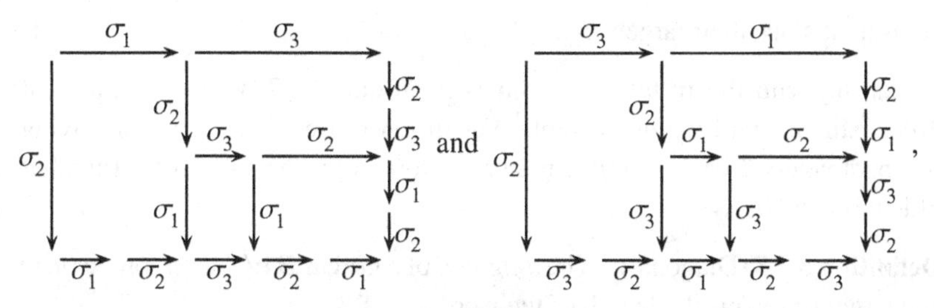

equivalent since $\sigma_2\sigma_3\sigma_1\sigma_2 \equiv^+ \sigma_2\sigma_1\sigma_3\sigma_2$ and

$$\sigma_1\sigma_2\sigma_3\sigma_2\sigma_1 \equiv^+ \sigma_1\sigma_3\sigma_2\sigma_3\sigma_1 \equiv^+ \sigma_3\sigma_1\sigma_2\sigma_1\sigma_3 \equiv^+ \sigma_3\sigma_2\sigma_1\sigma_2\sigma_3. \qquad \square$$

We treat similarly the braid relations of length 3.

Lemma 3.2.7 *For any i, j, k with $|j - k| = 1$, the pairs $(\sigma_i, \sigma_j\sigma_k\sigma_j)$ and $(\sigma_i, \sigma_k\sigma_j\sigma_k)$ are stable.*

Proof The verification is the same as in Lemma 3.2.6. We suppose $j = 1$ and $k = 2$. This time there are only three families to consider.

Case 1: $|i - j| \geqslant 2$ and $|i - k| \geqslant 2$, say $i = 4$. The grids are

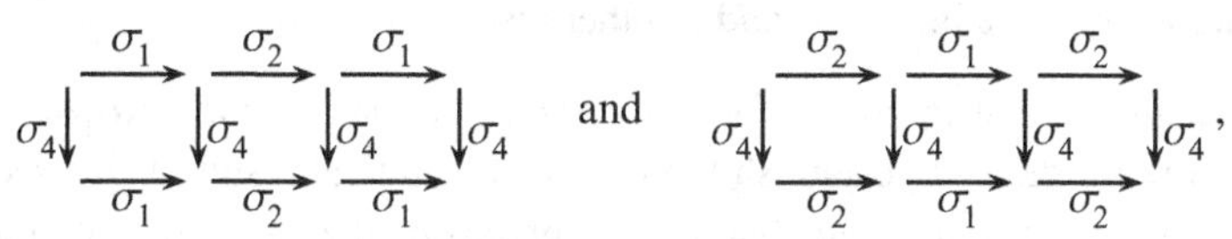

equivalent since $\sigma_1\sigma_2\sigma_1 \equiv^+ \sigma_2\sigma_1\sigma_2$.

Case 2: $|i - j| \geqslant 2$ and $|i - k| = 1$, so $i = 3$. The grids are

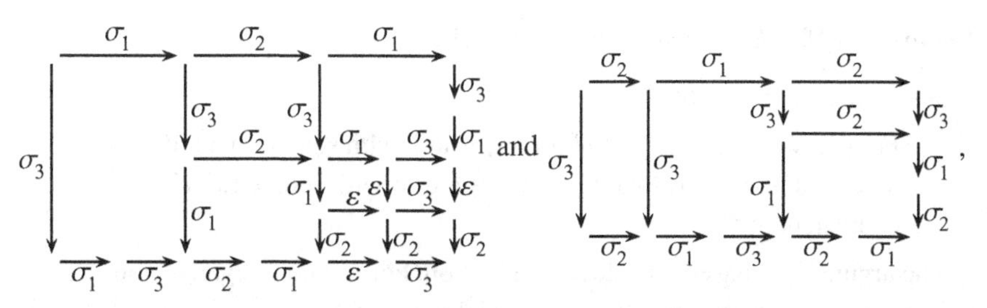

equivalent since $\sigma_3\sigma_1\sigma_2 = \sigma_3\sigma_1\varepsilon\sigma_2$ and

$$\sigma_2\sigma_1\sigma_3\sigma_2\sigma_1 \equiv^+ \sigma_2\sigma_1\sigma_2\sigma_3\sigma_1 \equiv^+ \sigma_1\sigma_2\sigma_1\sigma_3\sigma_1$$
$$\equiv^+ \sigma_1\sigma_2\sigma_3\sigma_1\sigma_3 \equiv^+ \sigma_1\sigma_3\sigma_2\sigma_1\sigma_3 = \sigma_1\sigma_3\sigma_2\sigma_1\varepsilon\sigma_3.$$

Case 3: $i = k$, suppose $i = 2$. The grids are

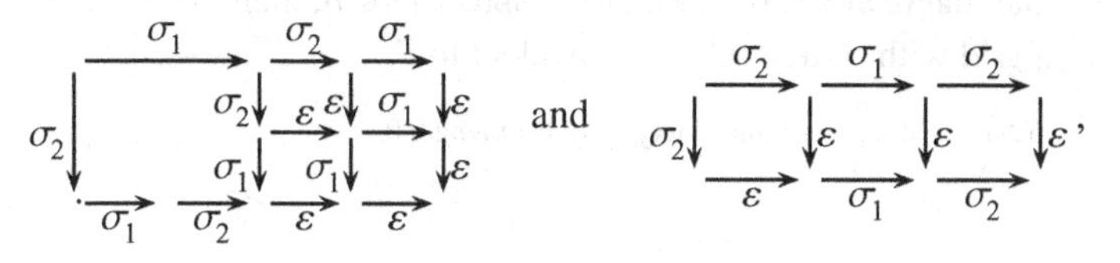

equivalent since their targets coincide. □

Starting with the results of Lemmas 3.2.6 and 3.2.7, we now show that every pair of braid words is stable. For this, we use an induction involving two parameters. Recall that, if w is a braid word, $|w|$ is the length of w (number of letters).

Definition 3.2.8 (Diagonal) The *diagonal* of a grid in $\mathcal{BW}_\infty^+$ with source (u, v) and target (u_1, v_1) is the length of the word uv_1.

For example, the diagonal of the grid of (3.17) is $|\sigma_1\sigma_2\sigma_1\sigma_3\sigma_2\varepsilon\sigma_1|$, or 6[6]. Note that in virtue of Lemma 3.2.3, the diagonal of a grid with source (u, v) and target (u_1, v_1) is also the length of vu_1, as $\equiv^+$-equivalent words have the same length.

Then, by Lemma 2.2.20, if two words w, w' are $\equiv^+$-equivalent, there exists a derivation (at least one) linking w to w'.

Definition 3.2.9 (Distance) The *(combinatorial) distance* between two words w, w', denoted $\mathrm{dist}(w, w')$, is the minimal length of a derivation linking w to w' if w and w' are $\equiv^+$-equivalent, and ∞ otherwise.

For example, $\mathrm{dist}(w, w) = 0$ for any word w, and $\mathrm{dist}(\sigma_1\sigma_2\sigma_1, \sigma_2\sigma_1\sigma_2) = 1$.

We are now ready to establish the principal technical result of this section. The proof is a bit delicate, but this is precisely the price to pay for the cancellativity in B_n^+: all the observations up to now have been quite simple, as are those to follow.

Lemma 3.2.10 *Every pair of words of $\mathcal{BW}_\infty^+$ is stable.*

Proof We must prove the following:

$\diamond$: For every u, v in $\mathcal{BW}_\infty^+$, and for any grid Γ with source (u, v), if $u' \equiv^+ u$ and $v' \equiv^+ v$, then there exists a grid with source (u', v') equivalent to Γ.

The argument is based on a double induction. For ℓ, d non-negative integers, we introduce two statements that are special cases of $\diamond$:

$\diamond_\ell$: For every u, v in $\mathcal{BW}_\infty^+$, and for any grid Γ with source (u, v) and diagonal $\leqslant \ell$, if $u' \equiv^+ u$ and $v' \equiv^+ v$, then there exists a grid with source (u', v') equivalent to Γ.

$\diamond_{\ell,d}$: For every u, v in $\mathcal{BW}_\infty^+$, and for any grid Γ with source (u, v) and diagonal $\leqslant \ell$, if $\mathrm{dist}(u', u) + \mathrm{dist}(v', v) \leqslant d$, then there exists a grid with source (u', v') equivalent to Γ.

[6] The contribution of ε is zero: the empty word is of length 0.

By definition, $\diamond$ is satisfied if and only if $\diamond_\ell$ is for every ℓ, and $\diamond_\ell$ is satisfied if and only if $\diamond_{\ell,d}$ is for every ℓ and d. We will establish $\diamond_{\ell,d}$ for every ℓ, d by induction on ℓ and, for each ℓ, by induction on d.

First consider $\ell = 0$. Suppose u, v are in $\mathcal{BW}^+_\infty$, Γ is a grid with source (u, v) and diagonal 0, and u' and v' in $\mathcal{BW}^+_\infty$ satisfy $u' \equiv^+ u$ and $v' \equiv^+ v$. If (u_1, v_1) is the target of Γ, then by definition $|uv_1| = |vu_1| = 0$, meaning u, v, u_1, and v_1 must be empty. Then, $u' \equiv^+ u$ and $v' \equiv^+ v$ impose that u' and v' are also empty. Thus Γ is the last type of grid of Definition 3.2.1, and it suffices to take $\Gamma' := \Gamma$. Hence, $\diamond_0$ is satisfied.

Consider now $\ell > 0$ and $d = 0$. Suppose u, v are in $\mathcal{BW}^+_\infty$, Γ is a grid with source (u, v) and diagonal ℓ, and u' and v' in $\mathcal{BW}^+_\infty$ satisfy $\mathsf{dist}(u', u) + \mathsf{dist}(v, v') = 0$. By definition, $u' = u$ and $v' = v$, and it suffices to choose $\Gamma' := \Gamma$. Hence, $\diamond_{\ell,0}$ is satisfied for any ℓ.

Next consider $\ell > 0$ and $d = 1$. Suppose u, v are in $\mathcal{BW}^+_\infty$, Γ is a grid with source (u, v) and diagonal ℓ, and u' and v' are words of $\mathcal{BW}^+_\infty$ satisfying $\mathsf{dist}(u', u) + \mathsf{dist}(v, v') = 1$. Up to a symmetry, we can suppose $u' = u$ and $\mathsf{dist}(v', v) = 1$. Then v' is obtained starting from v by applying exactly a braid relation, that is, there exist two words v_0, v_2 and a braid relation $w = w'$ satisfying $v = v_0 w v_2$ and $v' = v_0 w' v_2$. Applying Lemma 3.2.4, we can decompose Γ into the following three subgrids according to a schema of the type

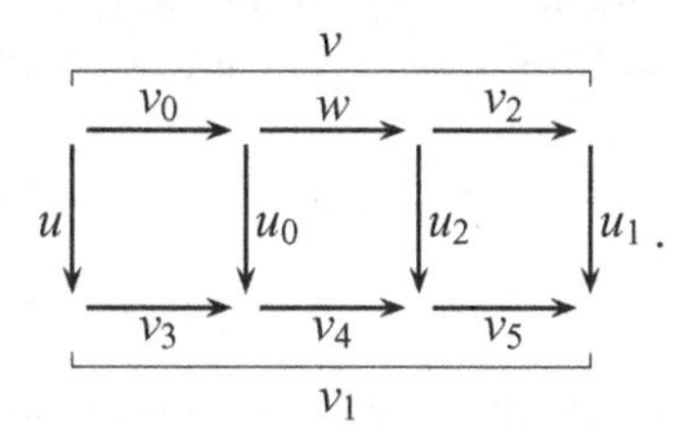

First suppose u_0 is empty. Then necessarily, u_2 and u_1 are empty, and $v_4 = w$ and $v_5 = v_2$. The grid Γ is thus as in the left image below, and taking for Γ' the grid on the right gives an equivalent grid with source (u, v').

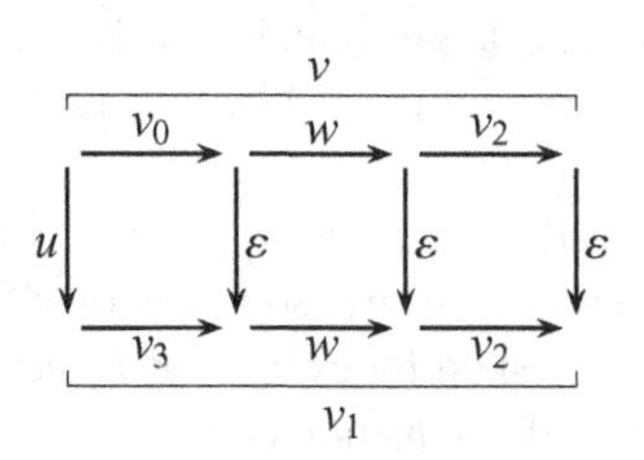

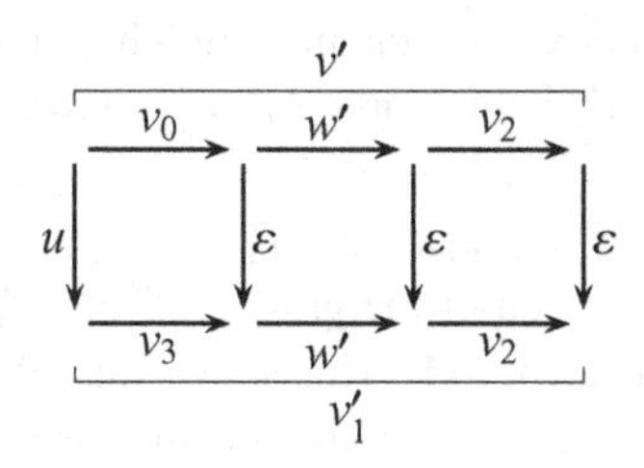

Now suppose u_0 nonempty. Write $u_0 = \sigma_i u_3$. Once again decomposing Γ, we obtain words $u_4, ..., u_7$ and v_6, v_7 such that Γ is the juxtaposition of five grids $\Gamma_0, ..., \Gamma_4$ as below on the left:

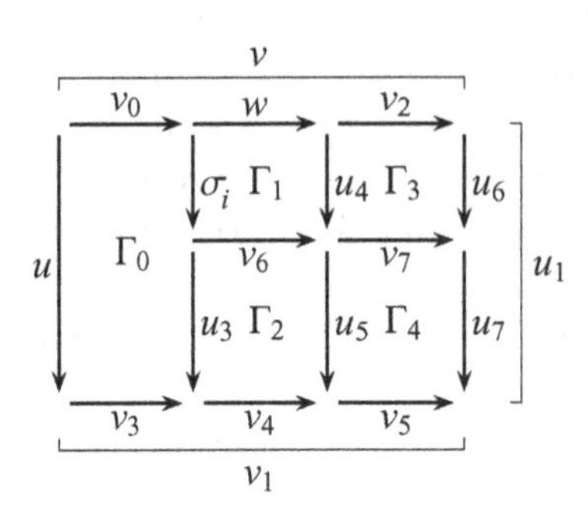

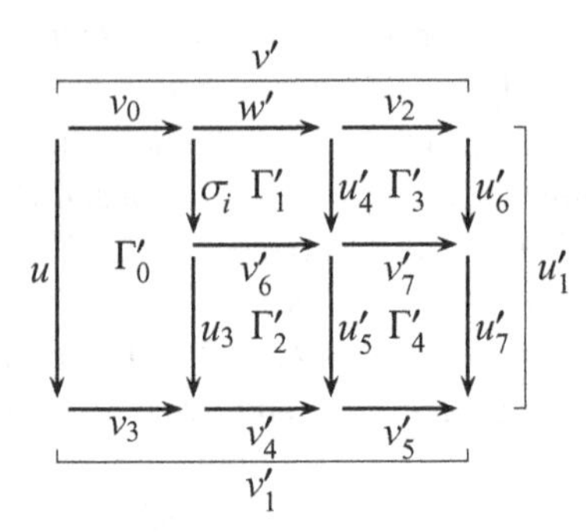

We establish the existence of a grid as above on the right, equivalent to Γ and with source (u', v'). We start with Γ_1. By Lemma 3.2.6 or 3.2.7, the pair (σ_i, w) is stable, hence there exists a grid Γ'_1 with source (σ_i, w') equivalent to Γ_1. Let (u'_4, v'_6) be its target; we thus have $u'_4 \equiv^+ u_4$ and $v'_6 \equiv^+ v_6$.

Next is Γ_2. We have $v'_6 \equiv^+ v_6$ and, in addition,

$$|u_3 v_4| < |\sigma_i u_3 v_4| \leqslant |v_0 \sigma_i u_3 v_4| \leqslant |v_0 s u_3 v_4 v_5| = |u v_1| \leqslant \ell,$$

thus $|u_3 v_4| < \ell$. By the induction hypothesis, $\diamond_{|u_3 v_4|}$ is satisfied, hence there exists a grid Γ'_2 with source (u_3, v'_6) equivalent to Γ_2. Let (u'_5, v'_4) be its target; then $u'_5 \equiv^+ u_5$ and $v'_4 \equiv^+ v_4$.

Now consider Γ_3. We have $u'_4 \equiv^+ u_4$ and, since $|w|$ is 2 or 3,

$$|u_4 v_7| < |w u_4 v_7| \leqslant |v_0 w u_4 v_7| \leqslant |v_0 w u_4 v_7 u_7| = |u v_1| \leqslant \ell,$$

so $|u_4 v_7| < \ell$. By the induction hypothesis, $\diamond_{|u_4 v_7|}$ is satisfied, hence there exists a grid Γ'_3 with source (u'_4, v_2) equivalent to Γ_3. Let (u'_6, v'_7) be its target; then $u'_6 \equiv^+ u_6$ and $v'_7 \equiv^+ v_7$.

We finish with Γ_4. We have $u'_5 \equiv^+ u_5$, $v'_7 \equiv^+ v_7$ and moreover,

$$|u_5 v_5| \leqslant |v_6 u_5 v_5| < |\sigma_i v_6 u_5 v_5| \leqslant |v_0 \sigma_i v_6 u_5 v_5| = |u v_1| \leqslant \ell,$$

so $|u_5 v_5| < \ell$. By the induction hypothesis, $\diamond_{|u_5 v_5|}$ is satisfied, hence there exists a grid Γ'_4 with source (u'_5, v'_4) equivalent to Γ_4. Let (u'_7, v'_5) be its target; then $u'_7 \equiv^+ u_7$ and $v'_5 \equiv^+ v_5$.

Set $u'_1 := u'_6 u'_7$ and $v'_1 := v_3 v'_4 v'_5$. Then $u'_1 \equiv^+ u_1$ and $v'_1 \equiv^+ v_1$, and the juxtaposition of the grids $\Gamma_0, \Gamma'_1, ..., \Gamma'_4$ is a grid Γ' whose source is (u, v') and is equivalent to Γ. This shows that, if $\diamond_{\ell'}$ is satisfied for every $\ell' < \ell$, then $\diamond_{\ell,1}$ is satisfied, thus completing the case $d = 1$ of the induction on ℓ.

Finally suppose $\ell > 0$ and $d \geqslant 2$. Let u, v be in $\mathcal{BW}_\infty^+$, Γ a grid with source (u, v) and diagonal ℓ, and u' and v' words of $\mathcal{BW}_\infty^+$ satisfying $\mathsf{dist}(u', u) + \mathsf{dist}(v', v) = d$. By definition, there exist u'', v'' satisfying

$$\mathsf{dist}(u'', u) + \mathsf{dist}(v'', v) = d - 1 \text{ and } \mathsf{dist}(u', u'') + \mathsf{dist}(v', v'') = 1.$$

If (u_1, v_1) is the target of Γ, then $|uv_1| \leqslant \ell$ and, by the induction hypothesis, $\diamond_{\ell,d-1}$ is satisfied. Hence there exists a grid Γ'' with source (u'', v'') equivalent to Γ. Let (u''_1, v''_1) be its target. Then $u'' \equiv^+ u$ and $v''_1 \equiv^+ v_1$ imply $|u''v''_1| = |uv_1| \leqslant \ell$. By the induction hypothesis, $\diamond_{\ell,1}$ is satisfied, hence there exists a grid Γ' with source (u', v') equivalent to Γ, thus also to Γ by the transitivity of $\equiv^+$. Hence $\diamond_{\ell,d}$ is satisfied, completing the induction. □

3.2.3 Application to the Simplification

We have seen in Lemma 3.2.3 that the existence of a grid with source (u, v) and target $(\varepsilon, \varepsilon)$ is a sufficient condition for u and v to be equivalent for $\equiv^+$. We are now in a position to establish the converse implication.

Proposition 3.2.11 (Equivalence) *Words u and v in $\mathcal{BW}_\infty^+$ are $\equiv^+$-equivalent if and only if $(u, v) \searrow (\varepsilon, \varepsilon)$.*

Proof

A simple induction on the length of u shows that, for any braid word u, there exists a grid with source (u, u) and target $(\varepsilon, \varepsilon)$: the result is trivial if u is empty. If not, we write $u = \sigma_i v$, and, applying the induction hypothesis, we obtain the desired grid for (u, u) by completing a grid for (v, v) according the schema to the right.

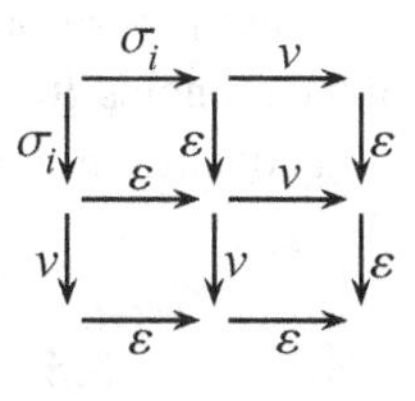

Now suppose $v \equiv^+ u$. Since (u, u) is stable and there exists a grid Γ with source (u, u) and target $(\varepsilon, \varepsilon)$, there exists a grid Γ' with source (u, v) equivalent to Γ. Let (u', v') be the target of Γ'. Hence $u' \equiv^+ \varepsilon$ and $v' \equiv^+ \varepsilon$: the only possibility is that u' and v' are empty. Hence the target of Γ' is $(\varepsilon, \varepsilon)$. □

In other words, we have established that

$$(u, v) \searrow (\varepsilon, \varepsilon) \quad \text{is equivalent to} \quad u \equiv^+ v. \tag{3.19}$$

The desired cancellativity result is now easy.

Proposition 3.2.12 (Left-cancellativity) *The monoid B^+_∞ is left-cancellative.*

Proof
Since the elements σ_i generate B^+_∞, it suffices to show that $\sigma_i a = \sigma_i b$ implies $a = b$ for any a, b in B^+_∞, or, equivalently, that $\sigma_i u \equiv^+ \sigma_i v$ implies $u \equiv^+ v$ for any u, v in $\mathcal{BW}^+_\infty$. Suppose $\sigma_i u \equiv^+ \sigma_i v$. By Proposition 3.2.11, there exists a grid Γ going from $(\sigma_i u, \sigma_i v)$ to $(\varepsilon, \varepsilon)$. By Lemma 3.2.4, the grid Γ can be decomposed into four as shown to the right:

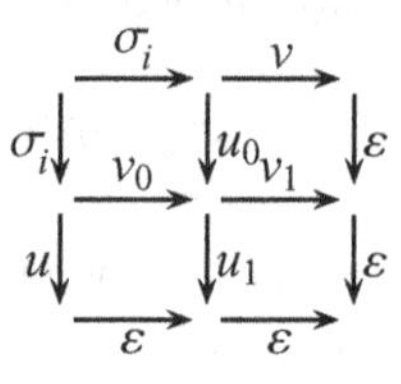

As there does not exist any braid relation of the type $\sigma_i... = \sigma_i...$, the only possibility is to have $u_0 = v_0 = \varepsilon$, hence $u_1 = u$ and $v_1 = v$. But then, the grid on the bottom right goes from (u, v) to $(\varepsilon, \varepsilon)$, and thus witnesses the equivalence $u \equiv^+ v$. Hence B^+_∞ is left-cancellative. □

Since, by 3.1.7, each monoid B^+_n is a submonoid of B^+_∞, we deduce that B^+_n is left-cancellative for every n.

For right simplification, we can exploit the palindromic character of the relations.

Proposition 3.2.13 (Right-cancellativity) *The monoid B^+_∞ is right-cancellative.*

Proof We use the antiautomorphism of Lemma 3.1.12. Suppose a, b are in B^+_∞ and $a\sigma_i = b\sigma_i$. By Lemma 3.1.12, we deduce $(a\sigma_i)^\sim = (b\sigma_i)^\sim$, or $\sigma_i\widetilde{a} = \sigma_i\widetilde{b}$. By left-cancellativity, we obtain $\widetilde{a} = \widetilde{b}$, and finally $a = b$. Thus B^+_∞ is right-cancellative.[7] □

As in Proposition 3.2.12, the right-cancellativity of B^+_∞ implies that of each of the monoids B^+_n.

In conclusion, for every $n \leqslant \infty$, the monoid B^+_n is cancellative.

3.3 Applications to the Study of Braids

3.3.1 The Group B_n as a Group of Fractions of B^+_n

Thanks to the results of Sections 3.1.3 and 3.2, we now know that each of the monoids B^+_n is cancellative and admits common right multiples. The conditions mentioned in Proposition 3.1.22 are thus fulfilled, and, supposing the validity

[7] Alternatively, we could develop a direct approach starting with the notion of a 'co-grid' where, guarding the same orientation of the arrows, we go from the bottom right towards the top left with elementary cells symmetric with those of Definition 3.2.1. Note that a co-grid is not a grid: the counterpart of $\begin{array}{c} \xrightarrow{\sigma_i} \\ \sigma_i\downarrow \quad \downarrow\varepsilon \\ \xrightarrow[\varepsilon]{} \end{array}$ is $\begin{array}{c} \xrightarrow{\varepsilon} \\ \varepsilon\downarrow \quad \downarrow\sigma_i \\ \xrightarrow[\sigma_i]{} \end{array}$, illegitimate in a grid.

of Ore's theorem, for which we refer to Appendix C, we can directly state the following fundamental result.

Proposition 3.3.1 (Embedding) *For every n, the homomorphism ι of Lemma 3.1.15 is an embedding of B_n^+ into B_n; any element of B_n can be written in the form $\iota(a)\iota(b)^{-1}$ with a, b in B_n^+.*

Exercise 3.3.2 (Shift) Show the compatibility of the endomorphism dec^+ of the monoid B_∞^+ and the endomorphism shift of the group B_∞.

We have thus responded positively to the embedding problem of Question 3.1.17. Since the homomorphism ι is injective, we can safely identify each element of B_n^+ and its image in B_n.

Convention 3.3.3 (Embedding) For every i, the generator σ_i of B_∞^+ is identified with the braid σ_i of B_∞.

With this, B_∞^+ can be identified with the submonoid of B_∞ generated by the braids σ_i, and its elements can from now on be called 'braids', like those of B_∞. Seen as a set of braids, B_∞^+ is formed by the braids that can be written in at least one way as products of the generators σ_i (without inverses); these braids are naturally called *positive*.[8]

As in Section 3.1.2, the injectivity of the homomorphism ι is expressed by the fact that the implication of (3.6) is an equivalence: for any w, w' in $\mathcal{BW}_\infty^+$,

$$w \equiv^+ w' \quad \Longleftrightarrow \quad w \equiv w'. \tag{3.20}$$

In other words, the relation $\equiv^+$ is the restriction of $\equiv$ to the positive words. Thus there is no longer any need to use distinct symbols, and from now on we will use $\equiv$ everywhere.

Now that the elements of B_∞^+ are (identified as) braids, we can verify that the equivalences of Section 3.1.3 correspond to isotopies. For example, in the braid diagram coded by $\underline{\sigma}_{i,j}$ in Lemma 3.1.23, the ith strand moves to the position j by passing under (case $i < j$) or over (case $i > j$) the intermediate strands:

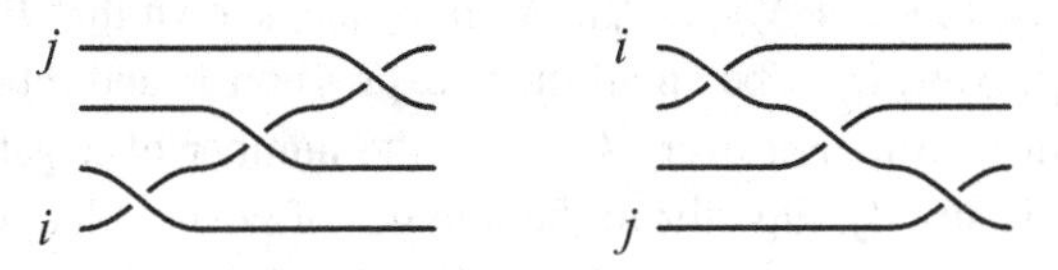

[8] Note that a positive braid also admits expressions that are not positive words: for example, the braid $\sigma_1\sigma_2$ is positive, while $\sigma_1\sigma_2\sigma_1^{-1}\sigma_1$ is a non-positive expression. Observe also that the mapping 'length', well-defined on B_n^+, does not extend to B_n: the braid $\sigma_1\sigma_2$, of length 2, is also represented by the word $\sigma_1\sigma_2\sigma_1^{-1}\sigma_1$, of length 4.

and the relations (3.9) and (3.10) correspond to migrating a supplementary crossing σ_k across the diagram.

Similarly, the braid diagram coded by $\underline{\Delta}_n$ is a half-turn of an n-strand braid, for example, for $\underline{\Delta}_4$,

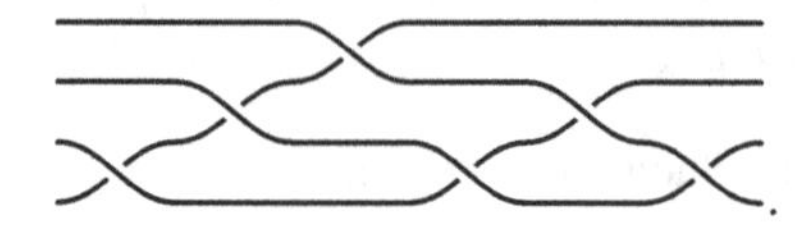

We conclude with two additional applications. The first again exploits the symmetries of the relations.

Proposition 3.3.4 (Left fractions) *The group B_n is a group of left fractions for the monoid B_n^+: any element of B_n can be expressed as $a^{-1}b$ with a and b in B_n^+.*

Proof Let g be arbitrary in B_n. Since B_n is a group of right fractions of B_n^+, there exists c and d in B_n^+ satisfying $\widetilde{g} = cd^{-1}$, where we recall that $\widetilde{g}$ is the image of g by the reversing of Proposition 2.1.26. Hence $g = (cd^{-1})^{\sim} = \widetilde{d^{-1}}\,\widetilde{c} = \widetilde{d}^{-1}\widetilde{c}$, or $g = a^{-1}b$ with $a := \widetilde{d}$ and $b := \widetilde{c}$. □

We next exploit the properties of the elements Δ_n of Definition 3.1.24 to make the powers Δ_n^m into universal denominators.[9]

Proposition 3.3.5 (Delta-fraction) *Any element g of B_n having an expression with p negative letters can be expressed as $\Delta_n^{-m}b$ with m in $\mathbb{Z}$ satisfying $m \leqslant p$ and b in B_n^+; the pair (m, b) is unique if in addition we require $\Delta_n \not\preccurlyeq b$.*

Proof We first show by induction on $p \geqslant 0$ that the braid $\Delta_n^p g$ belongs to B_n^+. For $p = 0$, the result is trivial. Suppose $p \geqslant 1$, and that an expression of g is $u\sigma_i^{-1}v$, where u contains at most $p-1$ negative letters and v is positive. Then $\Delta_n^p g = \Delta_n^p[u]\sigma_i^{-1}[v] = \Delta_n(\Delta_n^{p-1}[u])\sigma_i^{-1}[v]$. By the induction hypothesis, $\Delta_n^{p-1}[u]$ belongs to B_n^+, and consequently, $\Delta_n^p g = \phi_n(\Delta_n^{p-1}[u])(\Delta_n\sigma_i^{-1})[v]$. From this, since $\Delta_n\sigma_i^{-1}$ is positive by Lemma 3.1.28, $\Delta_n^p g$ is positive as the product of positive braids.

Next, let $P := \{k \in \mathbb{Z} \mid \Delta_n^k g \in B_n^+\}$. We have just shown that P is nonempty and contains p. Now, let w be an arbitrary expression g, and q be the number of positive letters in w. For every $k \leqslant -q$, the number of negative letters in the word $\underline{\Delta}_n^k w$ is strictly superior to the number of positive letters, hence this word does not represent a positive braid. We thus have $\Delta_n^k g \notin B_n^+$, so $k \notin P$. The set P is hence bounded below, so there exists a minimal integer m in P.

[9] Think of the decimal rationals, which can be expressed as fractions whose denominators are powers of 10.

Set $b := \Delta_n^m g$. By construction, b is positive, and $g = \Delta_n^{-m} b$. Moreover, if we had $\Delta_n \preccurlyeq b$, we could write $g = \Delta_n^{-(m-1)}(\Delta_n^{-1} b)$, contradicting the minimality of m in P. Thus there exists (m, b) in $\mathbb{Z} \times B_n^+$ satisfying $g = \Delta_n^{-m} b$ and $\Delta_n \not\preccurlyeq b$.

Finally, suppose $g = \Delta_n^m b = \Delta_n^{m'} b'$ with b, b' in B_n^+ and $\Delta_n \not\preccurlyeq b$ and $\Delta_n \not\preccurlyeq b'$. If $m \leqslant m'$, we can write $g = \Delta_n^{m'} b' = \Delta_n^m \Delta_n^{m'-m} b'$, hence $b = \Delta_n^{m'-m} b'$. The hypothesis $\Delta_n \not\preccurlyeq b$ then implies $m' = m$, and from this, $b' = b$. □

Example 3.3.6 (Delta-fraction) For $g := \sigma_2^{-2} \sigma_1^{-2} \sigma_2^2 \sigma_1^2$, the result predicts the existence of a unique pair (m, b) satisfying $g = \Delta_3^{-m} b$ with $m \leqslant 4$, $b \in B_3^+$, and $\Delta_3 \not\preccurlyeq b$. In this case, the values are $m = 3$ and $b = \sigma_2 \sigma_1^3 \sigma_2^3 \sigma_1^2$.

Exercise 3.3.7 (Centre)
(i) Show that Δ_n^2 is central in the monoid B_n^+ (i.e. commutes with every element).
(ii) Show that an element g of B_n is central if and only if it can be written $\Delta_n^{2m} a$ with a central in B_n^+.

A consequence of the above is the decidability of the word problem for B_n. The solution obtained here is just as calamitous in practice as that of Proposition 3.1.10, on which it is based, and a much better solution will be described in Section 3.3.3.

Corollary 3.3.8 (Decidability) *For every $n \leqslant \infty$, the word problem of the group B_n is decidable.*

Proof Let w be an arbitrary element of $\mathcal{BW}_\infty$. Suppose that w belongs to $\mathcal{BW}_n$ and contains m letters σ_i^{-1}. By Proposition 3.3.5, we can effectively find a word u in $\mathcal{BW}_n^+$ satisfying $w \equiv \underline{\Delta}_n^{-m} u$. Then $w \equiv \varepsilon$ is equivalent to $\underline{\Delta}_n^{-m} u \equiv \varepsilon$, hence to $u \equiv \underline{\Delta}_n^m$, or again to $u \equiv^+ \underline{\Delta}_n^m$, a relation which, by Proposition 3.1.10, is decidable. □

Exercise 3.3.9 (Automorphism) Show that $\sigma_i \mapsto \sigma_{n-i}$ can be extended to an automorphism of the monoid B_n^+, then to an automorphism of the group B_n, in fact the inner automorphism associated with conjugation by Δ_n.

3.3.2 A Solution of the Word Problem for B_n^+

We will now use the results of Section 3.2 to describe solutions for the word problems in the monoid B_n^+ and the group B_n much less distressing than the calamitous solutions of Proposition 3.1.10 and Corollary 3.3.8. We begin with the case of B_n^+.

First, we establish two additional properties of reversing grids for the presentation of the monoid B_∞^+. The first is a uniqueness result.

Lemma 3.3.10 *For any words u, v in $\mathcal{BW}^+_\infty$, there exists at most one grid with source (u, v).*

Proof An inspection of the braid relations shows that, for every pair of generators $\{\sigma_i, \sigma_j\}$, there does not exist any relation $\sigma_i \ldots = \sigma_j \ldots$ for $i = j$, and there exists exactly one for $i \neq j$. Hence, if s, t are either a letter σ_i, or ε, there exists in Definition 3.2.1 exactly one cell of the type $s\downarrow \overset{t}{\longrightarrow} \downarrow$. Hence, given a pair (u, v), there exists at most one way to construct a grid with source (u, v) starting from the top left corner and proceeding down and to the right: at no time is there a blockage, nor a multiplicity of choices. □

The preceding argument does not guarantee the existence of a grid: the inductive process described above might never terminate, with smaller and smaller arrows appearing without end (see Exercise 3.3.12). Such a phenomenon is impossible for the grids in B^+_∞.

Lemma 3.3.11 *For any words u, v in $\mathcal{BW}^+_\infty$, there exists exactly one grid with source (u, v). Moreover, if (u', v') is its target, then every common right-multiple of $[u]^+$ and $[v]^+$ in B^+_∞ is a right-multiple of the element $[uv']^+$.*

Proof
Select arbitrary u, v in $\mathcal{BW}^+_\infty$. By Equation (3.7), the elements $[u]^+$ and $[v]^+$ of B^+_∞ admit at least one common right-multiple, hence there exist positive words u'' and v'' satisfying $uv'' \equiv^+ vu''$. By Proposition 3.2.11, there exists a grid Γ leading from (uv'', vu'') to $(\varepsilon, \varepsilon)$. By Lemma 3.2.4, Γ can be decomposed in four as shown to the right.

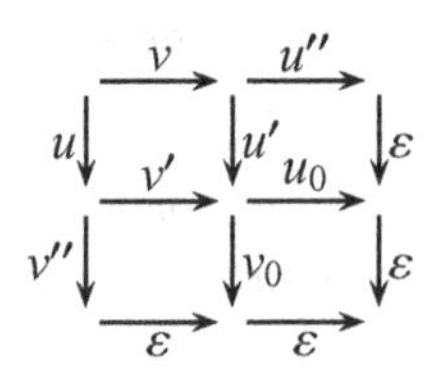

The subgrid on the top left is a grid with source (u, v). Moreover, we can read on the diagram the equality $[uv'']^+ = [uv']^+[v_0]^+$ in B^+_n, showing that $[uv'']^+$ is necessarily a right-multiple of $[uv']^+$. □

Exercise 3.3.12 (No such grid) Show that there does not exist a grid with source $(\mathsf{a}, \mathsf{bc})$ with respect to the presentation

$$\langle \mathsf{a}, \mathsf{b}, \mathsf{c} \mid \mathsf{aba} = \mathsf{bab}, \mathsf{bcb} = \mathsf{cbc}, \mathsf{cac} = \mathsf{aca} \rangle^+.$$

The preceding results of existence and unicity will allow us to define an effective algorithm. To ease the implementation, we begin with an alternative description of the inductive construction of Lemma 3.3.10 based on signed braid words, hence words[10] of $\mathcal{BW}_\infty$. The main ingredient here is a system of rewriting for such signed words.

Definition 3.3.13 (Reversing) For w, w' in $\mathcal{BW}_\infty$, w is said to be *reversible to* w' *in one step*, denoted $w \curvearrowright^1 w'$, if it is possible to write $w = uw_0v$ and $w' = uw'_0w$ with one of the following:
- $w_0 = \overline{\sigma}_i\sigma_i$ and $w'_0 = \varepsilon$,
- $w_0 = \overline{\sigma}_i\sigma_j$ with $|i-j| \geqslant 2$ and $w'_0 = \sigma_j\overline{\sigma}_i$, or
- $w_0 = \overline{\sigma}_i\sigma_j$ with $|i-j| = 1$ and $w'_0 = \sigma_j\sigma_i\overline{\sigma}_j\overline{\sigma}_i$.

The word w is said to *reduce* to w', denoted $w \curvearrowright w'$, if there exists a finite sequence $w_0, ..., w_m$ with $w_0 = w$, $w_m = w'$, and $w_k \curvearrowright^1 w_{k+1}$ for every k.

The terminology is natural: carrying out a reversing step in w consists in isolating a negative–positive factor of length 2, and substituting for it a positive–negative factor of length 0, 2, or 4: in short, we reverse $-+$ to $+-$.

Example 3.3.14 (Reversing) We start with $\overline{\sigma}_2^2\overline{\sigma}_1^2\sigma_2^2\sigma_1^2$. To simplify the expressions, we use a, b, ... for $\sigma_1, \sigma_2, ...$ and A, B, ... for $\overline{\sigma}_1, \overline{\sigma}_2, ...$, whereby the starting point is $w :=$ BBAAbbaa. We seek a positive–negative factor of length 2 to transform. Here, the only choice is the central factor Ab, which the definition prescribes replacing by baBA, leaving BBAbaBAbaa. Iterating the procedure and seeking each time the first reversible factor, we obtain the following stages:

0: BBAAbbaa
1: BBAbaBAbaa
2: BBbaBAaBAbaa
3: BaBAaBAbaa
4: abABBAaBAbaa
5: abABBBAbaa
6: abABBBbaBAaa
7: abABBaBAaa
8: abABabABBAaa
9: abAabABbABBAaa
10: abbABbABBAaa
11: abbAABBAaa
12: abbAABBa
13: abbAABabAB
14: abbAAabABbAB
15: abbAbABbAB
16: abbbaBAABbAB
17: abbbaBAAAB

and the reversing stops since, in the last word obtained, all the positive letters come before all the negative letters, hence there are no more factors that can be reversed.

[10] It is preferable here to use $\overline{\sigma}_i$ rather than σ_i^{-1} for the formal inverse of σ_i: the notation $\overline{\sigma}_i$ is more compact and has the advantage of reminding us that we are considering words, and not the elements they represent in the group B_∞.

Clearly the following result holds.

Lemma 3.3.15 *For any w, w', the relation $w \curvearrowright w'$ implies $w \equiv w'$.*

Proof Each stage of the reversing consists in replacing a factor by an $\equiv$-equivalent word. □

Constructing the grid with source (u, v) is equivalent to reversing the signed word $\overline{u}v$.

Lemma 3.3.16 *For any u, v, u', v' in $\mathcal{B}W_{\infty}^{+}$, the unique grid with source (u, v) admits (u', v') for target if and only if $\overline{u}v \curvearrowright v'\overline{u'}$.*

Proof Let Γ be a grid with source (u, v) and target (u', v'). We consider the signed words coding the paths of Γ starting from the bottom left corner and arriving at the top right corner and traversing the vertical arrows only from top to bottom and the horizontal arrows from left to right. The coding consists in concatenating the labels of the arrows traversed with the convention that a horizontal (*resp.*, vertical) arrow labelled σ_i contributes σ_i (*resp.*, $\overline{\sigma_i}$). For example, the left and top boundaries provide a path labelled $\overline{u}v$, and those on the bottom and right provide a path labelled $v'\overline{u'}$. The point is that, if γ is an elementary cell of Γ, and if w codes a path passing by the left and top boundaries of γ while w' codes the analogous path passing by those on the bottom and right, then w reverses to w' in one step. For example, if γ is of the type $\begin{array}{ccc} & \xrightarrow{\sigma_j} & \\ \sigma_i\big\downarrow & & \big\downarrow\sigma_i \\ & & \big\downarrow\sigma_j \\ & \xrightarrow[\sigma_j]{}\xrightarrow[\sigma_i]{} & \end{array}$ with $|i - j| = 1$, the contribution of the left and top boundaries is $\overline{\sigma_i}\sigma_j$, whereas that of the bottom and right boundaries is $\sigma_j\sigma_i\overline{\sigma_j}\,\overline{\sigma_i}$, and this is a typical case of reversing. It is easy to see that this is the same for all the elementary cells of Definition 3.2.1 (show this!). We conclude that the relation $\overline{u}v \curvearrowright v'\overline{u'}$ is satisfied.

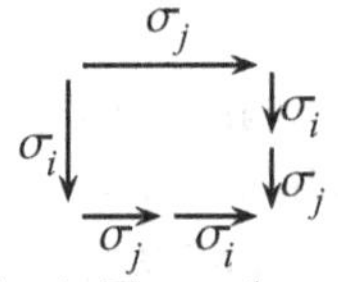

Inversely, when $\overline{u}v \curvearrowright v'\overline{u'}$, consider a finite sequence of reversing steps leading from $\overline{u}v$ to $v'\overline{u'}$. Let Γ_0 be the diagram obtained by writing vertically and from top to bottom the letters of u and horizontally from left to right those of v starting from a common summit. Then following the sequence of elementary reversings starting with $\overline{u}v$ corresponds to constructing step by step from Γ_0 a grid Γ with source (u, v) and target (u', v'). □

Trying a few examples quickly shows that by hand, it is more convenient to construct grids, whereas reversing is a better choice for computer algorithms, as it operates only on character strings.

From the above, we can derive a first algorithm resolving the word problem of B_{∞}^{+} with respect to $\{\sigma_1, \sigma_2, \ldots\}$.

Algorithm 3.3.17 (Word problem of B_∞^+ by reversing)
Input: two words u, v in $\mathcal{BW}_\infty^+$
Output: **yes** if u, v represent the same element of B_∞^+,
no otherwise

```
1: w ← ūv
2: REVERSE(w)
3: if w = ε then
4:     return yes
5: else
6:     return no
7: end if
```

```
 1: function REVERSE(w: word of BW∞)
 2:     while w contains a factor σ̄ᵢσⱼ do
 3:         w₀ ← the leftmost factor σ̄ᵢσⱼ in w
 4:         if w₀ = σ̄ᵢσᵢ then
 5:             replace w₀ by ε in w
 6:         else if w₀ = σ̄ᵢσⱼ with |i − j| = 1 then
 7:             replace w₀ by σⱼσᵢσ̄ⱼσ̄ᵢ in w
 8:         else if w₀ = σ̄ᵢσⱼ with |i − j| ⩾ 2 then
 9:             replace w₀ by σⱼσ̄ᵢ in w
10:         end if
11:     end while
12:     return w
13: end function
```

It is easy to prove that Algorithm 3.3.17 effectively resolves the word problem of B_∞^+ with respect to $\{\sigma_1, \sigma_2, ...\}$, hence, by specialization, that of B_n^+ with respect to $\{\sigma_1, ..., \sigma_{n-1}\}$.

Lemma 3.3.18 *Algorithm 3.3.17 always terminates and is correct: the initial words represent the same element of B_∞^+ if and only if the final word is empty.*

Proof The termination is guaranteed by Lemma 3.3.11: for any initial words u, v, there exists a grid with source (u, v), hence, by Lemma 3.3.16, the reversing of the word $\overline{u}v$ terminates with a positive–negative word $v'\overline{u'}$, the word w of step 3. By Proposition 3.2.11, the word w is empty if and only if the initial words u and v are equivalent. □

Example 3.3.19 (An instance of Algorithm 3.3.17) Starting with $u =: \sigma_1^2\sigma_2^2$ and $v := \sigma_2^2\sigma_1^2$, we again write these as aabb and bbaa. At step 1 of the algorithm, we form[11] $w := \overline{u}v$, or $w =$ BBAAbbaa. At step 2, we reverse this word, as in the example after Definition 3.3.13. The final word is abbbaBAAAB, obviously not empty. We conclude that u and v are not equivalent, and hence do not represent the same element in B_3^+, nor, consequently, in B_3.

If we would like to draw the grid corresponding to the above reversing, the appearance of smaller and smaller arrows ε renders the diagram unreadable. We can obtain simplified grids by authorizing new cells, regrouping several cells of Definition 3.2.1, typically

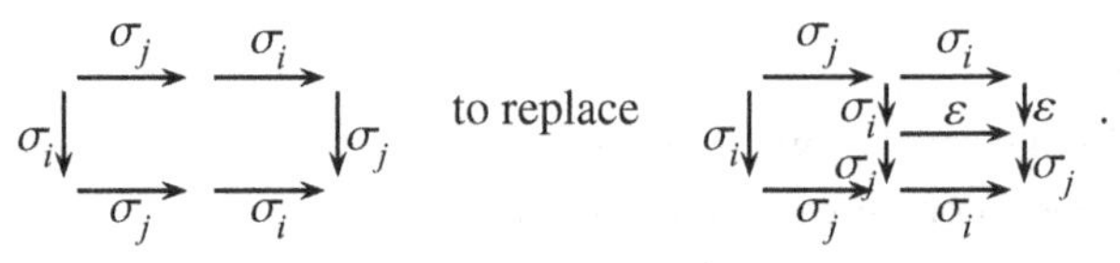

It is easy to verify that, in doing so, we do not change the final words obtained. With this convention, a simplified grid for the example of Definition 3.3.13 becomes

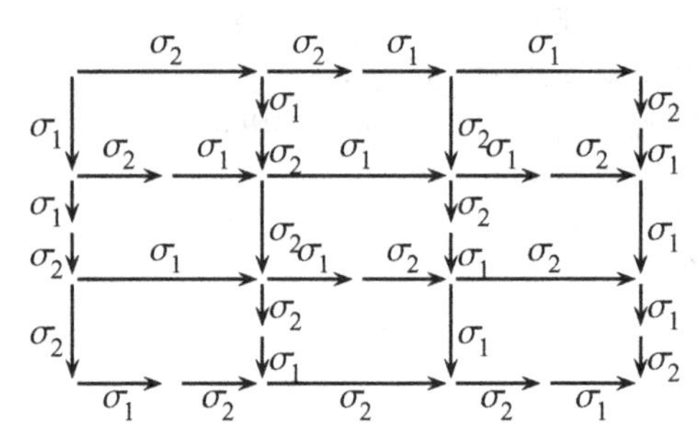

We have (finally!) responded to the question of Examples 1.3.11 and 2.1.23.

3.3.3 A Solution of the Word Problem for B_n

Thanks to the fact that the group B_n is the group of right fractions of the monoid B_n^+ and to the functioning of the reversing grids, little supplementary effort is required to go from the solution of the word problem of the monoid B_n^+ described above to a solution for the group B_n.

While, by definition, the source of a grid is a pair of positive words, the starting point of the reversing procedure described in Definition 3.3.13 is an arbitrary signed word, negative–positive or not. By Lemmas 3.3.16 and 3.3.11, the reversing of a word $\overline{u}v$ with positive u and v always terminates. In fact this is a general result.

[11] Recall that $\overline{u}$ is obtained from u by exchanging σ_i and $\overline{\sigma}_i$ and by reversing the order of the letters.

Lemma 3.3.20 *For any word w in $\mathcal{BW}_\infty$, the reversing of w terminates: there exist positive words u' and v' satisfying $w \curvearrowright v'\overline{u'}$.*

Proof Any nonempty signed word w can be written $\overline{u_1}v_1\overline{u_2}v_2 \cdots \overline{u_p}v_p$ with positive $u_1, ..., v_p$. We prove the result by induction on p. For $p=1$, the result is Lemma 3.3.11. Suppose $p \geqslant 2$, and set $w' := \overline{u_2}v_2 \cdots \overline{v_p}u_p$. Then $w = \overline{u_1}v_1w'$. First, by Lemma 3.3.11, there exist positive u_1' and v_1' satisfying $\overline{u_1}v_1 \curvearrowright v_1'\overline{u_1'}$. Next, by the induction hypothesis, there exist positive u_2' and v_2' satisfying $w' \curvearrowright v_2'\overline{u_2'}$, hence $w \curvearrowright v_1'\overline{u_1'}v_2'\overline{u_2'}$. Finally, again by Lemma 3.3.11, there exist positive u_3' and v_3' satisfying $\overline{u_1'}v_2' \curvearrowright v_3'\overline{u_3'}$. Putting this all together, we deduce

$$w = \overline{u_1}v_1w' \curvearrowright v_1'\overline{u_1'}w' \curvearrowright v_1'\overline{u_1'}v_2'\overline{u_2'} \curvearrowright v_1'v_3'\overline{u_3'}\,\overline{u_2'}.$$

This is the desired result, with $u' := u_2'u_3'$ and $v' := v_1'v_3'$. □

We are now ready to introduce a new algorithm based on the reversing relation on signed words. As usual, we provide the statement for the general case of B_∞; the case for each group B_n is deduced by specialization.

Algorithm 3.3.21 (Word problem of B_∞ by reversing)
Input: a signed braid word w in $\mathcal{BW}_\infty$
Output: **yes** if w represents 1 in B_∞,
no otherwise

1: REVERSE(w) ▹ at this stage, w is a word of type positive-negative
2: EXCHANGE(the positive and negative parts of w) ▹ if w was $v\overline{u}$, it is now $u\overline{v}$
3: REVERSE(w)
4: **if** $w = \varepsilon$ **then**
5: **return yes**
6: **else**
7: **return no**
8: **end if**

As was the case for Algorithm 3.3.17, it is easy to show that Algorithm 3.3.21 solves the word problem of the group B_∞ with respect to the generators σ_i, hence also solves the braid isotopy problem.

Lemma 3.3.22 *Algorithm 3.3.21 always terminates and is correct: the initial word represents* 1 *in B_∞ if and only if the final word is empty.*

Proof The procedure consists of two successive reversings, and its termination is thus guaranteed by Lemma 3.3.20. By Lemma 3.3.15 the positive–negative word $v\overline{u}$ obtained after step 1 is equivalent to the initial word w.

The relation $w \equiv \varepsilon$ is thus equivalent to $v\overline{u} \equiv \varepsilon$, hence to $u \equiv v$, then by embedding to $u \equiv^+ v$, and finally, to $\overline{u}v \curvearrowright \varepsilon$. Hence the algorithm is correct. □

Example 3.3.23 (An instance of Algorithm 3.3.21) We return to our favourite word $w := \overline{\sigma}_2^2\overline{\sigma}_1^2 o\sigma_2^2\sigma_1^2$ (alias $w := \sigma_2^{-2}\sigma_1^{-2}\sigma_2^2\sigma_1^2$), otherwise written $w :=$ BBAAbbaa (cf. Lemma 3.3.18). As seen in Definition 3.3.13, a first reversing gives a positive–negative word abbbaBAAAB. The exchange of the positive and negative factors leads to BAAABabbba, which then reverses to aabbAABB (the initial word backwards). This word is not empty, and we conclude that the initial word w does not represent 1 in B_∞, thus not in B_3. We have thus obtained[12] a definitive proof that the geometric braids $\sigma_1^2\sigma_2^2$ and $\sigma_2^2\sigma_1^2$ are *not* isotopic, a result beyond the power of the naive methods of Section 1.3.2 to establish.

In order to evaluate the efficiency of an algorithm, and notably to compare it to others, we count the number of elementary computation steps leading to the result, often a good approximation to the execution time on a computer. The result of course depends on what are considered the elementary steps, typically operations on a single character, letter or number. The total number of steps depends on the size of the input data, and the complexity is expressed as the order of magnitude of the number of steps as a function of the total length of the data. We speak then of algorithms in $O(\ell)$, $O(\ell^2)$, etc.[13] according to whether the total number of elementary steps is a function that is linear, quadratic, etc. of the cumulative length ℓ of the input data.

In the case of Algorithms 3.3.17 and 3.3.21, evaluating the complexity is difficult. It is easy to show that, for input words of length ℓ, a number of reversing steps of the type $\overline{u}v \curvearrowright v'\overline{u'}$ in $O(\ell^2)$ is sufficient, but results shown so far do not provide any bound on $|u'| + |v'|$ as a function of $|u| + |v|$, and hence no bound on the global complexity follows.

In fact, it would be possible, with a bit more work, to establish a global bound in $O(\ell^2)$ for the complete procedure in the group B_n, but we do not present this here, as better solutions will be shown in later chapters. We note here the extreme simplicity of the implementation of the reversing algorithms – mime Definition 3.3.13 and iterate – and the fact that they provide the most economical solution to the braid isotopy problem in terms of the underlying theoretical results, the only somewhat delicate point being the proof of the cancellativity of the monoids B_n^+.

[12] Finally, and after almost a hundred pages of effort! But on the way we have learned many other interesting results...

[13] The notation $O(f(\ell))$ signifies the existence of a constant C such that the number in question is at most $Cf(\ell)$.

3.4 Appendix C: Ore's Theorem

3.4.1 Statement of the Theorem

The general result – no more difficult to establish than for the special case of the monoids B_n^+ – concerns the embedding of a monoid in a group of fractions, and is an extension of the construction of the integers from the natural numbers, or the rational numbers from the integers.

Proposition 3.4.1 (Ore's theorem) *If M is a cancellative monoid and if every pair of elements of M admit a common right-multiple, then there exists an embedding ι of M into a group G whose elements can be written in the form $\iota(a)\iota(b)^{-1}$ with a and b in M.*

In the context of presented monoids, the preceding statement can be completed as follows.

Proposition 3.4.2 *In the context of Proposition 3.4.1, if the monoid M admits the presentation $\langle S \mid R \rangle^+$, the associated fraction group G admits the presentation $\langle S \mid R \rangle$.*

Clearly the statement of Proposition 3.1.22 can be deduced from these two results.

3.4.2 Proof of Ore's Theorem

The method is to construct a group structure on a quotient of the set of pairs of elements of M; the proof is decomposed into a sequence of verifications, none of which is difficult.

In the remainder of this appendix, we fix a monoid M satisfying the hypotheses of the theorem. We define a binary relation $\sim$ on $M \times M$ by

$$(a, b) \sim (a', b') \quad \Leftrightarrow \quad \exists x, x' \in M \ (ax = a'x' \text{ and } bx = b'x'). \tag{3.21}$$

If $ax = a'x'$ and $bx = b'x'$, so that the two equalities of (3.21) are satisfied, then (x, x') is said to *witness* for $(a, b) \sim (a', b')$.

Lemma 3.4.3 *The relation $(a, b) \sim (a', b')$ is equivalent to*

$$\forall y, y' \in M \ (ay = a'y' \ \Leftrightarrow \ by = b'y'). \tag{3.22}$$

Proof Suppose (x, x') witnesses for $(a, b) \sim (a', b')$, and $ay = a'y'$. By hypothesis, x and y have a common right-multiple, say $xd = yc$. Then $a'x'd = axd = ayc = a'y'c$, hence $x'd = y'c$ when simplifying by a' on the left.

We deduce $byc = bxd = b'x'd = b'y'c$, so that $by = b'y'$ when simplifying by c on the right. Hence $(a, b) \sim (a', b')$ implies (3.22).

Conversely, suppose (3.22). By hypothesis, a and a' have a common right-multiple, hence there exist x and x' in M satisfying $ax = a'x'$. By (3.22), we deduce $bx = b'x'$, and hence (x, x') witnesses for $(a, b) \sim (a', b')$. □

Lemma 3.4.4 *The relation $\sim$ is an equivalence relation.*

Proof Indeed, $(1, 1)$ witnesses for $(a, b) \sim (a, b)$, so $\sim$ is reflexive. Next, if (x, x') witnesses for $(a, b) \sim (a', b')$, then (x', x) witnesses for $(a', b') \sim (a, b)$, hence $\sim$ is symmetric. Finally, let $(a, b) \sim (a', b') \sim (a'', b'')$. By hypothesis, a, a', a'' have a common right-multiple, $ax = a'x' = a''x''$. By (3.22), the conjunction of the hypothesis $(a, b) \sim (a', b')$ and the equality $ax = a'x'$ imply $bx = b'x'$. Moreover, the hypothesis $(a', b') \sim (a'', b'')$ plus the equality $a'x' = a''x''$ imply $b'x' = b''x''$. Thus $ax = a''x''$ and $bx = b''x''$, so $(a, b) \sim (a'', b'')$, hence $\sim$ is transitive. □

For a and b in M, denote a/b the equivalence class of (a, b) with respect to $\sim$, and G the quotient set $M^2/\sim$.

Lemma 3.4.5 *For any a, b, c, d, x, y in M satisfying $bx = cy$, the class $(ax)/(dy)$ is independent of x and y.*

Proof Suppose $bx = cy$ and $bx' = cy'$. The elements y and y' have a common right-multiple, say $yt = y't'$. Then $bxt = cyt = cy't' = bx't'$, or $xt = x't'$ when simplifying by b on the left. Hence $dyt = dy't'$ and $axt = ax't'$, giving $(ax, dy) \sim (ax', dy')$. □

By Lemma 3.4.5, we can without ambiguity define a mapping $*$ from M^2 into G by setting

$$(a, b) * (c, d) = (ax)/(dy) \tag{3.23}$$

where $bx = cy$ is an arbitrary common multiple of b and c.

Lemma 3.4.6 *The operation $*$ induces a well-defined operation*[14] *on G.*

Proof Suppose $(a, b) \sim (a', b')$ and $(c, d) \sim (c', d')$. By hypothesis, b, b', c, and c' have a common right-multiple, say $bx = b'x' = cy = c'y'$. Then, by definition, we have

$$(a, b) * (c, d) = (ax)/(dy) \quad \text{and} \quad (a', b') * (c', d') = (a'x')/(d'y').$$

[14] i.e. the value of $(ax)/(dy)$ in (3.23) depends only on the classes a/b and c/d, and not on the pairs (a, b) and (c, d).

However, by Lemma 3.4.3, $bx = b'x'$ implies $ax = a'x'$, and $cy = c'y'$ implies $dy = d'y'$, hence $(ax)/(dy) = (a'x')/(d'y')$. □

Let $\cdot$ denote the binary operation on G induced by $*$.

Lemma 3.4.7 *The operation $\cdot$ on G is associative.*

Proof Let a, b, c, d, e, f be arbitrary elements of M. Select x, y, z, t, u, and v satisfying $bx = cy$, $dz = et$, and $yu = zv$ – corresponding to the schema below where an arrow is associated with an element of M and where a sequence of arrows represents a product:

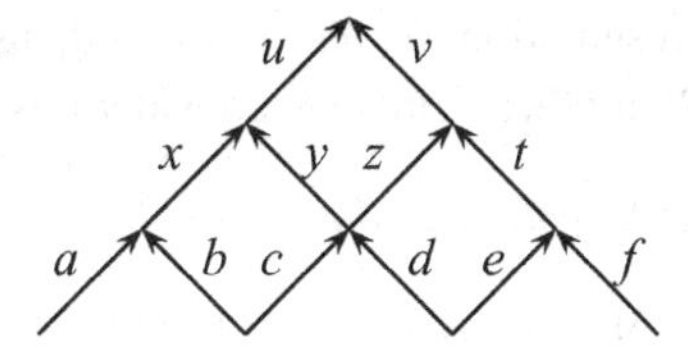

Then, by definition, we have

$$a/b \cdot (c/d \cdot e/f) = a/b \cdot (cz)/(ft) = (axu)/(ftv), \tag{3.24}$$

$$(a/b \cdot c/d) \cdot e/f = (ax)/(dy) \cdot e/f = (axu)/(ftv). \tag{3.25}$$

Indeed, the second equality in (3.24) results from the equalities $b(xu) = cyu = (cz)v$, and that of (3.25) from $(dy)u = dzv = e(tv)$. □

Lemma 3.4.8 *The structure $(G, \cdot)$ is a group whose unity is the class $1/1$, and where the inverse of a/b is b/a.*

Proof Let a and b be elements of M. Then $b\,1 = 1\,b$, thus $a/b * 1/1 = (a\,1)/(1\,b) = a/b$. Similarly, $1\,a = a\,1$, hence $1/1 * a/b = (1\,a)/(b\,1) = a/b$, and $1/1$ is the neutral element of $\cdot$. Moreover, $(x, x) \sim (1, 1)$ for any x, so $x/x = 1/1$. Finally, $b1 = b1$ implies $a/b * b/a = (a1)/(a1) = 1/1$ and $a1 = a1$ implies $b/a * a/b = (b1)/(b1) = 1/1$, hence b/a is the inverse of a/b. □

For a in M, we set $\iota(a) := a/1$.

Lemma 3.4.9 *The mapping ι is an injective homomorphism of M into G. Moreover, for any a/b in G, $a/b = \iota(a) \cdot \iota(b)^{-1}$.*

Proof For any a and b in M, we have $1\,b = b\,1$, hence

$$\iota(a) \cdot \iota(b) = a/1 \cdot b/1 = (a\,b)/(1\,1) = (ab)/1 = \iota(ab).$$

Moreover, suppose $\iota(a) = \iota(a')$, that is, $(a, 1) \sim (a', 1)$. By definition, there exist x and x' satisfying $ax = a'x'$ and $1x = 1x'$. We deduce $x = x'$, then $a = a'$ when simplifying by x on the right.

Finally, for a, b in M, we have $a/b = a/1 \cdot 1/b = a/1 \cdot (b/1)^{-1} = \iota(a) \cdot \iota(b)^{-1}$, as stated. □

The proof of Proposition 3.4.1 is complete: we have obtained a group G and an embedding ι of M into G such that any element of G is a right quotient of two elements in the image of ι.

3.4.3 Proof of Proposition 3.4.2

We continue with the notations from above, and suppose in addition that the monoid M admits the presentation $\langle S \mid R \rangle^{+}$. Without danger, we can identify M with its image by ι in G, in other words we consider ι as the identity.

Lemma 3.4.10 *The set S generates the group G.*

Proof Any element of M other than 1 is a product of elements of S. Since every element of the fraction group G can be written as ab^{-1} with a and b in M, every element of G is a product of elements of S and their inverses. Consequently, S generates G as a group. □

We denote $\equiv^{+}$ the congruence on S^* generated by R, and $\equiv$ the congruence on $(S \cup \overline{S})^*$ generated by R and the free group relations $\mathsf{Sym}(S)$. The evaluation of a signed S-word w in G is denoted eval_G. Saying that G admits the presentation $\langle S \mid R \rangle$ means showing that, for any word w in $(S \cup \overline{S})^*$, the relation $w \equiv \varepsilon$ is equivalent to $\mathsf{eval}_G(w) = 1$.

Lemma 3.4.11 *The relation $w \equiv \varepsilon$ implies* $\mathsf{eval}_G(w) = 1$.

Proof If $u = v$ is a relation of R, then u and v have the same evaluation in M, hence in G. Moreover, for any s in S, we have $\mathsf{eval}_G(s\overline{s}) = \mathsf{eval}_G(\overline{s}s) = 1$. Since the congruence $\equiv$ is generated by the pairs $\{u, v\}$ with $u = v$ a relation of R, plus the pairs $\{s\overline{s}, \varepsilon\}$ and $\{\overline{s}s, \varepsilon\}$, we conclude that $w \equiv \varepsilon$ implies $\mathsf{eval}_G(w) = 1$. □

Lemma 3.4.12 *The relation* $\mathsf{eval}_G(w) = 1$ *implies $w \equiv \varepsilon$.*

Proof First note that every signed S-word is $\equiv$-equivalent to a word of the form $u\overline{v}$ with positive u, v. For this, remark that any signed S-word can be written $w = u_1\overline{v_1}u_2\overline{v_2}\cdots u_p\overline{v_p}$ with positive $u_1, \ldots, u_p, v_1, \ldots, v_p$, and we reason by induction on $p \geqslant 1$. For $p = 1$, there is nothing to show. Suppose $p \geqslant 2$, and let $w' := u_2\overline{v_2}\cdots u_p\overline{v_p}$. By the induction hypothesis, there exist positive u'_2, v'_2 satisfying $w' \equiv u'_2\overline{v'_2}$, hence $w \equiv u_1\overline{v_1}u'_2\overline{v'_2}$. Since, in the monoid M, the elements represented by v_1 and u'_2 have a common right-multiple, there exist u'_3 and v'_3 in S^* satisfying $v_1u'_3 \equiv^{+} u'_2v'_3$, thus, a fortiori, $v_1u'_3 \equiv u'_2v'_3$. Using the free group relations, we deduce $\overline{v_1}u'_2 \equiv u'_3\overline{v'_3}$, and obtain $w \equiv u_1\overline{v_1}u'_2\overline{v'_2} \equiv u_1u'_3\overline{v'_3}\,\overline{v'_2}$, hence $w \equiv u\overline{v}$ by setting $u := u_1u'_3$ and $v := v'_2v'_3$.

Suppose then $\mathrm{eval}_G(w) = 1$. There exist u, v in S^* satisfying $w \equiv u\overline{v}$. Then $\mathsf{eval}_G(w) = 1$ implies $\mathsf{eval}_G(u) = \mathsf{eval}_G(v)$, hence $u \equiv^+ v$ as M is embedded in G and $\langle S \mid R \rangle^+$ is a presentation of M. We conclude a fortiori $u \equiv v$, so $u\overline{v} \equiv \varepsilon$, and finally, $w \equiv \varepsilon$. □

The proof of Proposition 3.4.2 is complete.

There of course exists a symmetric version of Ore's theorem where we suppose the existence of common left-multiples, and where the resulting group is a group of left-fractions for the monoid. Note that by Proposition 3.4.2, if a monoid possesses at the same time common left- and right-multiples, the two associated fraction groups are isomorphic and can without danger be identified. This is notably the case for a commutative monoid, such as $(\mathbb{N}, +)$ or $(\mathbb{Z}, \times)$.

4

The Greedy Normal Form

This chapter is a direct follow-on of the last, and it continues to exploit the properties of the monoids B_n^+ to study the groups B_n. The specific new ingredient here is the fact that the left-divisibility relation of the monoid B_n^+ has a lattice structure: any two elements of B_n^+ always admit a least common right-multiple ('right-lcm'), and a greatest common left-divisor ('left-gcd'). This result, which at first glance seems rather technical, is important as it allows us to construct unique normal forms, first for positive braids, then for arbitrary braids. This allows us to define, for any braid, a distinguished decomposition in elementary fragments that essentially turn out to be permutations of $\{1, ..., n\}$. Thus, a braid appears as a finite sequence of permutations. One of the (multiple) interests of such a decomposition is to provide solutions to the isotopy problem both theoretically efficient and simple in practice.

4.1 The Lattice Structure of B_∞^+

4.1.1 The Left-Divisibility Relation

We return here to the left-divisibility relation of the monoid B_∞^+ introduced in Definition 3.1.21 and study it in more detail. Recall that, for a, b in a monoid M, a is said to *left-divide* b, or equivalently, b is a *right-multiple* of a, denoted $a \preccurlyeq b$, if there exists x in M satisfying $ax = b$.

Example 4.1.1 (Divisibility) The braid σ_1 left-divides Δ_3 in B_∞^+, since we can write $\sigma_1 \cdot \sigma_2\sigma_1 = \Delta_3$. However, σ_1 does not left-divide $\sigma_2\sigma_1$, as the only expression for the braid $\sigma_2\sigma_1$ is the word $\sigma_2\sigma_1$, which does not begin with σ_1.

We know, from Proposition 3.1.7, that for every n, the monoid B_n^+ is the submonoid of B_∞^+ generated by $\sigma_1, ..., \sigma_{n-1}$. Clearly, for a, b in B_n^+, the relation '$a \preccurlyeq b$ in B_n^+' implies '$a \preccurlyeq b$ in B_∞^+', however, a priori, it could be that the converse is not valid. In fact, it is.

Lemma 4.1.2 *For a, b in B_n^+, the relation $a \preccurlyeq b$ is satisfied in B_n^+ if and only if this holds in B_∞^+.*

Proof Only the converse direction could pose a problem. Suppose a left-divides b in B_∞^+. Write $a = [u]$ and $b = [v]$, with u and v in $\mathcal{BW}_n^+$. The hypothesis that $a \preccurlyeq b$ is satisfied in B_∞^+ means that there exists a word w in $\mathcal{BW}_\infty^+$ satisfying $uw \equiv^+ v$. Then, by Lemma 3.1.5, the word w must belong to $\mathcal{BW}_n^+$, and hence $[w]$ belongs to B_n^+. Hence $a[w] = b$ in B_n^+, and $a \preccurlyeq b$ is satisfied in B_n^+. □

The divisibility relation of B_n^+ is thus the restriction of the relation of B_∞^+, and there is not the slightest risk in using the symbol $\preccurlyeq$ without an index.

From the definition, it is immediate that $a \preccurlyeq b$ implies $|a| \leqslant |b|$; this suggests an order relation.

Lemma 4.1.3 *The relation $\preccurlyeq$ is a partial order*[1] *with minimum* 1 *on B_∞^+.*

Proof The relation $\preccurlyeq$ is reflexive, as we can always write $a \cdot 1 = a$. It is antisymmetric, since, if we have $ax = b$ and $by = a$, then $a(xy) = a$, thus $xy = 1$ by simplifying by a on the left. However, the existence of the length function on B_∞^+ implies that 1 is the only invertible element of B_n^+ (Exercise 3.1.9). Hence $x = y = 1$, and $a = b$. Moreover, $\preccurlyeq$ is transitive, as the conjunction of $ax = b$ and $by = c$ implies $a(xy) = c$, so $a \preccurlyeq c$. Thus $\preccurlyeq$ is an equivalence relation. Finally, for any a, we have $1 \cdot a = a$, hence $1 \preccurlyeq a$. □

For $n \geqslant 3$, the relation $\preccurlyeq$ is not a total order on B_n^+: for example, we have neither $\sigma_1 \preccurlyeq \sigma_2$, nor $\sigma_2 \preccurlyeq \sigma_1$ (why?). However, we shall see that, in $(B_n^+, \preccurlyeq)$, two elements always admit a greatest common lower bound and a smallest common upper bound. It is usual here to speak of the 'greatest common divisor (gcd)' and 'least common multiple (lcm)'.

Definition 4.1.4 (Gcd, lcm) Let a, b, c be elements of a monoid M. The element c is said to be the *left-gcd* of a and b if, in M, $c \preccurlyeq a$ and $c \preccurlyeq b$, and $x \preccurlyeq c$ for every x satisfying $x \preccurlyeq a$ and $x \preccurlyeq b$. Similarly, c is said to be the *right-lcm* of a and b if, in M, $a \preccurlyeq c$ and $b \preccurlyeq c$, and $c \preccurlyeq x$ for every x satisfying $a \preccurlyeq x$ and $b \preccurlyeq x$.

Note that, in the case where $a \preccurlyeq b$ is satisfied, then a is a left-gcd and b a right-lcm of a and b.

[1] i.e. a reflexive, antisymmetric, and transitive relation.

We begin with the common multiples.[2]

Proposition 4.1.5 (Lcm) *For every $n \leqslant \infty$, any two elements of B_n^+ admit a unique right-lcm.*

Proof Let a, b be arbitrary in B_n^+, and u and v braid words representing respectively a and b. By Lemma 3.3.11, there exists exactly one grid Γ with source (u, v). Let (u', v') be the target of Γ. By definition, we have $[u]^+[v']^+ = [v]^+[u']^+$, thus $[uv']^+$ is a right-multiple of $[u]^+$ and $[v]^+$, that is, of a and b. Moreover, still by Lemma 3.3.11, every common right-multiple of $[u]^+$ and $[v]^+$ is a right-multiple of $[uv']^+$: this means $[uv']^+$ is a right-lcm of $[u]^+$ and $[v]^+$.

For the unicity, remark that, if c and c' are two right-lcms of a and b, then, by definition, we have at the same time $c \preccurlyeq c'$ and $c' \preccurlyeq c$, implying $c = c'$ by the antisymmetry of $\preccurlyeq$. □

We next move to the common divisors.[3]

Proposition 4.1.6 (Gcd) *For every $n \leqslant \infty$, any two elements of B_n^+ admit a unique left-gcd.*

Proof Suppose a, b in B_n^+. Let X be the set of common left-divisors of a and b. First, X is not empty since it contains at least 1. Let c be an element of X of maximal length: such an element exists since, for any x in X, we have $|x| \leqslant |a|$. Take any x in X. By Proposition 4.1.5, c and x admit a right-lcm, say y. Then we have at the same time $c \preccurlyeq a$ and $x \preccurlyeq a$, so $y \preccurlyeq a$ by the definition of the lcm. Similarly, $y \preccurlyeq b$, hence $y \in X$. However, by hypothesis, $c \preccurlyeq y$, hence $|c| \leqslant |y|$. The choice of c imposes $|c| = |y|$, hence $c = y$ and then $x \preccurlyeq c$. Any common left-divisor of a and b is thus a left-divisor of c. Consequently, c is a left-gcd of a and b. The argument for the unicity is the same as for the lcm. □

Exercise 4.1.7 (Gcd) Show that every set (finite or infinite) of positive braids admits a left-gcd.

By replacing $ax = b$ by $xa = b$ in Proposition 3.1.32, we introduce the notions of right-divisor and left-multiple, symmetric to those of left-divisor and right-multiple.

Proposition 4.1.8 (Left-lcm, right-gcd) *For every $n \leqslant \infty$, any two elements of B_n^+ admit a unique left-lcm, and a unique right-gcd.*

[2] Here, distinguishing between B_n^+ and B_∞^+ makes sense since, even if the divisibility relation of the one is the restriction of that of the other, the multiples are not the same: an element of B_n^+ admits many more right-multiples in B_∞^+ than in B_n^+.

[3] For symmetry, we adopt a statement similar to Proposition 4.1.5, but here, the case of B_∞^+ is sufficient, since a divisor of an element of B_n^+ is necessarily in B_n^+.

Exercise 4.1.9 Prove the result. [Hint: Use the antiautomorphism of Lemma 3.1.12.]

Exercise 4.1.10 (Computation of the gcd)
(i) Let $a, b \in B_\infty^+$. Suppose $ad = bc$ is the right-lcm of a and b, and $ed = fc$ the left-lcm of c and d. Show that there exists g satisfying $a = ge$ and $b = gf$, and that g is the left-gcd of a and b.
(ii) Deduce an algorithm based on the reversing procedure of Definition 3.3.13 to compute the gcd in B_∞^+. [Hint: Use the antiautomorphism $\widetilde{\ }$ for the computation of the left-lcm.]

Exercise 4.1.11 (Centre)
(i) An element $a \in B_n^+$ is said to be *quasi-central* if it satisfies

$$\forall b \in B_n^+ \, \exists c \in B_n^+ \, (ba = ac).$$

Show that every element Δ_n^m is quasi-central.
(ii) Conversely, show that, if a is quasi-central in B_n^+ and $\sigma_i \preccurlyeq a$, then also $\sigma_j \preccurlyeq a$ for $|i - j| = 1$.
(iii) Deduce that, if a is quasi-central, then it is a right-multiple Δ_n, and then show that a is a power of Δ_n.
(iv) Deduce that, for $n \geqslant 3$, the centre of B_n^+ is the submonoid generated by Δ_n^2, then, by using Exercise 3.3.6, that the centre of B_n is the subgroup generated by Δ_n^2.

4.1.2 The Permutation Braids

We have seen in Proposition 2.1.13 that the mapping 'perm' induces a homomorphism of B_n onto $\mathfrak{S}_n$, which is surjective, but not injective, since, for example, the identity is the image of both σ_1^2 and 1. We will describe here a (set-theoretic) section for perm, that is, a selection, for each f in $\mathfrak{S}_n$, of a distinguished braid whose permutation is f. We will see that the family formed by the $n!$ braids thus selected has numerous remarkable properties.

In what follows, the transposition exchanging i and $i{+}1$ is denoted s_i. In particular, s_i is the permutation of the braid σ_i.

The construction of the permutation braids is inductive: for $f \neq \mathrm{id}$, in order to define a permutation braid f, we consider the largest integer m moved by f, in which case necessarily $f(m) < m$; we ensure then that the mth strand starts from the position $f(m)$, and then iterate with a new permutation that only moves integers $< m$.

Definition 4.1.12 (Permutation braid) Let $f \in \mathfrak{S}_n$. We define $\underline{\sigma}_f$ in $\mathcal{BW}_n^+$ by $\underline{\sigma}_{\mathrm{id}} := \varepsilon$ and $\underline{\sigma}_f := \underline{\sigma}_{f(m),m}\underline{\sigma}_g$, with m the largest integer moved by f and g in $\mathfrak{S}_{n-1}$ defined by

$$g(i) := \begin{cases} f(i) & \text{for } i < m \text{ and } f(i) < f(m), \\ f(i) - 1 & \text{for } i < m \text{ and } f(i) > f(m), \\ i & \text{for } i \geqslant m. \end{cases} \tag{4.1}$$

We denote σ_f the braid[4] represented by $\underline{\sigma}_f$, and call it the *permutation braid* associated with f.

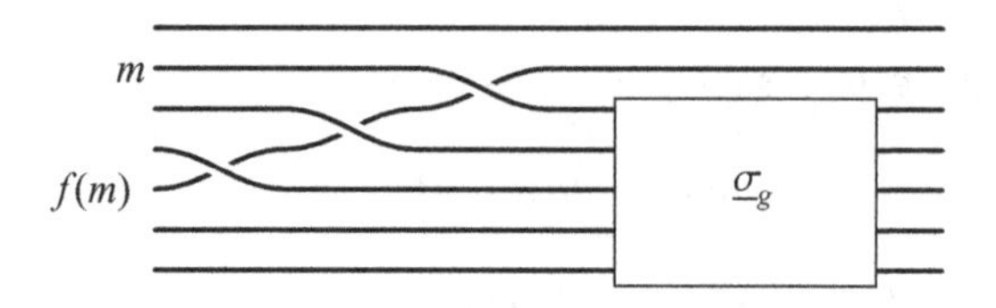

Figure 4.1 Inductive definition of $\underline{\sigma}_f$: we ensure that, if m is the largest integer moved by f, then the strand finishing at the position m starts from the position $f(m)$, and, then, there remains a permutation g moving only $m-1$ positions.

The definition implies that $\underline{\sigma}_{s_i}$ is equal to σ_i. It is easy to treat more complicated cases.

Example 4.1.13 (Permutation braid) Let f be the permutation[5] $\binom{1\,2\,3\,4}{3\,4\,1\,2}$. The integer '$m$' is 4, which is sent to 2, hence $\underline{\sigma}_f = \underline{\sigma}_{2,4}\underline{\sigma}_g$, with $g = \binom{1\,2\,3\,4}{2\,3\,1\,4}$. This time, '$m$' is 3, sent to 1, so $\underline{\sigma}_g = \underline{\sigma}_{1,3}\underline{\sigma}_h$, with $h = \binom{1\,2\,3\,4}{1\,2\,3\,4}$. However $\underline{\sigma}_h$ is the empty word, and thus $\underline{\sigma}_f = \underline{\sigma}_{2,4}\underline{\sigma}_{1,3} = \sigma_2\sigma_3\sigma_1\sigma_2$.

As expected, the mapping $f \mapsto \sigma_f$ from $\mathfrak{S}_n$ into B_n^+ is a set-theoretic section of perm.

Lemma 4.1.14 *For any permutation f in $\mathfrak{S}_n$,* $\mathrm{perm}(\sigma_f) = f$.

Proof Induction on n: we can see in Figure 4.1 that, if the permutation of σ_g is g, that of σ_f is f. □

It immediately follows that the mapping $f \mapsto \sigma_f$ is injective, and hence there exist exactly $n!$ permutation braids in B_n^+.

[4] You might be surprised that the same notation of type σ_x is employed for such distinct uses, but, in practice, there is little risk of confusion between the cases where x is an integer, a pair of integers, or a permutation.

[5] The value of $f(i)$ is indicated underneath i.

A specific permutation of $\mathfrak{S}_n$ plays an important role, namely the permutation ω_n, often called the *flip*, which exchanges i and $n - i$ for every i. We see the braid Δ_n of Definition 3.1.24 reappearing here.

Lemma 4.1.15 *For any $n \geqslant 1$, $\underline{\Delta}_n = \underline{\sigma}_{\omega_n}$, and hence $\Delta_n = \sigma_{\omega_n}$.*

Proof Again an induction on n. The result is clear for $n = 1$ and $n = 2$. For $n \geqslant 3$, the definition gives $\underline{\sigma}_{\omega_n} = \sigma_{1,n}\underline{\sigma}_{\omega_{n-1}}$, thus, applying the induction hypothesis, $\underline{\sigma}_{\omega_n} = \sigma_{1,n}\underline{\Delta}_{n-1} = \underline{\Delta}_n$. □

4.1.3 The Simple Braids

We will now study a new family of positive braids, that in the end will turn out to coincide with the permutation braids. However, it is more convenient to define them here using the number of inversions of a permutation, a sort of measure of its complexity.

Definition 4.1.16 (Inversion number) A pair (i, j) is an *inversion* of a permutation f of $\{1, ..., n\}$ if $1 \leqslant i < j \leqslant n$ and $f(i) > f(j)$. The *inversion number* of f, denoted $\mathsf{inv}(f)$, is the total number of pairs (i, j) that are inversions of f.

For example, id is not an inversion, hence $\mathsf{inv}(\mathrm{id}) = 0$. Next, the only inversion of s_i is the pair $(i, i{+}1)$, and $\mathsf{inv}(s_i) = 1$. Finally, every pair (i, j) satisfying $1 \leqslant i < j \leqslant n$ is an inversion of the flip ω_n, hence $\mathsf{inv}(\omega_n) = n(n{-}1)/2$. Note that, by definition, $\mathsf{inv}(f) \leqslant n(n{-}1)/2$ for any f in $\mathfrak{S}_n$. Moreover, if (i, j) is an inversion of f, then $(f(j), f(i))$ is an inversion of f^{-1}, otherwise $(f(i), f(j))$ is not an inversion of f^{-1}. Hence we always have

$$\mathsf{inv}(f^{-1}) = \mathsf{inv}(f). \tag{4.2}$$

We are now ready to introduce the simple braids. Recall, if a is a positive braid, $|a|$, the *length* of a, is the common length of all positive braid words representing a.

Definition 4.1.17 (Simple braid) A braid a is said to be *simple* if it is positive and satisfies $|a| = \mathsf{inv}(\mathsf{perm}(a))$.

For example, every braid σ_i is simple, since $|\sigma_i| = 1$ and $\mathsf{inv}(\mathsf{perm}(\sigma_i)) = \mathsf{inv}(s_i) = 1$. The unit braid is also simple, as $|1| = \mathsf{inv}(\mathrm{id}) = 0$. However, σ_1^2 is not simple: $|\sigma_1^2| = 2$, and $\mathsf{inv}(\mathsf{perm}(\sigma_1^2)) = \mathsf{inv}(\mathrm{id}) = 0$.

Note immediately that the inequality $\mathsf{inv}(f) \leqslant n(n{+}1)/2$, valid for any f in $\mathfrak{S}_n$, implies $|a| \leqslant n(n{+}1)/2$ for any simple braid in B_n^+. Hence there exists only a finite number of simple braids in B_n^+.

To study the simple braids, we need to control the growth of the inversion number.

Lemma 4.1.18 *If f in $\mathfrak{S}_n$ and $1 \leqslant k < n$, then*

$$\mathrm{inv}(f s_k) = \begin{cases} \mathrm{inv}(f) + 1 & \text{if } f(k) < f(k+1), \\ \mathrm{inv}(f) - 1 & \text{if } f(k) > f(k+1), \end{cases} \tag{4.3}$$

$$\mathrm{inv}(s_k f) = \begin{cases} \mathrm{inv}(f) + 1 & \text{if } f^{-1}(k) < f^{-1}(k+1), \\ \mathrm{inv}(f) - 1 & \text{if } f^{-1}(k) > f^{-1}(k+1). \end{cases} \tag{4.4}$$

Proof We compare the inversions of $f s_k$ with those of f. If (i, j) is not $(k, k{+}1)$, then (i, j) is an inversion of $f s_k$ if and only if it is an inversion of f, since an initial permutation of k and $k{+}1$ does not change their positions with respect to the other integers. On the other hand, $(k, k{+}1)$ is an inversion of $f s_k$ if and only if it is not an inversion of f, that is, if $f(k) < f(k+1)$. Thus (4.3) follows by summing.

The argument is similar for (4.4), but what counts here is to know if the integers sent by f onto k and $k{+}1$ are in increasing or decreasing order, and thus whether or not $(k, k{+}1)$ is an inversion of f^{-1}. □

An essential property of simple braids is that any braid dividing a simple braid to the left or to the right is simple.

Lemma 4.1.19 *Suppose that a and b are positive braids and the product ab is simple. Then a and b are simple.*

Proof By virtue of (4.3), multiplying on the right by a transposition augments the number of inversions by at most one. Hence, if g is the product of ℓ transpositions, then, for any permutation f, $\mathrm{inv}(fg) \leqslant \mathrm{inv}(f) + \ell$. Passing to braids, we deduce

$$\mathrm{inv}(\mathrm{perm}(ab)) \leqslant \mathrm{inv}(\mathrm{perm}(a)) + |b| \tag{4.5}$$

for any positive braid a, b. In particular (taking $a = 1$)

$$\mathrm{inv}(\mathrm{perm}(b)) \leqslant |b|.$$

Now suppose a is not simple: then $\mathrm{inv}(\mathrm{perm}(a)) < |a|$, hence, by (4.5),

$$\mathrm{inv}(\mathrm{perm}(ab)) \leqslant \mathrm{inv}(\mathrm{perm}(a)) + |b| < |ab|$$

for any b, implying that ab is not simple. By contraposition, if ab is simple, a must be simple.

Symmetrically, (4.4) leads to

$$\mathrm{inv}(\mathrm{perm}(ab)) \leqslant |a| + \mathrm{inv}(\mathrm{perm}(b)), \tag{4.6}$$

and, as above, we conclude that, if b is not simple, then neither is ab. Hence, if ab is simple, so must be b. □

We now study the simplicity of permutation braids. The principal technical step consists in comparing[6] σ_f and $\sigma_{s_k f}$.

Lemma 4.1.20 *Let f be a permutation of $\{1, ..., n\}$ such that σ_f is a simple braid. Then, for $1 \leqslant k \leqslant n-1$, two cases are possible;*
(i) *either $f^{-1}(k) < f^{-1}(k+1)$, and then the braid $\sigma_k\sigma_f$ is simple, and equal to $\sigma_{s_k f}$,*
(ii) *or $f^{-1}(k) > f^{-1}(k+1)$, and then $\sigma_k\sigma_f$ is not simple.*

Proof By hypothesis, $|\sigma_f| = \mathsf{inv}(f)$. If $(k, k+1)$ is not an inversion of f^{-1}, (4.3) implies $\mathsf{inv}(s_k f) = \mathsf{inv}(f) + 1$, thus $\mathsf{inv}(s_k f) = |\sigma_k\sigma_f|$, hence $\sigma_k\sigma_f$ is simple. If $(k, k+1)$ is an inversion of f^{-1}, (4.3) implies $\mathsf{inv}(s_k f) = \mathsf{inv}(f) - 1$, hence $\mathsf{inv}(s_k f) < |\sigma_k\sigma_f|$, and $\sigma_k\sigma_f$ is not simple.

It remains only to establish the equality $\sigma_k\sigma_f = \sigma_{s_k f}$ in case (i). If f is the identity, the result reduces to the equality $\sigma_k = \sigma_{s_k}$. We now consider the case $f \neq \mathsf{id}$, and reason by induction on the largest integer m moved by f, hence satisfying $f(m) < m$. By Definition 4.1.12, we have $\sigma_f = \sigma_{f(m),m}\sigma_g$ with g as in Definition (4.1). By Lemma 4.1.19, the hypothesis that σ_f is simple implies that σ_g is as well. Setting $f' := \sigma_k f$, we are going to determine $\sigma_{f'}$; for this, we distinguish several cases according to the position of k with respect to m and $f(m)$.

First suppose $k < f(m) - 1$. The largest integer moved by f' is m, which is sent to $f(m)$. Hence $\sigma_{f'} = \sigma_{f(m),m}\sigma_{g'}$, with g' defined by (4.1), giving, in this case, $g' = g\sigma_k$. The induction hypothesis then can be applied to g, implying $\sigma_{s_k g} = \sigma_k\sigma_g$. Moreover, σ_k and $\sigma_{f(m),m}$ commute, since $k \leqslant i - 2$ for each generator σ_i appearing in $\underline{\sigma}_{f(m),m}$, and, as desired, we obtain

$$\sigma_{f'} = \sigma_{f(m),m}\sigma_k\sigma_g = \sigma_k\sigma_{f(m),m}\sigma_g = \sigma_k\sigma_f.$$

Next suppose $k = f(m) - 1$. The largest integer moved by f' is again m, but this time it is sent to $f(m) - 1$. Consequently we obtain $\sigma_{f'} = \sigma_{f(m)-1,m}\sigma_{g'}$, where, this time, (4.1) gives $g' = g$. With the same notation as above, we thus obtain

$$\sigma_{f'} = \sigma_{f(m)-1,m}\sigma_g = \sigma_k\sigma_{f(m),m}\sigma_g = \sigma_k\sigma_f.$$

[6] It might seem more natural to compare σ_f and $\sigma_{f s_k}$, to separate according to the inversions of f, rather than of f^{-1}, however, because of the definition of σ_f 'on the left', the argument does not work (at least, not directly).

Now suppose $k = f(m)$. By definition, $m = f^{-1}(f(m))$ and $f^{-1}(f(m)+1) = g^{-1}(f(m)) < m$, hence $f^{-1}(k) > f^{-1}(k+1)$. By (ii), it follows that $\sigma_k\sigma_f$ is not simple, thus this case is excluded here.

Suppose then that $f(m) + 1 \leqslant k < m$. The largest integer moved by f' is again m, sent to $f(m)$. Hence $\sigma_{f'} = \sigma_{f(m),m}\sigma_{g'}$, where (4.1) gives here $g' = s_{k-1}g$. By applying the induction hypothesis to g and s_{k-1}, then using the equivalence (3.9), we obtain

$$\sigma_{f'} = \sigma_{f(m),m}\sigma_{s_{k-1}g} = \sigma_{f(m),m}\sigma_{k-1}\sigma_g = \sigma_k\sigma_{f(m),m}\sigma_g = \sigma_k\sigma_f.$$

Finally suppose $k \geqslant m$. Then the largest integer moved by f' is $k+1$, sent to k, thus $\sigma_{f'} = \sigma_{k,k+1}\sigma_{g'}$, and (4.1) gives $g' = f$. We thus find directly $\sigma_{f'} = \sigma_k\sigma_f$.

All cases have been treated successfully. □

We now come to the desired equivalence result.

Proposition 4.1.21 (Simple braid) *For every $n \geqslant 2$ and for any braid a in B_n^+, there is equivalence between:*

(i) *a is a permutation braid,*

(ii) *a is simple,*

(iii) *a is a right-divisor of Δ_n,*

(iv) *a is a left-divisor of Δ_n.*

Proof We show by induction on the integer $\mathsf{inv}(f)$ that, for any permutation f in $\mathfrak{S}_n$, the braid σ_f is simple. If $\mathsf{inv}(f)$ is zero, then f is the identity, and σ_f is the unit braid, which is simple. Otherwise, neither f nor f^{-1} are the identity, and hence there exists an integer k satisfying $f^{-1}(k) > f^{-1}(k+1)$. Set $g := s_k f$. Since s_k^2 is the identity, we also have $f = s_k g$, and, by (4.4), $\mathsf{inv}(g) = \mathsf{inv}(f) - 1$. By the induction hypothesis, σ_g is simple. The equality $\mathsf{inv}(f) = \mathsf{inv}(g) + 1$ implies $g^{-1}(k) < g^{-1}(k+1)$. Hence, σ_f, which is also $\sigma_k\sigma_g$, is simple by Lemma 4.1.20. Thus, (i) implies (ii).

Conversely, we show by induction on $\mathsf{inv}(\mathsf{perm}(a))$ that, if a is a simple braid, then $a = \sigma_{\mathsf{perm}(a)}$. For $\mathsf{inv}(\mathsf{perm}(a)) = 0$, meaning $\mathsf{perm}(a) = \mathsf{id}$, the hypothesis that a is simple implies $|a| = 1$, hence $a = 1 = \sigma_{\mathsf{id}}$. Otherwise, write $a = \sigma_k b$ with b positive. Then $\mathsf{perm}(a) = s_k\mathsf{perm}(b)$, and $|a| = 1 + |b|$. The formula (4.4) implies $\mathsf{inv}(\mathsf{perm}(a)) \leqslant \mathsf{inv}(\mathsf{perm}(b)) + 1$, hence the only possibility is that b is simple with $\mathsf{inv}(\mathsf{perm}(b)) = \mathsf{inv}(\mathsf{perm}(a)) - 1$. The induction hypothesis applied to b gives $b = \sigma_{\mathsf{perm}(b)}$. Then Lemma 4.1.20 implies $a = \sigma_{s_k\mathsf{perm}(b)}$, or $a = \sigma_{\mathsf{perm}(a)}$. Thus, (ii) implies (i).

Next, we show by reverse induction on $\mathsf{inv}(f) \leqslant n(n-1)/2$ that, for any permutation f in $\mathfrak{S}_n$, the braid σ_f right-divides Δ_n. In the case $\mathsf{inv}(f) = n(n-1)/2$, the only possibility is that f is the flip ω_n, hence $\sigma_f = \Delta_n$. Suppose $\mathsf{inv}(f) < n(n-1)/2$, so $\mathsf{inv}(f^{-1}) < n(n-1)/2$. Then there exists at least one integer k satisfying $f^{-1}(k) < f^{-1}(k+1)$, and, by (4.4), $\mathsf{inv}(s_k f) > \mathsf{inv}(f)$, and, by Lemma 4.1.20, $\sigma_{s_k f} = \sigma_k \sigma_f$. Hence σ_f right-divides $\sigma_{s_k f}$. Moreover, the induction hypothesis applied to $s_k f$ implies that $\sigma_{s_k f}$ right-divides Δ_n. By transitivity of right-divisibility, we deduce that σ_f right-divides Δ_n. Hence (i) and (ii) imply (iii).

The above results show that, for any simple braid a in B_n^+, there exists a braid b satisfying $ba = \Delta_n$. As the monoid B_n^+ is right simplifiable, the braid b is unique. Denote this braid $\phi(a)$. Lemma 4.1.19 implies that $\phi(a)$ is simple, hence ϕ is a mapping of the set $\mathcal{S}_n$ of simple braids of B_n^+ into itself. However, as B_n^+ is left simplifiable, the mapping ϕ is injective. By the equivalence of (i) and (ii), it follows that the set $\mathcal{S}_n$ is finite, with $n!$ elements. The injection ϕ is thus a bijection. This means that for any simple braid b, there exists a simple braid a satisfying $ba = \Delta_n$. In other words, every simple braid left-divides Δ_n. Hence (i) and (ii) imply (iv).

Finally, by Lemma 4.1.19, every right- or left-divisor of Δ_n is simple. Hence (iii) and (iv) each imply (ii). □

As an application, we can deduce that the family of simple braids is closed under several relations and operations.

Corollary 4.1.22 (Closure)
(i) *Every right- or left-divisor of a simple braid is simple.*
(ii) *The right-lcm and left-lcm of two simple braids are simple.*

Proof
(i) Let a be a simple braid belonging to B_n^+, and suppose a' left-divides a. By definition, there exists b satisfying $a'b = a$ and, by Lemma 4.1.19, the braids a' and b must be simple. Similarly, if a' right-divides a, there exists b satisfying $ba' = a$ and, again, b and a' must be simple.
(ii) Let a_1 and a_2 be two simple braids, and a their right-lcm (which exists and is unique by Proposition 4.1.5). By Proposition 4.1.21, Δ_n is a right-multiple of a_1 and of a_2 hence, by definition of the right-lcm, Δ_n is a right-multiple of a and hence a must be simple.

The argument is the same, mutatis mutandis, for the left-lcm. □

Even if it is superfluous for what follows, the following geometric characterization of simple braids is natural and welcome.

Proposition 4.1.23 (Geometry of simple braids) *For any positive braid a, the following properties are equivalent:*
(i) *The braid a is simple.*
(ii) *In at least one positive diagram representing a, any two strands cross at least once.*

(iii) *In every positive diagram representing a, any two strands cross at most once.*

Proof Denote $N_{i,j}(D)$ the number of crossings of the ith and jth strands in a (positive) braid diagram D. Suppose D is a positive diagram representing a and for every i, j, $N_{i,j}(D) \leqslant 1$. We evaluate the inversion number of $\mathsf{perm}(a)$. By hypothesis, with each new crossing σ_k, the strands that cross had not done so previously, so by (4.3), the inversion number increments by one. Consequently, at the end, necessarily $\mathsf{inv}(\mathsf{perm}(a)) = |a|$, and, by definition, a is simple. Hence (ii) implies (i).

Now suppose a simple, and let D be an arbitrary positive diagram representing a. This time we use the linking numbers. By Example 1.3.10, the contribution of σ_k is $+1/2$ for $\lambda_{k,k+1}$ and 0 otherwise. Then the formula (2.3) implies that the number $N_{i,j}(D)$ is equal[7] to twice the linking number $\lambda_{i,j}(a)$. Since a is simple, there exists by Proposition 4.1.21 a (simple) positive braid b satisfying $ab = \Delta_n$, and, for any i, j, we obtain $N_{i,j}(D) \leqslant 2\lambda_{i,j}(\Delta_n)$. However, for example because of the inductive definition of the diagram on page 82, this number is 1. Hence (i) implies (iii), which clearly implies (ii). □

4.2 Normal Form, the Case of Positive Braids

4.2.1 The Head of a Positive Braid

The first step of the construction of a normal form consists in identifying, for each positive braid, a maximal simple left-divisor.

Definition 4.2.1 (Head) If a is an n-strand positive braid, the *head* of a, denoted $H(a)$, is the left-gcd of a and Δ_n.

The head of the trivial braid is certainly 1, since 1 has no other left-divisor than itself. The head of σ_i is σ_i since, by Lemma 3.1.28, σ_i left-divides Δ_n and admits no other left-divisor than 1 and itself. The head of Δ_n is Δ_n.

Defined as such, the head depends on the element Δ_n, hence on the reference monoid B_n^+. In fact, this is not the case, as the head admits an alternative characterization independent of the index n.

Lemma 4.2.2 *The head of a positive braid a is the unique maximal simple braid left-dividing a.*

Proof Let $a \in B_n^+$. By definition, $H(a)$ left-divides both a and Δ_n, thus is simple. Now, let b be a simple braid left-dividing a. Then b left-divides a

[7] and hence, in general, depends only on the isotopy class of D.

and Δ_n, hence b left-divides the left-gcd of these elements, which is $H(a)$. Hence $H(a)$ is right-multiple of every simple left-divisor of a. □

In other words, $H(a)$ is the unique simple braid satisfying

$$H(a) \preccurlyeq a \quad \text{and} \quad \forall b \text{ simple } (b \preccurlyeq a \Rightarrow b \preccurlyeq H(a)). \tag{4.7}$$

We begin with the direct consequences of the definition.

Lemma 4.2.3
(i) *The relation* $H(a) = 1$ *is equivalent to* $a = 1$.
(ii) *The property 'a is simple' is equivalent to* $H(a) = a$.
(iii) *The relation* $a \preccurlyeq b$ *implies* $H(a) \preccurlyeq H(b)$.

Proof
(i) If a is not 1, there exists at least one generator σ_i left-dividing a. However, σ_i is simple, thus if it divides a, it also divides $H(a)$, which cannot be 1.
(ii) If a is simple, then it is the largest simple divisor of a.
(iii) Suppose $a \preccurlyeq b$. Then $H(a)$ is simple and left-divides a, hence also b. By Lemma 4.2.2, it follows that $H(a)$ left-divides $H(b)$. □

In what follows, it is convenient to introduce a new binary operation on positive braids.

Definition 4.2.4 (Right-complement) If a, b are positive braids, the *right-complement* of a in b, denoted $a\backslash b$ ('a under b') is the unique braid b' satisfying $ab' = \mathsf{lcm}(a, b)$.

The existence of the right-complement is guaranteed by that of the right-lcm, seen in Proposition 4.1.5, and its unicity by the left-simplifiability of the monoid B_n^+, seen in Proposition 3.2.13.

The right-complement is linked with the divisibility and product by several relations.

Lemma 4.2.5 *For any positive braids* a, b, c,
(i) $a \preccurlyeq b$ *is equivalent to* $b\backslash a = 1$, *and* $a \preccurlyeq bc$ *is equivalent to* $b\backslash a \preccurlyeq c$;
(ii) $a\backslash(bc) = a\backslash b \cdot (b\backslash a)\backslash c$ *and* $(bc)\backslash a = c\backslash(b\backslash a)$.

Proof
(i) We begin with the second equivalence. In the presence of $b \preccurlyeq bc$, the relation $a \preccurlyeq bc$ implies $\mathsf{lcm}(a, b) \preccurlyeq bc$. Conversely, as $a \preccurlyeq \mathsf{lcm}(a, b)$, the relation $\mathsf{lcm}(a, b) \preccurlyeq bc$ implies $a \preccurlyeq bc$. Hence $a \preccurlyeq bc$ is equivalent to $\mathsf{lcm}(a, b) \preccurlyeq bc$, thus to $b(b\backslash a) \preccurlyeq bc$, and, from this, to $b\backslash a \preccurlyeq c$ by simplifying

on the left by b. Consequently, $a \preccurlyeq bc$ is equivalent to $b\backslash a \preccurlyeq c$. For the first equivalence, taking $c = 1$ in the above, we obtain that $a \preccurlyeq b$ is equivalent to $b\backslash a \preccurlyeq 1$, hence to $b\backslash a = 1$ since 1 is the only divisor of 1.
(ii) Observe first that for any braid words u, v, u', v',

$$\text{If there exists a grid with source } (u, v) \text{ and target } (u', v'), \text{ then } [v'] = [u]\backslash[v] \text{ and } [u'] = [v]\backslash[u]: \tag{4.8}$$

this is the case, since, by Proposition 4.1.5, uv' represents the right-lcm of $[u]$ and $[v]$. So let a, b, c be three positive braids, and u, v, w the braid words representing them. By Lemma 3.2.4, the grid Γ with source (u, vw) is the juxtaposition of the grid Γ' with source (u, v) and the grid Γ'' with source (u', w), where (u', v') is the target of Γ'. Let (u'', w') be the target of Γ''. Then the target of Γ is $(u'', v'w')$. By (4.8), we have $a\backslash(bc) = [v'w']$ and $(bc)\backslash a = [u'']$. However, (4.8) gives $[u'] = b\backslash a$ and $[v'] = a\backslash b$, then $[w'] = [u']\backslash[w] = (b\backslash a)\backslash c$, and $[u''] = [w]\backslash[u'] = c\backslash(b\backslash a)$. Bringing together these equalities, we obtain the stated formulas. □

This leads to a new closure property for the family of simple braids.

Lemma 4.2.6 *If a is simple, then for any positive braid b, the braid $b\backslash a$ is simple.*

Proof Every positive braid is a product of generators σ_i, hence a product of simple braids. We show by induction on $p \geqslant 0$ that, if a is simple and b is a product of at most p simple braids, then $b\backslash a$ is simple. For $p = 0$, the only possible choice is $b = 1$, so we have $1\backslash a = a$, which is simple. Suppose $p \geqslant 1$ and let b be a braid product of at most p simple braids. We write $b = b'c$, where b' is the product of at most $p-1$ simple braids and where c is simple. By Lemma 4.2.5(ii), $b\backslash a = c\backslash(b'\backslash a)$. By the induction hypothesis, $b'\backslash a$ is simple, as is c. By Corollary 4.1.22(ii), it follows that $\mathrm{lcm}(c, (b'\backslash a))$, that is, $c(b\backslash a)$, is simple, hence, by Corollary 4.1.22(i), $b\backslash a$, a right-divisor of the preceding braid, is simple. □

In view of future applications, also note that a translation of the formulas of Lemma 4.2.5(ii) in terms of the lcm directly gives a recipe for an iterated lcm of the type $\mathrm{lcm}(a, bc)$.

Lemma 4.2.7 *For $a, b, c \in B_\infty^+$, if $ab' = ba' = \mathrm{lcm}(a, b)$ and $a'c' = ca'' = \mathrm{lcm}(a', c)$, then $\mathrm{lcm}(a, bc) = ab'c' = bca''$.*

Exercise 4.2.8 (Iterated lcm) Give a direct proof of the lemma.

Thus equipped, we can now return to the heads of braids.

Lemma 4.2.9 *For any a, b in B_∞^+, the equality $H(ab) = H(aH(b))$ holds.*

Proof Let a, b be positive braids. First, $H(b) \preccurlyeq b$ implies $aH(b) \preccurlyeq ab$, thus $H(aH(b)) \preccurlyeq H(ab)$ by Lemma 4.2.3. Conversely, suppose c is simple and satisfies $c \preccurlyeq ab$. By Lemma 4.2.5(i), we deduce $a\backslash c \preccurlyeq b$. However, by Lemma 4.2.6, the braid $a\backslash c$ is simple because c is. By the characteristic property of $H(b)$, the relation $c\backslash a \preccurlyeq b$ is thus equivalent to $c\backslash a \preccurlyeq H(b)$, so, again by Lemma 4.2.5(i), to $c \preccurlyeq aH(b)$. Setting $c := H(ab)$, we obtain $H(ab) \preccurlyeq H(aH(b))$, hence the equality. □

4.2.2 The Normal Form

At this point, we have obtained, for any positive braid a, a distinguished decomposition

$$a = H(a) \cdot a', \tag{4.9}$$

where the first factor is a simple braid and, more precisely, the 'largest' simple braid possible here. By iterating the construction, we obtain a decomposition as a product of simple braids.

Definition 4.2.10 (Normal sequence) A sequence $(s_1, ..., s_d)$ of simple braids is said to be *normal* if either it is the empty sequence (case $d = 0$), or, for every $k \leqslant d$, $s_k = H(s_k s_{k+1} \cdots s_d)$.

Note that a sequence $(s_1, ..., s_d)$ is normal if and only if the extended sequence $(s_1, ..., s_d, 1)$ is as well: adding or deleting trivial terms to the right does not affect the normality.

The desired result is then easy. It is natural to say that a sequence $(s_1, ..., s_d)$ is a *decomposition* of a braid a if $a = s_1 \cdots s_d$. By convention the empty sequence is a decomposition of the unit braid.

Lemma 4.2.11 *Every positive braid admits a unique normal decomposition whose last term is not* 1.

Proof Let $a \in B_n^+$. We show the existence of a normal decomposition of a by induction on $|a|$. For $|a| = 0$, $a = 1$, and the empty sequence is a normal decomposition. Suppose $|a| \geqslant 1$. By Lemma 4.2.3, the simple braid $H(a)$ is not 1. From this, in the decomposition $a = H(a) \cdot a'$ of (4.9), $H(a)$ is not 1, and we conclude $|a'| < |a|$. If $a' = 1$, and thus $a = H(a)$, the braid a is simple and (a) is its normal decomposition. Otherwise, by the induction hypothesis, a' admits a normal decomposition $(s_2, ..., s_d)$. By definition, $(H(a), s_2, ..., s_d)$ is then a normal decomposition of a.

The unicity follows from a trivial induction on d since, by definition, the first factor of a normal decomposition of a must be $H(a)$. □

We thus have a normal form on B_n^+.

Definition 4.2.12 (Normal form, degree) In the context of Lemma 4.2.11, the sequence $(s_1, ..., s_d)$ is called the *normal form* of a. The integer d is called the *degree* of a, denoted $\mathsf{deg}(a)$.

We sometimes add to 'normal form' the adjective 'greedy' as a reminder that each term of the decomposition swallows the largest possible portion of the part not yet factored.

Exercise 4.2.13 (Normal forms) Determine the normal forms of σ_1^k, Δ_n^k, $\sigma_1\sigma_2$, and $\sigma_1^2\sigma_2^2$.

Exercise 4.2.14 (Algorithm) Write pseudocode for the computation of the normal form on B_n^+ based on the direct inductive definition.

As defined above, the normal form is not very useful, as we lack a simple criterion characterizing it: being the head of a braid a is a global property of a, requiring knowledge of the whole of a. The principal interest of the normal form is the existence of a *local* characterization of normal sequences.

We start with the following consequence of the definition and (4.7).

Lemma 4.2.15 *If s_1, s_2 are simple braids, then (s_1, s_2) is normal if and only if for any simple braid t,*

$$t \preccurlyeq s_1 s_2 \quad \textit{implies} \quad t \preccurlyeq s_1. \tag{4.10}$$

We thus obtain a local criterion, concerning only adjacent terms.

Lemma 4.2.16 *If $s_1, ..., s_d$ are simple braids, $(s_1, ..., s_d)$ is normal if and only if (s_k, s_{k+1}) is normal for $1 \leqslant k < d$.*

Proof Suppose $k < d$. For any sequence $(s_1, ..., s_d)$ of simple braids, by Lemma 4.2.3 we have

$$s_k = H(s_k) \preccurlyeq H(s_k s_{k+1}) \preccurlyeq H(s_k \cdots s_d). \tag{4.11}$$

If $(s_1, ..., s_d)$ is normal, then $s_k = H(s_k \cdots s_d)$, so by (4.11), the equality $s_k = H(s_k s_{k+1})$ holds, hence, by Lemma 4.2.15, (s_k, s_{k+1}) is normal.

The converse is the non-trivial point. We show by induction on $d \geqslant 2$ that, if (s_k, s_{k+1}) is normal for each k, then $(s_1, ..., s_d)$ is normal. For $d = 2$, there is nothing to show. Suppose $d \geqslant 3$. By the induction hypothesis the sequence $(s_2, ..., s_d)$ is normal, so to establish the normality of $(s_1, ..., s_d)$, it suffices to show that s_1 is the head of $s_1 \cdots s_d$. By Lemma 4.2.2, this means showing that any simple braid t dividing $s_1 \cdots s_d$ left-divides s_1. Suppose $t \preccurlyeq s_1 \cdots s_d$. By Lemma 4.2.5(i), we have $(s_1 s_2 \cdots s_d)\backslash t = 1$, hence, by Lemma 4.2.5(ii),

$(s_2 \cdots s_d)\backslash(s_1\backslash t) = 1$, which, by Lemma 4.2.5(i), implies $s_1\backslash t \preccurlyeq s_2 \cdots s_d$. However, by Corollary 4.1.22, $s_1\backslash t$ is simple, thus the normality of $s_2 \cdots s_d$ implies $s_1\backslash t \preccurlyeq s_2$, so, again by Lemma 4.2.5(i), $t \preccurlyeq s_1 s_2$. By hypothesis, (s_1, s_2) is normal, hence $t \preccurlyeq s_1$, and in conclusion $(s_1, ..., s_d)$ is normal. □

We finish with an alternative characterization of normality, to be used in what follows, hence not left merely as an exercise. For s simple, we introduce the two simple braids $\partial_n s$ and $\widetilde{\partial}_n s$ satisfying $s \cdot \partial_n s = \widetilde{\partial}_n s \cdot s = \Delta_n$. Thus $\partial_n s = s\backslash\Delta_n$ in B_n^+, or again $\partial_n s = s^{-1}\Delta_n$ in B_n. Note the coherence with the notation of Lemma 3.1.28.

Lemma 4.2.17 *For s_1, s_2 simple in B_n^+, the following are equivalent:*
(i) *the sequence (s_1, s_2) is normal,*
(ii) *the left-gcd of $\partial_n s_1$ and s_2 is 1.*

Proof Suppose (i), and suppose σ_i left-divides $\partial_n s_1$ and s_2. Then $s_1\sigma_i \preccurlyeq s_1 s_2$ and also $s_1\sigma_i \preccurlyeq s_1\partial_n s_1$, or $s_1\sigma_i \preccurlyeq \Delta_n$, hence $s_1\sigma_i$ is simple. This would give $s_1\sigma_i \preccurlyeq H(s_1 s_2)$, in contradiction with the normality of (s_1, s_2). Thus $\gcd(\partial_n s_1, s_2)$ is trivial.

Conversely, suppose $\gcd(\partial_n s_1, s_2) = 1$. If (s_1, s_2) is not normal, there exists a proper right-multiple of s_1 which is simple and divides $s_1 s_2$, so is certainly of the form $s_1\sigma_i$. The hypothesis that $s_1\sigma_i$ is simple implies $s_1\sigma_i \preccurlyeq s_1\partial_n s_1$, thus $\sigma_i \preccurlyeq \partial_n s_1$. Moreover, the hypothesis that $s_1\sigma_i$ left-divides $s_1 s_2$ implies $\sigma_i \preccurlyeq s_2$. Hence $\sigma_i \preccurlyeq \gcd(\partial_n s_1, s_2)$, and the latter cannot be trivial. Thus (s_1, s_2) is normal. □

Exercise 4.2.18 (Normality) Show that (s_1, s_2) is normal if and only if for any left-divisor σ_i of s_2, the braid $s_1\sigma_i$ is not simple.

4.2.3 Computation of the Normal Form

The greedy normal form gives for each positive braid a distinguished decomposition in terms of simple braids (or, if you wish, in terms of permutations), and, from this, a unique expression in terms of the generators σ_i as soon as we have fixed an expression for each of the simple braids (finite in number for a given index n). Nevertheless, the practical interest of such a decomposition depends on the possibility of computing it in an efficient algorithmic manner. We shall see here that this is indeed the case.

Convention 4.2.19 It will be convenient in what follows to associate with each positive braid a an arrow labelled a, and with the product of two braids the concatenation of the corresponding arrows: this allows us to visualize an equality $ab = cd$ as a 'commutative diagram' of the type $a\downarrow \overset{c}{\underset{b}{\square}} \downarrow d$. We agree to mark a normal sequence (s, t) using a small arc linking the head of s with the tail of t, as in $\xrightarrow{s}\frown\xrightarrow{t}$. By Lemma 4.2.16, a sequence of simple braids $(s_1, ..., s_d)$ is normal if and only if it corresponds to a diagram of the form

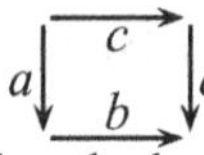

$$\xrightarrow{s_1}\frown\xrightarrow{s_2}\frown\cdots\cdots\frown\xrightarrow{s_d}.$$

All the procedures of normalization described in what follows correspond to diagrams obtained by juxtaposing elementary commutative squares, here called *tiles*. Several types will appear later on, but a single one suffices here.[8]

Definition 4.2.20 (Type A⁻/A tiles)
A *type* A⁻ *tile* is a quadruplet of simple braids (t, s, s', t') satisfying $ts = s't'$; it is said to be *of type* A if, in addition, (s', t') is normal.

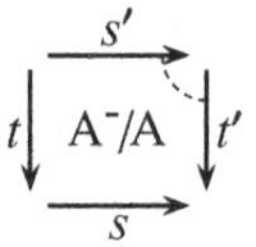

s' (top), t (left), A⁻/A, t' (right), s (bottom)

According to our conventions, a type A⁻ tile (t, s, s', t') corresponds to the above diagram, with the small normality arc for type A.

The first step of the normalization concerns the product of two simple braids, and can be stated as the existence of a tile.

Lemma 4.2.21 *For any simple braids s and t, there exists a unique type* A *tile (t, s, s', t').*

Proof For s, t simple in B_n^+, set

$$s' := H(ts) := \mathsf{gcd}(ts, \Delta_n) \quad \text{and} \quad t' := s'\backslash(ts). \tag{4.12}$$

By definition of the head, s' is simple and $s' \preccurlyeq ts$, so $s't' = ts$. Next, since t is simple and satisfies $t \preccurlyeq ts$, we have $t \preccurlyeq s'$, say $s' = ts''$. We thus find $ts = ts''t'$, so $s = s''t'$. By Lemma 4.1.19, we deduce that t' is simple. Finally, (s', t') is normal by construction, as s' is the head of ts, which is $s't'$. The unicity results from that of the normal form. □

What the above result says is that the normal form of the product of two simple braids is of length at most 2: for (s, t, s', t') as above, the normal form of ts is (s', t') for $t' \neq 1$, (s') for $t' = 1$ and $s' \neq 1$, and the empty sequence $()$ for $s' = t' = 1$.

[8] The 'diminishing' type A⁻ is not used here, but the subsequent use of the type C⁻ for Lemma 4.3.19 and the domino rule $\widetilde{\mathrm{AC}}$ seems indispensable. We thus chose the option of a unique framework highlighting the similarities.

By (4.12), the construction of a type A tile is easy given algorithms for computing the left-gcd and right-complement. By (4.8), the complement can be computed by the reversing procedure of Section 3.3.2. The left-gcd of two positive braids a and b can be computed by the composition of three reversings (see Exercise 4.1.10) or, more efficiently, by seeking (by reversing) a generator σ_i dividing a and b, and then, in the case of success, iterating with the quotients $\sigma_i\backslash a$ and $\sigma_i\backslash b$, up to exhaustion. For every n, we can thus pre-calculate a complete table of the type A tiles, see Table 4.1 for B_3^+.

type A: $t\downarrow$ $s\rightarrow$	1	a	b	ab	ba	Δ
1	1, 1	a, 1	b, 1	ab, 1	ba, 1	Δ, 1
a	a, 1	a, a	ab, 1	a, ab	Δ, 1	Δ, b
b	b, 1	ba, 1	b, b	Δ, 1	b, ba	Δ, a
ab	ab, 1	Δ, 1	ab, b	Δ, b	ab, ba	Δ, ba
ba	ba, 1	ba, a	Δ, 1	ba, ab	Δ, a	Δ, ab
Δ	Δ, 1	Δ, a	Δ, b	Δ, ab	Δ, ba	Δ, Δ

Table 4.1 *Type* A *tiles in* B_3^+. *We use* a, b, *and* Δ *for* σ_1, σ_2, *and* Δ_3. *For example, we read for* ab, a *the value* Δ, 1, *meaning with our conventions of notation that* $(\sigma_1\sigma_2, \sigma_1, \Delta_3, 1)$ *is a type* A *tile, hence a normal decomposition of* $\sigma_1\sigma_2 \cdot \sigma_1$ *is* $(\Delta_3, 1)$.

From this, a 'magic rule' will allow us to determine normal decompositions step by step.

Lemma 4.2.22 (Domino AA)
If (t_0, s_1, s'_1, t_1) *and* (t_1, s_2, s'_2, t_2) *are tiles of type* A *and* A$^-$, *and if* (s_1, s_2) *is normal, then* (s'_1, s'_2) *is also normal.*

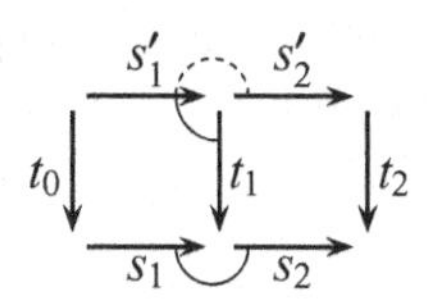

Proof Suppose x is simple and satisfies $x \preccurlyeq s'_1 s'_2$. A fortiori, $x \preccurlyeq s'_1 s'_2 t_2$, or $x \preccurlyeq t_0 s_1 s_2$ by using the commutativity of the diagram. By Lemma 4.2.5, we deduce $t_0\backslash x \preccurlyeq s_1 s_2$. However, by Lemma 4.2.6, $t_0\backslash x$ is simple, since x is by hypothesis. As (s_1, s_2) is normal, $t_0\backslash x \preccurlyeq s_1$, so $x \preccurlyeq t_0 s_1$, or $x \preccurlyeq s'_1 t_1$. Since (s'_1, t_1) is normal, we deduce $x \preccurlyeq s'_1$. Consequently, $x \preccurlyeq s'_1 s'_2$ implies $x \preccurlyeq s'_1$, and (s'_1, s'_2) is normal. □

We now present the principal computation rule relating the normal decompositions of a and ta for t simple.

Proposition 4.2.23 (Left-multiplication) *(See Figure 4.2) Suppose* $(s_1, ..., s_d)$ *is the normal decomposition of a positive braid a and t is simple. Let* $t_0 := t$ *and, for k increasing from* 1 *to d, let* $(t_{k-1}, s_k, s'_k, t_k)$ *be a type* A *tile. Then* $(s'_1, ..., s'_d, t_d)$ *is a normal decomposition of ta.*

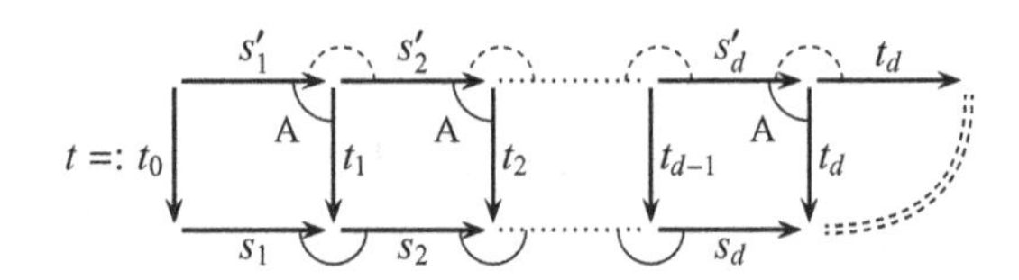

Figure 4.2 Passage from a normal decomposition of a to one of ta: starting with t, we construct the diagram from left to right with d type-A tiles; the sequence $(s'_1, ..., s'_d, t_d)$ is normal.

Proof The existence of the tiles results from Lemma 4.2.21. Then, the commutativity of the diagram of Figure 4.2 implies $s'_1 \cdots s'_d t_d = t_0 s_1 \cdots s_d$, hence $(s'_1, ..., s'_d, t_d)$ is a decomposition of ta. Next, by the domino rule AA, the normality of (s_k, s_{k+1}) implies that of (s'_k, s'_{k+1}) for $k < d$. Finally, (s'_d, t_d) is normal by construction. By Lemma 4.2.16, we conclude that the sequence $(s'_1, ..., s'_d, t_d)$ is normal. □

From this, we obtain by iteration a method to determine a normal decomposition of any positive braid. Recall that a sequence S of elements of a set X is considered as a mapping of an interval of $\mathbb{N}$ into X, so that $S(p)$ designates the pth term of S. In particular this applies to words, considered as sequences of letters.

Algorithm 4.2.24 (Normal form in B_n^+)
Input: a braid word w in $\mathcal{BW}_n^+$
Output: the normal form S of the braid $[w]$

```
1: S ← ()
2: for p decreasing from |w| to 1 do
3:     S ← LeftMult(S, w(p))
4: end for
5: while LastTerm(S) = 1 do
6:     REMOVE LastTerm(S)
7: end while
8: return S
```

```
1: function LeftMult(S: normal sequence, t: simple braid)
2:     S′ ← ()
3:     for k increasing from 1 to |S| do
4:         (s′, t) ← TileA(t, S(k))
5:         S′ ← Concat(S′, (s′))
6:     end for
7:     return Concat(S′, (t))
8: end function
```

```
1: function TileA(t, s: simple braids)
2:     s′ ← gcd(ts, Δ_n)
3:     t′ ← s′\(ts)
4:     return (s′, t′)
5: end function
```

The correctness of the algorithm follows directly from the results above: starting from an empty sequence, we successively multiply by each of the letters of w starting with the last, with the aid of Proposition 4.2.23, hence by using as building blocks the type A tiles, that is, the normalization of the product of two simple braids.

Example 4.2.25 (Normalization) As illustrated in Figure 4.3, the application of Algorithm 4.2.24 to $w := \sigma_2^2\sigma_1^2$ leads first to the normal decomposition $(\sigma_2, \sigma_2\sigma_1, \sigma_1, 1)$ then, after suppression of the final trivial term, to the normal form $(\sigma_2, \sigma_2\sigma_1, \sigma_1)$.

The complexity of Algorithm 4.2.24 can be read from the diagram of Figure 4.3.

Proposition 4.2.26 (Complexity) *For any n, the normal form of a braid in B_n^+ of length ℓ is calculable in time $O(\ell^2)$.*

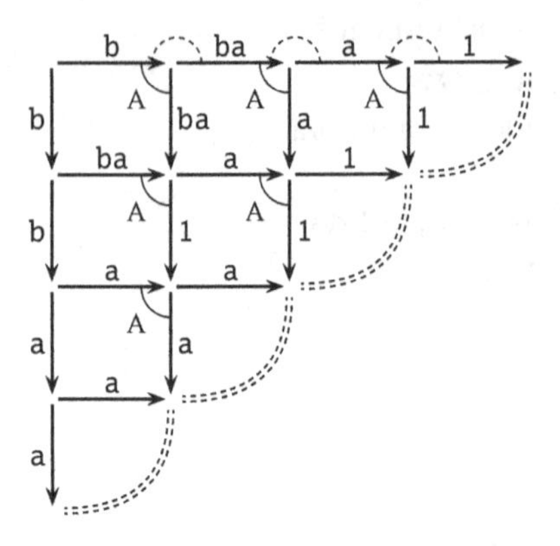

Figure 4.3 Normalization of $w := \sigma_2^2\sigma_1^2$ (alias bbaa) by Algorithm 4.2.24: starting from w written as a column, we construct the diagram with six type A tiles read from Table 4.1; at the top appears the normal decomposition $(\sigma_1, \sigma_1\sigma_2, \sigma_2, 1)$, then, by suppression of the final 1, the normal form $(\sigma_1, \sigma_1\sigma_2, \sigma_2)$.

Proof There exists in B_n^+ a finite number of pairs of simple braids, in fact $(n!)^2$. Hence, for fixed n, we can precalculate the normalization of any two simple braids (as in Table 4.1), so that the normalization of a product of two simple braids can be done in time $O(1)$. Then, by Proposition 4.2.23, determining a normal form of $\sigma_i a$ from that of a requires at most $O(|a|)$, since, by construction, the normal decomposition of a obtained is of length $|a|$. Then, applying ℓ times the method to braids of length increasing from 1 to ℓ takes time $O(\ell^2)$. □

We conclude with a more theoretical application.

Lemma 4.2.27 *For a in B_n^+ and $m \geqslant 1$, the following are equivalent:*
(i) *a left-divides Δ_n^m,*
(ii) *a can be expressed as the product of m simple braids,*
(iii) $\mathrm{deg}(a)$ *is at most m.*

Proof We show the equivalence of the three conditions by induction on m. For $m = 1$, the result is clear: (i) is equivalent to (ii) by definition, and, in this case, the normal form of a is either (a), or the empty sequence.

Suppose $m \geqslant 2$. First suppose (i), that is, $a \preccurlyeq \Delta_n^m$. Set $a' := \Delta_n \backslash a$ and $s := a \backslash \Delta_n$. Since Δ_n is simple, so is s by Lemma 4.2.6. By Lemma 4.2.5(i), the hypothesis $a \preccurlyeq \Delta_n^m$, written $a \preccurlyeq \Delta_n\Delta_n^{m-1}$, implies $a' \preccurlyeq \Delta_n^{m-1}$, and the induction hypothesis implies the existence of simple $s_1, ..., s_{m-1}$ satisfying $a' = s_1 \cdots s_{m-1}$, so $as = \Delta_n s_1 \cdots s_{m-1}$. Setting $s_k' := \phi_n(s_k)$ for $k = 1, ..., m-1$, where ϕ_n is the automorphism of B_∞ swapping σ_i and σ_{n-i} (cf. Exercises 3.1.27 and 3.3.9), and $s' := \widetilde{\partial}_n s$, we obtain $as = s'_1 \cdots s'_{m-1} \Delta_n s' s$. Simplifying on the right by s gives $a = s'_1 \cdots s'_{m-1} s'$, hence (i) ⇒ (ii).

Next, suppose (ii), say $a = s_1 \cdots s_m$ with $s_1, ..., s_m$ simple. Set $a' := s_1 \cdots s_{m-1}$. By the induction hypothesis, there exists b' satisfying $a'b' = \Delta_n^{m-1}$, and we find

$$\Delta_n^m = \Delta_n^{m-1} \Delta_n = a'b'\Delta_n = a'\Delta_n \phi_n(b') = a' s_m \cdot \partial_n s_m \phi_n(b') = ab,$$

with $b := \partial_n s_m \phi_n(b')$. Hence $a \preccurlyeq \Delta_n^m$, and (ii) ⇒ (i).

The implication (ii) ⇒ (iii) results directly from Proposition 4.2.23: applying m times the rule of left multiplication starting with the empty sequence provides a normal decomposition of length m. Finally, (iii) implies (ii) by definition. □

In particular, the above result implies that the degree of a braid a in B_n^+ is the smallest integer m satisfying $a \preccurlyeq \Delta_n^m$.

4.3 Normal Forms, Arbitrary Braids

The normal form of the positive braids can easily be extended to arbitrary braids. We construct here two normal forms, the first, non-symmetric, using the powers of Δ_n in the role of universal denominators, while the other, symmetric, is based on a notion of irreducible fractions analogue to that of the rational numbers.

4.3.1 The Delta-Normal Form

Combining Proposition 3.3.5 with the normal form on B_n^+, we obtain a distinguished decomposition for the elements of B_n.

Lemma 4.3.1 *Any braid g in B_n admits a unique decomposition $g = \Delta_n^m s_1 \cdots s_d$ with m in $\mathbb{Z}$ and $(s_1, ..., s_d)$ normal in B_n^+ satisfying $s_1 \neq \Delta_n$ and $s_d \neq 1$.*

Proof The only point to justify, according to Proposition 3.3.5, is that if $(s_1, ..., s_d)$ is a normal decomposition of a braid b in B_n^+, then the condition $\Delta_n \not\preccurlyeq b$ is equivalent to $s_1 \neq \Delta_n$. However, clearly $s_1 = \Delta_n$ implies $\Delta_n \preccurlyeq b$. Conversely, as Δ_n is simple, $\Delta_n \preccurlyeq b$ implies $\Delta_n \preccurlyeq H(b)$, hence $\Delta_n \preccurlyeq s_1$, but this is possible only for $s_1 = \Delta_n$. □

We thus obtain a normal form on B_n, where the powers of Δ_n play a fundamental role.

Definition 4.3.2 (Δ_n-normal form) In the context of Proposition 4.3.1, the sequence $(\Delta_n^m, s_1, ..., s_d)$ – more often denoted[9] $(\Delta_n^m \mid s_1, ..., s_d)$ – is called the Δ_n-*normal form* of g. In addition, we set[10] $\mathsf{inf}_n(g) := m$, $\mathsf{sup}_n(g) := m + d$, and $\mathsf{lc}_n(g) := d$, where this integer is called the *canonical length* of g.

It is easy to recognize whether a sequence $(\Delta_n^m \mid s_1, ..., s_d)$ is a Δ_n-normal form, that is, whether there exists a braid whose sequence is the Δ_n-normal form: it suffices to verify that s_1 is not Δ_n, s_d is not 1, and for every k, the pair (s_k, s_{k-1}) is normal.

Examples 4.3.3 (Δ_n-normal forms)

For $n \geqslant 3$, the Δ_n-normal form of σ_i is $(\Delta_n^0 \mid \sigma_i)$ for $n \geqslant i+1$; consequently, we find $\mathsf{inf}_n(\sigma_i) = 0$, $\mathsf{sup}_n(\sigma_i) = \mathsf{lc}_n(\sigma_1) = 1$. Indeed, σ_i is positive, simple, and not divisible by Δ_n.

However, the Δ_2-normal form of σ_1 is $(\Delta_2^1 \mid 1)$, since $\Delta_2 = \sigma_1$, so $\mathsf{inf}_2(\sigma_i) = 1$, $\mathsf{sup}_2(\sigma_i) = 1$, and $\mathsf{lc}_2(\sigma_1) = 0$.

More interestingly, for $n \geqslant \max(i + 1, 3)$, the Δ_n-normal form of σ_i^{-1} is $(\Delta_n^{-1} \mid b)$, where b is the unique positive braid satisfying $b\sigma_i = \Delta_n$, that is, $b = \widetilde{\partial}_n \sigma_i$. For example, for $n \geqslant 2$, the Δ_3-normal form of σ_1^{-1} is $(\Delta_3^{-1} \mid \sigma_1\sigma_2)$, and that of σ_2^{-1} is $(\Delta_3^{-1} \mid \sigma_2\sigma_1)$. Hence $\mathsf{inf}_n(\sigma_1^{-1}) = \mathsf{inf}_n(\sigma_2^{-1}) = -1$, $\mathsf{sup}_n(\sigma_1^{-1}) = \mathsf{sup}_n(\sigma_2^{-1}) = 0$, and $\mathsf{lc}_n(\sigma_1^{-1}) = \mathsf{lc}_n(\sigma_2^{-1}) = 1$.

The Δ_2-normal form of σ_1^{-1} is $(\Delta_2^{-1} \mid 1)$.

Exercise 4.3.4 (Multiplication by Δ_n^e) Show that, if the Δ_n-normal form of a braid g is $(\Delta_n^m \mid s_1, ..., s_d)$, then for any integer e, that of $\Delta_n^e g$ is $(\Delta_n^{m+e} \mid s_1, ..., s_d)$.

If we would like to obtain a unique distinguished braid word for each braid, we could use the permutation words of Definition 4.1.12 as representatives of the simple braids. Then, the Δ_3-normal word representing σ_1^{-1} is the word $\underline{\Delta}_3^{-1}\sigma_1\sigma_2$, or $\overline{\sigma}_2\overline{\sigma}_1\overline{\sigma}_2\sigma_1\sigma_2$, whereas the one representing σ_2^{-1} is $\overline{\sigma}_2\overline{\sigma}_1\overline{\sigma}_2\sigma_2\sigma_1$ (which is not freely reducible).

The Δ_n-normal form is easy to compute. The only new ingredient is the control of the conjugation by Δ_n.

Lemma 4.3.5 *For $n \geqslant 2$, let ϕ_n be the automorphism of B_n sending σ_i to σ_{n-i} for $1 \leqslant i < n$. Then, for any braid g in B_n,*

$$g \cdot \Delta_n = \Delta_n \cdot \phi_n(g). \tag{4.13}$$

[9] So as to clearly separate the portion 'power of Δ'; we could also use a compact form of the type $(m \mid s_1, ..., s_d)_n$, as only the integers m and n count, Δ itself being superfluous.

[10] When the reference group B_n is fixed, the index n is often omitted.

Proof By Lemma 3.1.26, the result is true when g is a generator σ_i. From this, it extends to all of B_n, since ϕ_n is the inner automorphism of conjugation by Δ_n. □

As in the case of positive braids, the principal computation rule links a Δ_n-normal decomposition[11] of a braid g to those of tg and $t^{-1}g$ when t is simple.

Proposition 4.3.6 (Left multiplication and division) *Suppose $(\Delta_n^m \mid s_1, ..., s_d)$ is a Δ_n-normal decomposition of a braid g in B_n and $t \in B_n^+$ is simple.*
(i) Set $t_0 := \phi_n^m(t)$, and let $(s'_1, ..., s'_d, t_d)$ be a normal decomposition of $t_0 s_1 \cdots s_d$. If s'_1 is not Δ_n, the sequence $(\Delta_n^m \mid s'_1, ..., s'_d, t_d)$ is a Δ_n-normal decomposition of tg. If s'_1 is Δ_n, the sequence $(\Delta_n^{m+1} \mid s'_2, ..., s'_d, t_d)$ is a Δ_n-normal decomposition (Figure 4.2, top).
(ii) Now set $t_0 := \phi_n^m(\widetilde{\partial}_n t)$, and let $(s'_1, ..., s'_d, t_d)$ be a normal decomposition of $t_0 s_1 \cdots s_d$. If s'_1 is not Δ_n, the sequence $(\Delta_n^{m-1} \mid s'_1, ..., s'_d, t_d)$ is a Δ_n-normal decomposition of $t^{-1}g$. If s'_1 is Δ_n, the sequence $(\Delta_n^m \mid s'_2, ..., s'_d, t_d)$ is a Δ_n-normal decomposition (Figure 4.2, bottom).

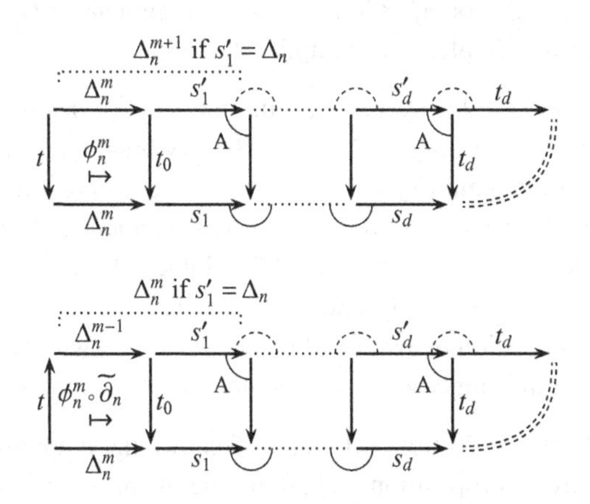

Figure 4.4 Passage from a Δ_n-normal decomposition of g to one of tg (at the top) and $t^{-1}g$ (at the bottom). On the top, we set $t_0 := \phi_n^m(t)$, and normalize from left to right as in Proposition 4.2.23 by juxtaposing tiles of type A; the Δ_n-normal decomposition obtained is $(\Delta_n^m \mid s'_1, ..., s'_d, t_d)$, except when $s'_1 = \Delta_n$, in which case we regroup Δ_n^m and s_1 into Δ_n^{m+1}. On the bottom, we introduce the simple element $\widetilde{\partial}_n t$ satisfying $\widetilde{\partial}_n t \cdot t = \Delta_n$; starting with $t_0 := \phi_n^m(\widetilde{\partial}_n t)$, the rest is the same.

[11] Here again, we avoid speaking of 'the Δ_n-normal form' in order to avoid distinguishing the eventual trivial terms at the end of the decompositions.

Proof
(i) By Lemma 4.3.5, $t\Delta_n^m = \Delta_n^m \phi_n^m(t)$, hence

$$tg = t\Delta_n^m s_0 \cdots s_d = \Delta_n^m \phi_n^m(t) s_0 \cdots s_d = \Delta_n^m t_0 s_0 \cdots s_d = \Delta_n^m s_1' \cdots s_d' t_d,$$

thus the sequence $(\Delta_n^m \mid s_1', ..., s_d', t_d)$ is a decomposition of tg. Moreover, the sequence $(s_1', ..., s_d', t_d)$ is normal. The only situation where neither $(\Delta_n^m \mid s_1', ..., s_d', t_d)$, nor $(\Delta_n^{m+1} \mid s_2', ..., s_d', t_d)$ would be Δ_n-normal is the case $s_1' = s_2' = \Delta_n$. So, suppose $s_1' = s_2' = \Delta_n$. Then $\Delta_n^2 \preccurlyeq s_1' \cdots s_d' t_d$, hence $\Delta_n^2 \preccurlyeq t_0 s_1 \cdots s_d$. As t is simple, so is t_0. Thus $\Delta_n = t_0 \cdot \partial_n t_0$, so $\Delta_n^2 = t_0 \Delta_n \phi_n(\partial_n t_0)$, and $\Delta_n^2 \preccurlyeq t_0 s_1 \cdots s_d$ implies $\Delta_n \preccurlyeq s_1 \cdots s_d$, thus $s_1 = \Delta_n$, contradicting the hypothesis that $(\Delta_n^m \mid s_1, ..., s_d)$ is Δ_n-normal.
(ii) The argument is similar. First $\Delta_n = \widetilde{\partial}_n t \cdot t$, or $t^{-1} = \Delta_n^{-1} \cdot \widetilde{\partial}_n t$, then $t^{-1}\Delta_n^m = \Delta_n^{-1} \widetilde{\partial}_n t \Delta_n^m = \Delta_n^{-1} \Delta_n^m \phi_n^m(\widetilde{\partial}_n t) = \Delta_n^{m-1} t_0$, and we find

$$t^{-1}g = t^{-1}\Delta_n^m s_1 \cdots s_d = \Delta_n^{m-1} t_0 s_1 \cdots s_d = \Delta_n^{m-1} s_1' \cdots s_d' t_d,$$

hence the sequence $(\Delta_n^{m-1} \mid s_1', ..., s_d', t_d)$ is a decomposition of $t^{-1}g$. Moreover, the sequence $(s_1', ..., s_d', t_d)$ is normal, and as in (i), it only remains to exclude the case $s_1' = s_2' = \Delta_n$, or $\Delta_n^2 \preccurlyeq t_0 s_1 \cdots s_d$. The argument of (i) remains valid since once again the simplicity of t implies that of t_0. □

As in Algorithm 4.2.24, by iterating the method of Proposition 4.3.6 we obtain an algorithm determining, from an arbitrary braid word w, a Δ_n-normal decomposition of the braid $[w]$, then, by eliminating eventual final terms 1, its Δ_n-normal form. As in the positive case, the basic element is the determination of the type A tiles – hence, in the case of B_3, Table 4.1 – plus the table of the values of the function $\widetilde{\partial}_n$, easily determined by reversing. Note that, when a Δ_n-normal decomposition of a suffix of the word w is known, we can start with it rather than with the empty word so as to not repeat the initial steps.

Example 4.3.7 (Normalization) Figure 4.5 shows the mapping of the algorithm derived from Proposition 4.3.20 to the word $w := \sigma_2^{-2}\sigma_1^{-2}\sigma_2^2\sigma_1^2$. As we have seen in Example 4.2.25 that the normal form of the suffix $\sigma_2^2\sigma_1^2$ of w is $(\sigma_2, \sigma_2\sigma_1, \sigma_1)$, we can start with this. The Δ_3-normal form obtained is $(\Delta_3^{-3} \mid \sigma_2\sigma_1, \sigma_1, \sigma_1\sigma_2, \sigma_2, \sigma_2\sigma_1, \sigma_1)$, from which we obtain the values $\mathsf{lc}_3(\sigma_2^{-2}\sigma_1^{-2}\sigma_2^2\sigma_1^2) = 3$, $\mathsf{inf}_3(\sigma_2^{-2}\sigma_1^{-2}\sigma_2^2\sigma_1^2) = -3$, and $\mathsf{sup}_3(\sigma_2^{-2}\sigma_1^{-2}\sigma_2^2\sigma_1^2) = 3$.

Exercise 4.3.8 (Algorithm) Adapt the plan of Algorithm 4.2.24 to write pseudocode for an algorithm determining the Δ_n-normal form of a braid in B_n, following the method of Proposition 4.3.6.

Since ϕ_n is involutive, applying ϕ_n^m changes nothing if m is even, and exchanges everywhere σ_i and σ_{n-i} if m is odd. It is in fact easy to pre-calculate

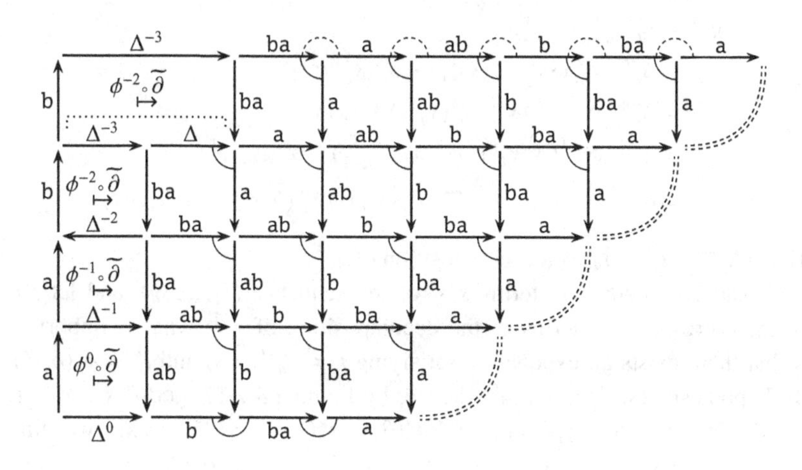

Figure 4.5 Determination of the Δ_3-normal form of $\sigma_2^{-2}\sigma_1^{-2}\sigma_2^2\sigma_1^2$ by the algorithm derived from Proposition 4.3.6: we start with the sequence $(\sigma_2, \sigma_2\sigma_1, \sigma_1)$ already computed as the (Δ_n)-normal form of the suffix $\sigma_2^2\sigma_1^2$ in Example 4.2.25, and apply the method of Proposition 4.3.6 to left multiply and divide by the successive letters starting with the last (in the present case, there are only divisions); the Δ_3-normal form can be read on the top line of the diagram. Note that after the third step, we are in the case '$s_1' = \Delta_3$', and thus can regroup Δ_3^{-3} and Δ_3 into Δ_3^{-2}.

the values $\widetilde{\partial}_n t$ for simple t in B_n^+. The cost of a left multiplication by a simple braid or its inverse following the method of Proposition 4.3.6 is thus linear, leading directly to the following result.

Proposition 4.3.9 (Complexity) *For any n, the Δ_n-normal form of a braid in B_n of length ℓ is computable in time $O(\ell^2)$.*

The unicity of the Δ-normal form implies a simple relation between the normal form of a braid and that of its inverse.

Proposition 4.3.10 (Inverse) *If $(\Delta_n^m \mid s_1, \ldots, s_d)$ is the Δ_n-normal form of a braid g in B_n, the Δ_n-normal form of the braid g^{-1} is $(\Delta_n^{-m-d} \mid t_1, \ldots, t_d)$ with $t_k := \phi_n^{m+d-k}(\widetilde{\partial}_n s_{d+1-k})$ for every k.*

Proof It suffices to verify that the stated sequence is a decomposition of g^{-1} and is Δ_n-normal. However, given the equalities $\phi_n^2 = \mathsf{id}$ and $\phi_n(\Delta_n) = \Delta_n$, and setting $e := d + m$, the equality $g = \Delta_n^m s_1 \cdots s_d$ implies

$$\begin{aligned}
g^{-1} &= s_d^{-1} s_{d-1}^{-1} \cdots s_1^{-1} \Delta_n^{-m} \\
&= \Delta_n^{-m} \cdot \phi_n^m(s_d^{-1}) \phi_n^m(s_{d-1}^{-1}) \cdots \phi_n^m(s_1^{-1}) \\
&= \Delta_n^{-e} \cdot \Delta_n^d \cdot \phi_n^m(s_d^{-1}) \phi_n^m(s_{d-1}^{-1}) \cdots \phi_n^m(s_1^{-1}) \\
&= \Delta_n^{-e} \cdot \phi_n^{e-1}(\Delta_n s_d^{-1}) \cdot \phi_n^{e-2}(\Delta_n s_{d-1}^{-1}) \cdots \phi_n^m(\Delta_n s_1^{-1}) \\
&= \Delta_n^{-e} \cdot \phi_n^{e-1}(\widetilde{\partial}_n s_d) \cdot \phi_n^{e-2}(\widetilde{\partial}_n s_{d-1}) \cdots \phi_n^m(\widetilde{\partial}_n s_1) = \Delta_n^{-e} t_1 \cdots t_d.
\end{aligned}$$

Thus $(\Delta_n^{-m-d} \,|\, t_1, ..., t_d)$ is a decomposition of g^{-1}.

Consider two adjacent terms s, s' of the sequence $(s_1, ..., s_d)$, and let t, t' be the corresponding terms of the decomposition of g^{-1}: what is important is that there exists an exponent e satisfying $t = \phi_n^{e+1}(\widetilde{\partial}_n s)$ and $t' = \phi_n^e(\widetilde{\partial}_n s')$. By hypothesis, (s, s') is normal, hence, by Lemma 4.2.17, $\mathsf{gcd}(\partial_n s, s') = 1$. As ϕ_n is an automorphism, $\mathsf{gcd}(\phi_n^e(\partial_n s), \phi_n^e(s')) = 1$. Next, we find $\Delta_n \partial_n s = \widetilde{\partial}_n s \cdot s \cdot \partial_n s = \widetilde{\partial}_n s \, \Delta_n$, so $\partial_n s = \Delta_n^{-1} \widetilde{\partial}_n s \Delta_n = \phi_n(\widetilde{\partial}_n s)$, or $\partial_n s = \phi_n(\widetilde{\partial}_n s)$. Moreover, since, by construction, ∂_n and $\widetilde{\partial}_n$ are inverses one of the other, we obtain $\phi_n^e(s') = \phi_n^e(\partial_n(\widetilde{\partial}_n s')) = \partial_n \phi_n^e(\widetilde{\partial}_n s')$. Hence the above equality can be rewritten $\mathsf{gcd}(\phi_n^{e+1}(\widetilde{\partial}_n s), \partial_n \phi_n^e(\widetilde{\partial}_n s')) = 1$, or $\mathsf{gcd}(\partial_n t, t') = 1$: with a new application of Lemma 4.2.17, we conclude that (t', t) is normal. The sequence $(t_1, ..., t_d)$ is thus normal.

Finally, $\widetilde{\partial}_n s_d$ cannot be Δ_n since s_d is not 1, and $\widetilde{\partial}_n s_1$ is not 1 as s_1 is not Δ_n. Hence $(\Delta_n^{-m-d} \,|\, t_1, ..., t_d)$ is Δ_n-normal. □

Note that the above result implies for any g in B_n

$$\mathsf{inf}_n(g^{-1}) = -\mathsf{sup}_n(g), \; \mathsf{sup}_n(g^{-1}) = -\mathsf{inf}_n(g), \; \mathsf{lc}_n(g^{-1}) = \mathsf{lc}_n(g).$$

4.3.2 The Symmetric Normal Form

As we will see in Section 4.4, the properties of the Δ_n-normal form of braids are interesting, but this form is not symmetric, and above all not intrinsic in the sense where, as seen in the Examples 4.3.3, the Δ_n-normal form of a braid in B_n does not coincide with its Δ_{n+1}-normal form considered as an element of B_{n+1}. With just a bit more effort, we can obtain a normal form avoiding these annoyances.

In what follows, there are many questions involving the gcd, so we begin with a small preparatory result. Remember that $\mathsf{gcd}(a, b)$ designates the left-gcd of two positive braids a, b.

Lemma 4.3.11 *For any positive braids a, b, c,*

$$\mathsf{gcd}(ab, ac) = a \cdot \mathsf{gcd}(b, c). \tag{4.14}$$

Proof Since $x \preccurlyeq b$ implies $ax \preccurlyeq ab$, we have $a \cdot \mathsf{gcd}(b, c) \preccurlyeq ab$ and $a \cdot \mathsf{gcd}(b, c) \preccurlyeq ac$, hence $a \cdot \mathsf{gcd}(b, c) \preccurlyeq \mathsf{gcd}(ab, ac)$.

Conversely, suppose $x \preccurlyeq \mathsf{gcd}(ab, ac)$. Then $x \preccurlyeq ab$ and $x \preccurlyeq ac$, which, by Lemma 4.2.5(i), implies $a\backslash x \preccurlyeq b$ and $a\backslash x \preccurlyeq c$, hence $a\backslash x \preccurlyeq \mathsf{gcd}(b, c)$. Again by Lemma 4.2.5(i), we conclude $x \preccurlyeq a \cdot \mathsf{gcd}(b, c)$, hence $\mathsf{gcd}(ab, ac) \preccurlyeq a \cdot \mathsf{gcd}(b, c)$. □

By Propositions 3.3.1 and 3.3.4, every braid can be expressed as both a right quotient and a left quotient of positive braids. Such fractionary expressions are not unique: for example, for $g = a^{-1}b$ with positive a, b, we also have $g = (ca)^{-1}(cb)$ for any positive braid c.[12] Nevertheless, as for rational numbers and fractions, unicity appears when we consider the irreducible fractions.

Lemma 4.3.12 *Every braid g in B_n admits a unique expression $a^{-1}b$ with a, b in B_n^+ satisfying* $\mathsf{gcd}(a, b) = 1$.

Proof Let $g \in B_n$. By Proposition 3.3.4, there exist a', b' in B_n^+ satisfying $g = a'^{-1}b'$. Let $c := \mathsf{gcd}(a', b')$, and define a and b by $a' = ca$ and $b' = cb$. Then $g = (ca)^{-1}(cb)$, hence $g = a^{-1}b$. Moreover, Lemma 4.3.11 implies $\mathsf{gcd}(a', b') = c \cdot \mathsf{gcd}(a, b)$, thus $\mathsf{gcd}(a, b) = 1$.

Now suppose $g = a^{-1}b = a'^{-1}b'$ with a, b, a', b' in B_n^+ satisfying $\mathsf{gcd}(a, b) = \mathsf{gcd}(a', b') = 1$. By Proposition 3.1.32, a and a' admit a common left multiple, say $ca = c'a'$. First of all, the equality $g = a^{-1}b = a'^{-1}b'$ implies by inversion $b^{-1}a = b'^{-1}a'$ in B_n, and we can write

$$cb \cdot b'^{-1}a' = cb \cdot b^{-1}a = ca = c'a' = c'b' \cdot b'^{-1}a',$$

hence $cb = c'b'$. Then let c'' be the right-lcm of c and c'. From the relations $c \preccurlyeq ca$ and $c' \preccurlyeq c'a' = ca$, we deduce $c'' \preccurlyeq ca$, then, similarly, from $c \preccurlyeq cb$ and $c' \preccurlyeq c'b' = cb$, we deduce $c'' \preccurlyeq cb$, then, finally, $c'' \preccurlyeq \mathsf{gcd}(ca, cb)$, hence $c'' \preccurlyeq c$ using the general law $\mathsf{gcd}(zx, zy) = z \cdot \mathsf{gcd}(x, y)$ valid for any x, y, z in B_n^+ (Lemma 4.3.11). We conclude $c' \preccurlyeq c$, and mutatis mutandis, $c \preccurlyeq c'$, so $c = c'$, therefore $a = a'$ and $b = b'$, hence the required unicity. □

By combining Lemma 4.3.12 with the normal form on B_n^+, we obtain a distinguished decomposition for the elements of B_n.

Lemma 4.3.13 *Every braid g in B_n admits a unique decomposition $g = r_e^{-1} \cdots r_1^{-1} s_1 \cdots s_d$ with $(r_1, ..., r_e)$ and $(s_1, ..., s_d)$ normal in B_n^+ satisfying* $\mathsf{gcd}(s_1, r_1) = 1$.

[12] This is the exact analogue to multiplying the numerator and denominator of a fraction in $\mathbb{Q}$ by the same quantity.

Proof The only point to justify for 4.3.12 is that if $(r_1, ..., r_e)$ and $(s_1, ..., s_d)$ are the normal decompositions of two positive braids a and b of B_n^+, then the condition $\mathsf{gcd}(a, b) = 1$ is equivalent to $\mathsf{gcd}(r_1, s_1) = 1$. Clearly, a left-divisor common to r_1 and s_1 divides $r_1 \cdots r_e$ and $s_1 \cdots s_d$, hence $\mathsf{gcd}(a, b) = 1$ implies $\mathsf{gcd}(r_1, s_1) = 1$. Conversely, if a and b admit a non-trivial common left-divisor, they admit at least one of the form σ_i, which is simple. By definition of the normal form, $\sigma_i \preccurlyeq a$ implies $\sigma_i \preccurlyeq r_1$, and $\sigma_i \preccurlyeq b$ implies $\sigma_i \preccurlyeq s_1$, hence $\sigma_i \preccurlyeq \mathsf{gcd}(r_1, s_1)$. Thus $\mathsf{gcd}(r_1, s_1) = 1$ implies $\mathsf{gcd}(a, b) = 1$. □

We thus have a normal form on B_n, where the numerator and denominator play symmetric roles.

Definition 4.3.14 (Symmetric normal form) In the context of Lemma 4.3.13, the sequence $(r_e^{-1}, ..., r_1^{-1}, s_1, ..., s_d)$ is called the *symmetric normal form* of g.

As in the positive case, it is easy to decide if a sequence of simple braids $(r_e^{-1}, ..., r_1^{-1}, s_1, ..., s_d)$ is a symmetric normal form, that is, if there exists a braid for which it is the symmetric normal form: it suffices to verify that $r_e \neq 1$, $s_d \neq 1$, $\mathsf{gcd}(r_1, s_1) = 1$ and that, for every appropriate k, (r_k, r_{k+1}) and (s_k, s_{k+1}) are normal. This characterization shows in particular that the symmetric normal form is independent of the reference group B_n.[13]

Examples 4.3.15 (Symmetric normal forms) For any i, the symmetric normal form of σ_i is (σ_i), and that of σ_i^{-1} is (σ_i^{-1}), of length 1. Next, for any positive braid, the symmetric normal form coincides with the normal form of Definition 4.2.12, whereas that of a negative braid a^{-1} is $(s_d^{-1}, ..., s_1^{-1})$ where $(s_1, ..., s_d)$ is the normal form of the positive braid a.

We now come to the computation of the symmetric normal form and the construction of an algorithm similar to Algorithm 4.2.24. The computations again bring into play the simple braids, and, as in Section 4.2.3, they will be adequately illustrated by diagrams following Convention 4.2.19, amended with the rule that an arrow $\xleftarrow{s}$ or $\downarrow s$ traversed backwards from its orientation represents s^{-1}. In what follows, we use a small arc between two arrows s, t with the same source as in $\xleftarrow{s}\frown\xrightarrow{t}$ to indicate the normality of (s^{-1}, t), that is, the $\mathsf{gcd}(s, t)$ is 1. In this way, a symmetric normal sequence $(r_e^{-1}, ..., r_1^{-1}, s_1, ..., s_d)$ corresponds to a diagram

$$\xleftarrow{r_e}\frown\cdots\cdots\frown\xleftarrow{r_2}\frown\xleftarrow{r_1}\frown\xrightarrow{s_1}\frown\xrightarrow{s_2}\frown\cdots\cdots\frown\xrightarrow{s_d}.$$

[13] It also shows that this normality is a *local* property in the sense where each condition only involves one term or two neighbouring terms. Hence the symmetric normal sequences form a regular language and can be recognized by a finite automaton.

Note that, when the arrows are oriented to the left, like the arrows r_i above, the normality is read from right to left: in other words, we declare (s^{-1}, t^{-1}) to be normal if (t, s) is.

The computation of the symmetric normal form is similar to that of the normal form of positive braids, with a base procedure which, from a normal decomposition of a braid g, provides one for tg and for $t^{-1}g$ when t is simple. For this, we again use tiles analogous to those of Definition 4.2.20.

Definition 4.3.16 (Tiles of type $\widetilde{\mathrm{A}}$, B, C)

(i) a *tile of type* $\widetilde{\mathrm{A}}^{-}$ is a quadruplet of simple braids (t, s, s', t') satisfying $t^{-1}s^{-1} = s'^{-1}t'^{-1}$; it is said to be *of type* $\widetilde{\mathrm{A}}$ if, as well, (s'^{-1}, t'^{-1}) is normal.

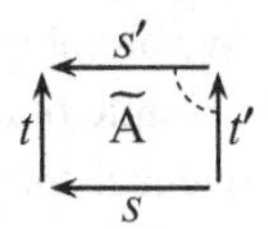

(ii) a *tile of type* B^{-} is a quadruplet of simple braids (t, s, s', t') satisfying $ts^{-1} = s'^{-1}t'$; it is said to be *of type* B if, as well, (s'^{-1}, t') is normal.

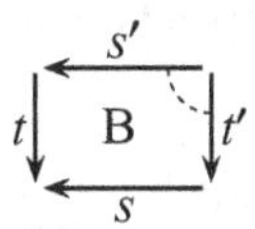

(iii) a *tile of type* C^{-} is a quadruplet of simple braids (t, s, s', t') satisfying $t^{-1}s = s'^{-1}t'$ and such that in addition, there exists r satisfying $t = rs'$ and $s = rt'$; it is said to be *of type* C if, as well, (s'^{-1}, t') is normal.

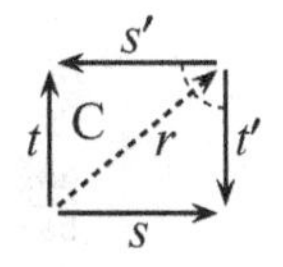

The four types of tiles correspond to the four combinations of signs ± in the normalization of a pair of simple braids (t^e, s^d): type A corresponds to ++, type $\widetilde{\mathrm{A}}$ to −−, type B to +−, and type C to −+. Note that a tile of type $\widetilde{\mathrm{A}}$ is a tile of type A read backwards: (t, s, s', t') is of type $\widetilde{\mathrm{A}}$ if and only if (s, t, t', s') is of type A. Also note that, for the type C, the existence of the diagonal arrow is guaranteed: by Lemma 4.3.12, when (s'^{-1}, t) is normal, $r := \mathsf{gcd}(s, t)$ gives the desired factorizations.[14]

Once again, we begin with a result on existence.

Lemma 4.3.17 *For any pair of simple braids s, t, there exists a unique tile (t, s, s', t') of type $\widetilde{\mathrm{A}}$ (resp., B, resp., C).*

Proof As in the case ++, treated in Lemma 4.2.21, we must show that, whatever the signs of e and d, the braid $t^e s^d$ possesses a symmetric normal decomposition of the form $s'^e t'^d$ with s' and t' simple, necessarily unique by the unicity of the symmetric normal form.

For the case −− (type $\widetilde{\mathrm{A}}$), we apply the case ++ to the pair (s, t).

For the case −+ (type C), let $r := \mathsf{gcd}(s, t)$ and s', t' determined by $t = rs'$ and $s = rt'$. Then s' and t' are simple as right divisors of simple braids, and by construction, $t^{-1}s = s'^{-1}t'$ in B_∞. Moreover, by 4.3.11, the gcd of s' and t' is 1, hence (s'^{-1}, t') is symmetric normal.

[14] Observe that, for the other types, A, B, $\widetilde{\mathrm{A}}$, the diagonal arrows also exist, but we will not use them here.

Finally, for the case $+-$ (type B), the simple braids s and t admit a left-lcm, say $s_0 t = t_0 s$, which is simple, and s_0 and t_0, left-dividing the simple braid $s_0 t$, are themselves simple. By construction, $ts^{-1} = s_0^{-1} t_0$ in B_∞. The case $-+$ thus implies the existence of a symmetric normal decomposition $s'^{-1}t'$ for $s_0^{-1}t_0$, thus for ts^{-1}.[15] □

The lcm and gcd are computed by reversings, thus, as with Table 4.2.3, for each n, we can pre-calculate the table of type B and C tiles for B_n^+, see Table 4.2 for B_3^+.

We are about to establish five (!) new domino rules, each affirming that, if the simple braids on the bottom of two adjacent tiles form a symmetric normal sequence, then it is the same for those on the top (with their signs, which depend on the tiles considered).

type B:

$t\downarrow$ $s\rightarrow$	1	a	b	ab	ba	Δ
1	1, 1	a, 1	b, 1	ab, 1	ba, 1	Δ, 1
a	1, a	1, 1	ab, ba	ab, b	b, 1	ab, 1
b	1, b	ba, ab	1, 1	a, 1	ba, a	ba, 1
ab	1, ab	b, ab	1, a	1, 1	b, a	1, b
ba	1, ba	1, b	a, ba	a, b	1, 1	a, 1
Δ	1, Δ	1, ab	1, ba	1, b	1, a	1, 1

type C:

$t\downarrow$ $s\rightarrow$	1	a	b	ab	ba	Δ
1	1, 1	1, a	1, b	1, ab	1, ba	1, Δ
a	a, 1	1, 1	a, b	1, b	a, ba	1, ba
b	b, 1	b, a	1, 1	b, ab	1, a	1, ab
ab	ab, 1	b, 1	ab, b	1, 1	ab, ba	1, a
ba	ba, 1	ba, a	a, 1	ba, ab	1, 1	b, 1
Δ	Δ, 1	ba, 1	ab, 1	a, 1	b, 1	1, 1

Table 4.2 *Tiles of type* B *(on the top) and* C *(on the bottom) in* B_3^+. *We use* a, b, *and* Δ *for* σ_1, σ_2, *and* Δ_3. *For example, we read on the top for* ab, a *the value* b, ab, *meaning that* $(\sigma_1\sigma_2, \sigma_1, \sigma_2, \sigma_1\sigma_2)$ *is a type* B *tile, hence a normal decomposition of* $\sigma_1\sigma_2 \cdot \sigma_1^{-1}$ *is* $(\sigma_2^{-1}, \sigma_1\sigma_2)$. *Similarly, for* ab, a, *we read on the bottom the value* b, 1, *meaning that* $(\sigma_1\sigma_2, \sigma_1, \sigma_2, 1)$ *is a type* C *tile, hence a normal decomposition of* $(\sigma_1\sigma_2)^{-1} \cdot \sigma_1$ *is* $(\sigma_2^{-1}, 1)$.

[15] Indeed, we can (easily) show that $\mathsf{gcd}(s_0, t_0) = 1$, hence $s' = s_0$ and $t' = t_0$, i.e. that (t, s, s_0, t_0) is the desired tile.

Definition 4.3.18 (Domino)
The *domino rule* XY is said to be valid if, with (t_0, s_1, s_1', t_1) and (t_1, s_2, s_2', t_2) as tiles of type X and Y˜, the normality of $(s_1^{\pm}, s_2^{\pm})$ implies that of $(s_1'^{\pm}, s_2'^{\pm})$.

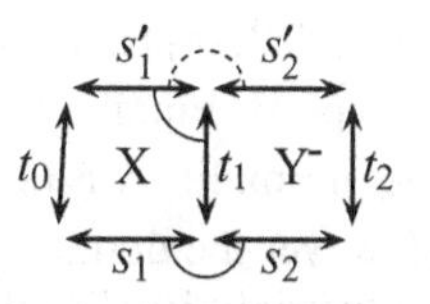

The domino rule AA of Lemma 4.2.22 plays in the above framework.

We present here the results required for what is to follow. The proofs are sometimes a bit delicate, but these are truly at the heart of the computations on braids, and we can also find a playful but elegant side to these arguments, all of which repose on the lattice properties of the simple braids...

Lemma 4.3.19 *The domino rules of* AA, BB, BA, $\widetilde{\mathrm{A}}\widetilde{\mathrm{A}}$, $\widetilde{\mathrm{A}}$C, *and* CA *are valid.*

Proof We will examine each case separately with its figure. The case AA was already handled in Lemma 4.2.22.

case BB. Suppose x is simple and satisfies $x \preccurlyeq s_2's_1'$. A fortiori, $x \preccurlyeq s_2's_1't_0$, or $x \preccurlyeq t_2s_2s_1$ by commutativity. By Lemma 4.2.5, we have $t_2\backslash x \preccurlyeq s_2s_1$. By Lemma 4.2.6, $t_2\backslash x$ is simple, and as (s_2, s_1) is normal, we deduce $t_2\backslash x \preccurlyeq s_2$, so $x \preccurlyeq t_2s_2$, or $x \preccurlyeq s_2't_1$.

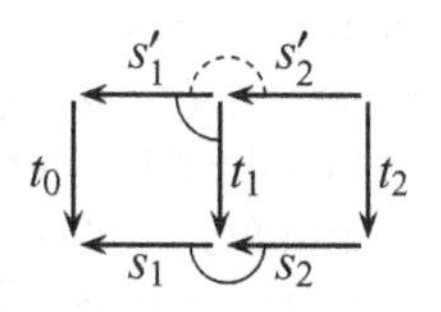

Moreover, we have supposed $x \preccurlyeq s_2's_1'$. Hence $x \preccurlyeq \mathsf{gcd}(s_2's_1', s_2't_1)$, or $x \preccurlyeq s_2' \cdot \mathsf{gcd}(s_1', t_1)$ by Lemma 4.3.11, giving $x \preccurlyeq s_2'$ since, by hypothesis, $\mathsf{gcd}(s_1', t_1)$ is trivial. Thus (s_2', s_1') is normal.

case BA. Suppose $x \preccurlyeq s_1'$ and $x \preccurlyeq s_2'$. Thus $x \preccurlyeq s_1't_0 = t_1s_1$, and $x \preccurlyeq s_2't_2 = t_1s_2$. Then, by Lemma 4.2.5, we deduce $t_1\backslash x \preccurlyeq s_1$ and $t_1\backslash x \preccurlyeq s_2$. The hypothesis $\mathsf{gcd}(s_1, s_2) = 1$ implies $t_1\backslash x = 1$, or $x \preccurlyeq t_1$. Since we also have $x \preccurlyeq s_1'$, we conclude $x \preccurlyeq \mathsf{gcd}(s_1', t_1) = 1$, so $x = 1$. Hence $\mathsf{gcd}(s_1', s_2')$ is trivial, and the sequence $(s_1'^{-1}, s_2')$ is symmetric normal.

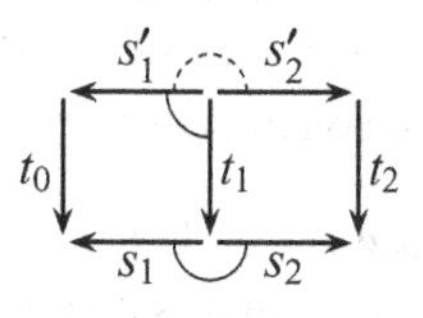

case $\widetilde{\mathrm{A}}\widetilde{\mathrm{A}}$. In addition to the diagram represented on the right, we will use an auxiliary diagram. For this, we suppose all the braids in play belong to B_n^+ and introduce the braids $\partial_n t_i$, simply denoted ∂t_i, and $\phi_n(s_i)$, denoted ϕs_i. By construction, the braids ∂t_i and ϕs_i are all simple. Let $i = 1$ or 2. By hypothesis, $t_is_i' = s_it_{i-1}$, hence $t_is_i'\partial t_{i-1} = s_it_{i-1}\partial t_{i-1} = s_i\Delta = \Delta\phi s_i = t_i\partial t_i\phi s_i$, and then $s_i'\partial t_{i-1} = \partial t_i\phi s_i$. Consequently, $(\partial t_0, \phi s_1, s_1', \partial t_1)$ and $(\partial t_1, \phi s_2, s_2', \partial t_2)$ are type B˜ tiles. Moreover, by hypothesis, (t_1, s_1') is normal.

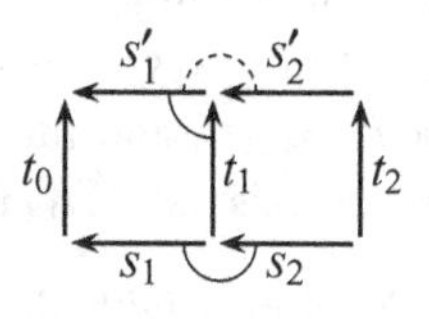

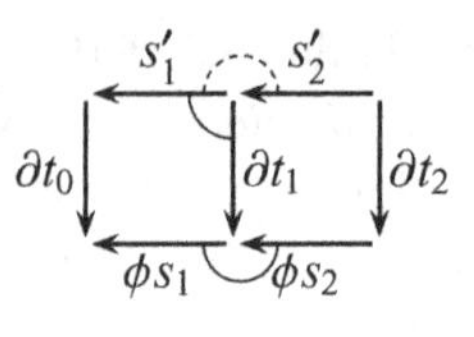

By Lemma 4.2.17, we deduce $\mathsf{gcd}(\partial t_1, s'_1) = 1$, thus $(s_1'^{-1}, \partial t_1)$ is normal, so the tile $(\partial t_0, \phi s_1, s'_1, \partial t_1)$ is of type B. Finally, since ϕ is an automorphism of B^+_∞, the normality of (s_2, s_1) implies that of $(\phi s_2, \phi s_1)$. However the domino rule BB applied to the diagram on the right implies that the sequence (s'_2, s'_1) is normal.[16]

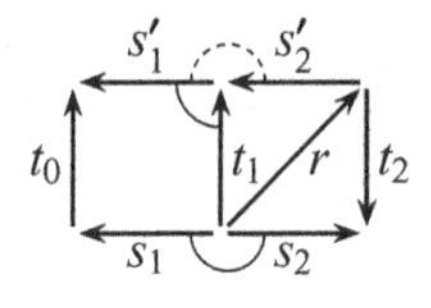

case $\widetilde{\text{A}}$C. We proceed as for $\widetilde{\text{A}}\widetilde{\text{A}}$ by passing to an auxiliary diagram. Again, suppose the braids are in B^+_n and write ∂ for ∂_n and ϕ for ϕ_n. By construction, the braids ∂t_i and ϕs_i are all simple. By hypothesis, $t_1 s'_1 = s_1 t_0$.

Multiplying on the right by ∂t_0, gives

$$t_1 s'_1 \partial t_0 = s_1 t_0 \partial t_0 = s_1 \Delta = \Delta \phi s_1 = t_1 \partial t_1 \phi s_1,$$

so $s'_1 \partial t_0 = \partial t_1 \phi s_1$. Moreover, by Lemma 4.2.17, the hypothesis that (t_1, s'_1) is normal implies $\mathsf{gcd}(\partial t_1, s'_1) = 1$, hence $(s_1'^{-1}, \partial t_1)$ is normal. Consequently, $(\partial t_0, \phi s_1, s'_1, \partial t_1)$ is a type B tile. Next, using the commutativity of the type C⁻ tiles and the relation $r = \mathsf{gcd}(t_1, s_2)$, we find

$$t_1 \partial t_1 \phi s_2 = \Delta \phi s_2 = s_2 \Delta = r t_2 \Delta = r \Delta \phi t_2 = r s'_2 \partial s'_2 \phi t_2 = t_1 \partial s'_2 \phi t_2,$$

so $\partial t_1 \phi s_2 = \partial s'_2 \phi_2$, hence $(\partial t_1, \phi s_2, \partial s'_2, \phi t_2)$ is a type A⁻ tile. Finally, the normality of (s_1^{-1}, s_2), that is, the condition $\mathsf{gcd}(s_1, s_2) = 1$, implies $\mathsf{gcd}(\phi s_1, \phi s_2) = 1$, hence the normality of $(\phi s_1^{-1}, \phi s_2)$. But then the domino rule BA applied to the diagram on the right shows that $(s_1'^{-1}, \partial s'_2)$ is normal, meaning $\mathsf{gcd}(\partial s'_2, s'_1)$ is 1. By Lemma 4.2.17, we conclude that (s'_2, s'_1) is normal.

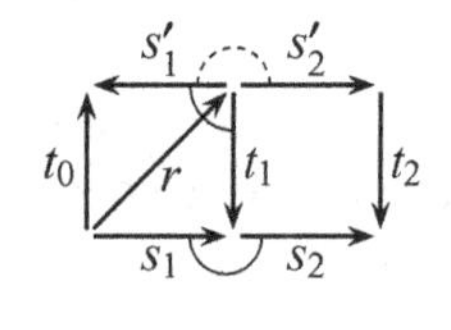

case CA. Let $r := \mathsf{gcd}(s'_1, t_1)$. Suppose x is simple and satisfies $x \preccurlyeq s'_1$ and $x \preccurlyeq s'_2$. We deduce $rx \preccurlyeq rs'_1 = t_0$, implying that rx is simple since t_0 is. Similarly, $x \preccurlyeq s'_2$ implies $rx \preccurlyeq rs'_2$ and, a fortiori, $rx \preccurlyeq rs'_2 t_2$, or $rx \preccurlyeq r t_1 s_2$, or again $rx \preccurlyeq s_1 s_2$. Since (s_1, s_2) is normal and rx is simple, we deduce $rx \preccurlyeq s_1$, or $rx \preccurlyeq r t_1$, and finally, $x \preccurlyeq t_1$. Thus $x \preccurlyeq \mathsf{gcd}(s_1, t_0)$, and $x = 1$ since, by hypothesis, $(s_1'^{-1}, t_1)$ is normal. □

Equipped with the domino rules, it is easy to obtain for the symmetric normal form of arbitrary braids the equivalent of Proposition 4.2.23 for the normal form of positive braids.

[16] Attention! the rule $\widetilde{\text{A}}\widetilde{\text{A}}$ is not the rule AA since the inversion of the sense of the arrows exchanges the hypothesis with the conclusion: the rule AA affirms here that, if (s'_2, s'_1) and (s_2, t_1) are normal, then so is (s_1, s_2).

Proposition 4.3.20 (Left-multiplication and division)
Suppose $(r_e^{-1}, ..., r_1^{-1}, s_1, ..., s_d)$ *is a symmetric normal decomposition of a braid* g *and* t *is simple. Set* $t_{-e} := t$.
(i) *For* k *decreasing from* e *to* 1, *let* $(t_{-k},\ rr_k, r'_k, t_{-k-1})$ *be a type* B *tile, then, for* k *increasing from* 1 *to* d, *let* $(t_{k-1}, s_k, s'_k, t_k)$ *be a type* A *tile. Then* $(r_e'^{-1}, ..., r_1'^{-1}, s'_1, ..., s'_d, t_d)$ *is a symmetric normal decomposition of* tg *(Figure 4.6).*
(ii) *For* k *decreasing from* e *to* 1, *let* $(t_{-k}, r_k, r'_k, t_{-k-1})$ *be a type* $\widetilde{\mathrm{A}}$ *tile, let* (t_0, s_1, r'_0, t_1) *be a type* C *tile, then, for* k *increasing from* 2 *to* d, *let* $(t_{k-1}, s_k, s'_{k-1}, t_k)$ *be a type* A *tile. Then* $(r_e'^{-1}, ..., r_1'^{-1}, r_0'^{-1}, s'_1, ..., s'_{d-1}, t_d)$ *is a symmetric normal decomposition of* $t^{-1}g$ *(Figure 4.7).*

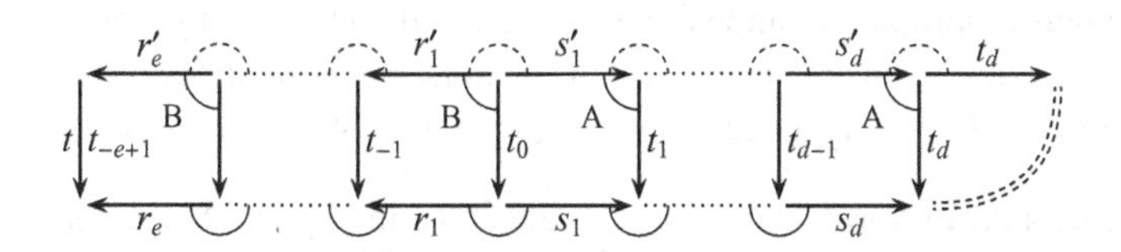

Figure 4.6 Passage from a symmetric normal decomposition of g to one of tg: starting from t, we fill the diagram from left to right with e type B tiles, then d type A tiles. The domino rules BB, BA, and AA assure that the sequence $(r_e'^{-1}, ..., r_1'^{-1}, s'_1, ..., s'_d, t_d)$ is symmetric normal.

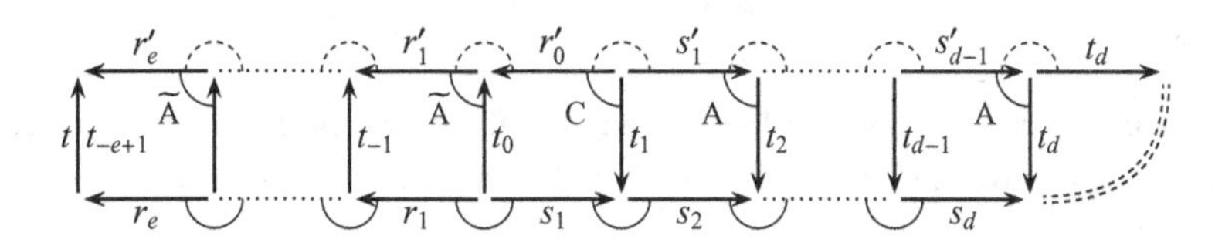

Figure 4.7 Passage from a symmetric normal decomposition of g to one of $t^{-1}g$: starting from t, we fill the diagram from left to right with e type $\widetilde{\mathrm{A}}$ tiles, a type C tile, then $d-1$ type A tiles. The domino rules $\widetilde{\mathrm{A}}\widetilde{\mathrm{A}}$, $\widetilde{\mathrm{A}}$C, CA, and AA assure that the sequence $(r_e'^{-1}, ..., r_1'^{-1}, r_0'^{-1}, s'_1, ..., s'_{d-1}, t_d)$ is symmetric normal.

Proof
(i) The existence of the type B and A tiles follows from Lemmas 4.3.17 and 4.2.21.Then, the commutativity of the diagram of Figure 4.6 implies

$$r_e'^{-1} \cdots r_1'^{-1} s'_1 \cdots s'_d t_d = r_e'^{-1} \cdots r_1'^{-1} t_0 s_1 \cdots s_d = t_{-e} r_e^{-1} \cdots r_1^{-1} s_1 \cdots s_d = tg,$$

so the sequence $(r_e'^{-1}, ..., r_1'^{-1}, s'_1, ..., s'_d, t_d)$ is a decomposition of tg. Finally, the normality of this sequence follows from the domino rules: $e-1$ times the domino BB, once the domino BA, and then $d-1$ times the domino AA.

(ii) The existence of the type $\widetilde{\mathrm{A}}$, C, and A tiles follows from Lemmas 4.2.21 and 4.3.16. Then, the commutativity of the diagram of Figure 4.7 implies

$$\begin{aligned} r_e'^{-1} \cdots r_1'^{-1} r_0'^{-1} s_1' \cdots s_{d-1}' t_d &= r_e'^{-1} \cdots r_1'^{-1} r_0'^{-1} t_1 s_2 \cdots s_d \\ &= r_e'^{-1} \cdots r_1'^{-1} t_0^{-1} s_1 s_2 \cdots s_d \\ &= t_{-e} r_e^{-1} \cdots r_1^{-1} s_1 \cdots s_d = t^{-1} g, \end{aligned}$$

hence $(r_e'^{-1}, ..., r_1'^{-1}, r_0'^{-1}, s_1', ..., s_{d-1}', t_d)$ is a decomposition of $t^{-1}g$. The normality follows from the domino rules: e–1 the domino $\widetilde{\mathrm{A}}\widetilde{\mathrm{A}}$, once the dominoes $\widetilde{\mathrm{A}}$C then CA, and finally d–2 times the domino AA. □

Equipped with the preceding results, it is must be clear that, as in Algorithm 4.2.24, we obtain an algorithm determining for an arbitrary braid word w, a symmetric normal decomposition of the braid $[w]$, then, by removing the eventual terms 1 on the two ends, its symmetric normal form. The basic elements are the four types of tiles – so, in the case of B_3, Tables 4.1 and 4.2.

Example 4.3.21 (Normalization) As illustrated in Figure 4.8, the application of the algorithm derived from Proposition 4.3.20 to the word $w := \sigma_2^{-2}\sigma_1^{-2}\sigma_2^2\sigma_1^2$ leads to the symmetric normal form $(\sigma_2^{-1}, (\sigma_1\sigma_2)^{-1}, \sigma_1^{-1}, \sigma_2, \sigma_2\sigma_1, \sigma_1)$. Note that, when a symmetric normal decomposition of a suffix of the input word is known, we can start with this rather than with the empty word, thus avoiding the duplication of the initial steps. For w, we have seen in Example 4.2.25 that the normal form of the suffix $\sigma_2^2\sigma_1^2$ is $(\sigma_2, \sigma_2\sigma_1, \sigma_1)$, and thus we can start from here.

As in the case of positive braids, the complexity of the algorithm derived from Proposition 4.3.20 can be read from the diagram of Figure 4.8.

Proposition 4.3.22 (Complexity) *For any n, the symmetric normal form of a braid in B_n given by an expression of length ℓ is computable in time $O(\ell^2)$.*

Proof Once the normalization tables for the product and quotient of two simple braids have been pre-calculated, determining the symmetric normal form of $\sigma_i^{\pm 1}[w]$ from that of $[w]$ requires at most $O(|w|)$ steps. Then, applying ℓ times the method to the words of length increasing from 1 to ℓ can be done in time $O(\ell^2)$. □

When the index n is not bounded, the global complexity increases to $O(\ell^2 n \log n)$ as the normalization of the words of length 2 can no longer be exhaustively pre-calculated and must be determined dynamically. As explained in (Epstein et al., 1992, Chapter IX), this is equivalent to a sort procedure, whose elementary cost is $O(n \log n)$.

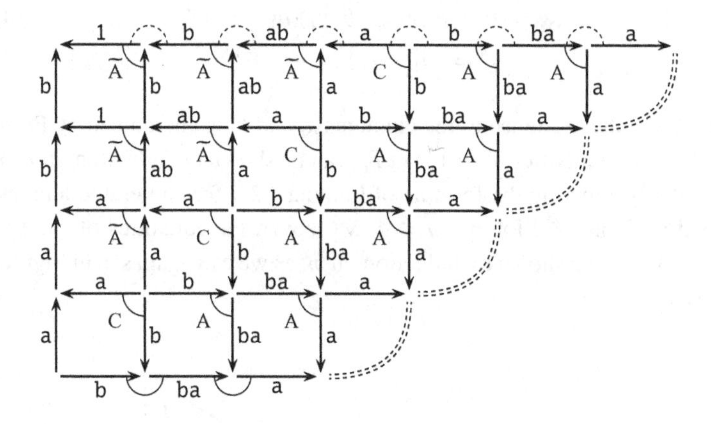

Figure 4.8 With the algorithm derived from Proposition 4.3.20, we determine the symmetric normal form of $w := \sigma_2^{-2}\sigma_1^{-2}\sigma_2^2\sigma_1^2$: starting with w written as a column, we fill the diagram with type A, $\widetilde{\text{A}}$, B, and C tiles according to the values read in Tables 4.1 and 4.2. The bottom four lines, corresponding to the normalization of bbaa, were determined in Figure 4.3 and are not duplicated here. We finish on the first row with the symmetric normal decomposition $(1, \sigma_2^{-1}, (\sigma_1\sigma_2)^{-1}, \sigma_1^{-1}, \sigma_2, \sigma_2\sigma_1, \sigma_1)$, then, after suppression of the initial 1, with the normal form $(\sigma_2^{-1}, (\sigma_1\sigma_2)^{-1}, \sigma_1^{-1}, \sigma_2, \sigma_2\sigma_1, \sigma_1)$; we gained time by starting with the normal form of the suffix $\sigma_2^2\sigma_1^2$, calculated in Example 4.2.25.

4.4 Applications

4.4.1 Torsion in B_n

A natural question for an infinite group is to know if it has torsion elements, that is, if there exist elements g distinct from 1 satisfying $g^m = 1$ for at least one non-zero integer m. We shall see that, as B_∞ is a group of right-fractions for B_∞^+ and as two elements in B_∞^+ always have a right-lcm, the group B_∞, and hence each of the groups B_n, has no torsion elements.

We begin with a general result linking the powers of an element of B_∞ to iterated lcms of the numerator and denominator of a fractionary expression.

Lemma 4.4.1 *Suppose $g = a_1 b_1^{-1}$ with a_1, b_1 in B_∞^+. Let $(a_i)_{i\geqslant 1}$ and $(b_i)_{i\geqslant 1}$ be the sequences in B_∞^+ inductively defined for $i \geqslant 1$ by*

$$a_i b_{i+1} = b_i a_{i+1} = \mathrm{lcm}(a_i, b_i).$$

Then, for any integers p, q, m,

$$a_1 \cdots a_p b_{p+1} \cdots b_{p+q} = b_1 \cdots b_q a_{q+1} \cdots a_{p+q} = \mathrm{lcm}(a_1 \cdots a_p, b_1 \cdots b_q), \tag{4.15}$$

$$g = (a_1 \cdots a_m)(a_{m+1} b_{m+1}^{-1})(a_1 \cdots a_m)^{-1}, \tag{4.16}$$

$$g^m = (a_1 \cdots a_m)(b_1 \cdots b_m)^{-1}. \tag{4.17}$$

Proof First, the existence of the elements a_i and b_i is guaranteed by Proposition 4.1.5. The equalities of (4.15) in B_∞^+ can be shown by induction on p and q by repeatedly applying the formula of Lemma 4.2.7 for an iterated lcm, as we can read in Figure 4.9 for $p + q \leqslant 4$. Moreover, the equalities of (4.16) and (4.17) in B_∞ can be shown by induction on m, as we can again see in Figure 4.9 for $m \leqslant 4$. □

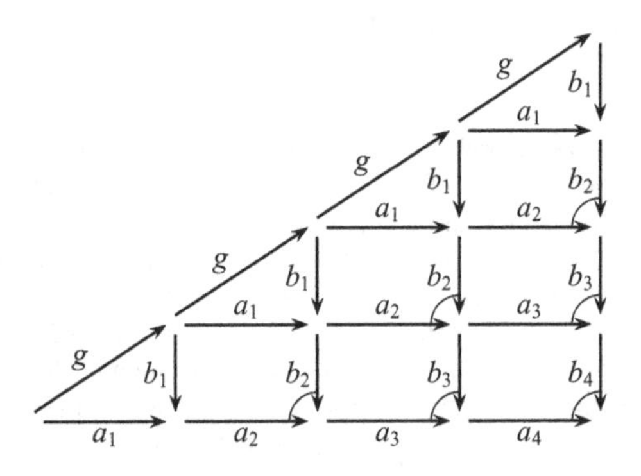

Figure 4.9 Decomposition of the powers of a fraction g in terms of iterated lcms of the numerator a_1 and the denominator b_1; the diagram is commutative, and the small arcs indicate that the rectangle concerned is a right-lcm. We read at the same time the equalities such as $a_1a_2b_1b_2 = b_1b_2a_3a_4 = \mathrm{lcm}(a_1a_2, b_1b_2)$, an instance of (4.15), $g = (a_1a_2a_3)a_4b_4^{-1}(a_1a_2a_3)^{-1}$, an instance of (4.16), and $g^4 = (a_1a_2a_3a_4)(b_1b_2b_3b_4)^{-1}$, an instance of (4.17).

From this, the result on the absence of torsion is easy.

Proposition 4.4.2 (Torsion-free) *The group B_∞ is torsion-free: any equality $g^m = 1$ in B_∞ implies $g = 1$.*

Proof Suppose g is a braid satisfying $g^m = 1$. We express g as $a_1b_1^{-1}$ with a_1, b_1 in B_∞^+, and introduce the sequences $(a_i)_{i\geqslant 1}$ and $(b_i)_{i\geqslant 1}$ as in Lemma 4.4.1. Set $a := a_1 \cdots a_m$ and $b := b_1 \cdots b_m$. Then (4.17) implies $a = b$, hence certainly $\mathrm{lcm}(a, b) = a$. However, by (4.15), we also have $\mathrm{lcm}(a, b) = a \cdot b_{m+1} \cdots b_{2m}$. Comparing, we deduce $b_{m+1} \cdots b_{2m} = 1$ hence, in particular, $b_{m+1} = 1$. Thus, (4.16) implies $g = aa_{m+1}a^{-1}$, so $g^m = 1$ implies $a_{m+1}^m = 1$, hence $a_{m+1} = 1$ since a_{m+1} is in B_∞^+. From this, $g = aa_{m+1}a^{-1}$ implies $g = 1$. □

4.4.2 The Conjugation Problem

Along with the word problem, another fundamental question for a presented group G is that of conjugacy: rather the determining if two words w, w' represent the same element of G, hence satisfying $[w'] = [w]$, we wish to know whether they represent *conjugate* elements, that is, whether there exists h in G satisfying $[w'] = h[w]h^{-1}$. This problem is in general more difficult than the word problem; the decidability of the one does not assure that of the other.

Notation 4.4.3 (Conjugation) For g, g' elements of a group G, denote $g \simeq g'$ for $\exists h{\in}G\,(g' = hgh^{-1})$, meaning g and g' are *conjugates* in G, and define $\mathsf{Conj}(g) := \{g'{\in}G \mid g' \simeq g\}$.

In the case of the group B_n, we will establish the decidability of the conjugation problem as a direct application of the Δ_n-normal form of Section 4.3.1. We begin with an observation on the conjugate elements.

Lemma 4.4.4 *Two braids g, g' of B_n are conjugate if and only if they are positively conjugate, that is, if there exists c in B_n^+ satisfying $g' = cgc^{-1}$.*

Proof Suppose $g' = hgh^{-1}$. By Proposition 3.3.5, there exists an integer m, that we can suppose even, and a in B_n^+ satisfying $\Delta_n^m h = a$. Since m is even, the braid Δ_n^m is in the centre of B_n, hence $g' = hgh^{-1} = a\Delta_n^{-m} g \Delta_n^m a^{-1} = aga^{-1}$. Thus g and g' are positively conjugate. The converse implication is trivial. □

The importance of this easy technical result is that by restricting ourselves to conjugations by positive braids, we can use their normal forms, decomposing an arbitrary conjugation into a product of conjugations by simple braids.[17]

It is convenient to extend the relation of left-divisibility of the monoid B_∞^+ to the group B_∞.

Notation 4.4.5 (Relation $\preccurlyeq$ on B_∞) For g, g' in B_∞, define

$$g \preccurlyeq g' \quad \Longleftrightarrow \quad \exists b{\in}B_\infty^+\,(gb = g'). \tag{4.18}$$

The relation $\preccurlyeq$ is then a partial order on B_∞ (prove it!), as is its restriction to B_n for every n. Note that the definition of the latter is not ambiguous: for g, g' in B_n, any relation $gb = g'$ with $b \in B_\infty^+$ implies $b \in B_n^+$, as we can see by expressing g and g' as fractions.

It is easy to characterize $\mathsf{inf}(g)$ and $\mathsf{sup}(g)$ in terms of $\preccurlyeq$.

[17] Conjugation is a symmetric relation, hence the existence of positive c satisfying $g' = cgc^{-1}$ implies that of positive c' satisfying $g = c'g'c'^{-1}$; however, there is no reason in general for c and c' to coincide.

Lemma 4.4.6 *For any braid g in B_n, the following are equivalent:*

$$p \leqslant \mathsf{inf}_n(g) \iff \Delta_n^p \preccurlyeq g, \tag{4.19}$$

$$\mathsf{sup}_n(g) \leqslant q \iff g \preccurlyeq \Delta_n^q. \tag{4.20}$$

Proof Let $(\Delta_n^m \,|\, s_1, ..., s_d)$ be the Δ_n-normal form of g. If $p \leqslant \mathsf{inf}_n(g)$, in other words $p \leqslant m$, we can write $g = \Delta_n^p \cdot \Delta_n^{m-p} s_1 \cdots s_d$, hence $\Delta_n^p \preccurlyeq g$. Conversely, suppose $\Delta_n^p \preccurlyeq g$, say $g = \Delta_n^p b$. If $(\Delta_n^r \,|\, t_1, ..., t_e)$ is a Δ_n-normal decomposition of b, with $r \geqslant 0$, then $(\Delta_n^{p+r} \,|\, t_1, ..., t_e)$ is a Δ_n-normal, sequence and a decomposition of g, hence it is the Δ_n-normal form of g, thus $\mathsf{inf}_n(g) = p + r$. This shows (4.19).

Symmetrically, suppose $\mathsf{sup}_n(g) \leqslant q$, or $m + d \leqslant q$. First, by Lemma 4.2.27, we must have $s_1 \cdots s_d \preccurlyeq \Delta_n^d$. From this, we deduce $g = \Delta_n^m s_1 \cdots s_d \preccurlyeq \Delta_n^{m+d} \preccurlyeq \Delta_n^q$. Conversely, suppose $g \preccurlyeq \Delta_n^q$, or $\Delta_n^m s_1 \cdots s_d \preccurlyeq \Delta_n^q$. Dividing by Δ_n^m gives $s_1 \cdots s_d \preccurlyeq \Delta_n^{q-m}$ in B_∞, hence in B_∞^+. However, by hypothesis, the sequence $(s_1, ..., s_d)$ is normal, thus by Lemma 4.2.27, we must have $q - m \geqslant d$, so $m + d \leqslant q$, or $\mathsf{sup}_n(g) \leqslant q$. □

Next, we introduce a notation of a kind of 'interval' describing the position of a braid with respect to the powers of the element Δ_n.

Notation 4.4.7 (Interval $[p, q]_n$) For $p \leqslant q$ in $\mathbb{Z}$, define

$$[p, q]_n := \{g \in B_n \mid \Delta_n^p \preccurlyeq g \preccurlyeq \Delta_n^q\}. \tag{4.21}$$

As with $\mathsf{inf}(g)$ and $\mathsf{sup}(g)$, we omit the index n when the reference group is fixed. The equivalences of Lemma 4.4.6 imply that for any g in B_n, the interval $[\mathsf{inf}(g), \mathsf{sup}(g)]$ is the smallest interval containing g.

We now arrive at the crucial observation.

Lemma 4.4.8 *Suppose $g' = cgc^{-1}$ in B_n with c positive. Let s be the head of c. Then $\mathsf{inf}(s^{-1}g's) \geqslant \min(\mathsf{inf}(g), \mathsf{inf}(g'))$.*

Proof Let $m := \min(\mathsf{inf}(g), \mathsf{inf}(g'))$. Then there exist a and a' in B_n^+ satisfying $g = \Delta_n^m a$ and $g' = \Delta_n^m a'$. Write $c = sb$, and let $t := \widetilde{\partial}_n s$, guaranteeing $ts = \Delta_n$. The hypothesis $g' = cgc^{-1}$ implies $cg = g'c$ in B_n^+, hence

$$sb\Delta_n^m a = \Delta_n^m a'c, \tag{4.22}$$

so from now on we reason in the monoid B_n^+. Multiplying (4.22) by t on the left, we obtain $\Delta_n b \Delta_n^m a = t\Delta_n^m a'c$. Moving back Δ_n^m with the automorphism ϕ_n^m, we deduce $\Delta_n^{m+1}\phi_n^m(b)a = \Delta_n^m \phi_n^m(t)a'c$, hence $\Delta_n^{m+1} \preccurlyeq \Delta_n^m \phi_n^m(t)a'c$, then, simplifying on the left by Δ_n^m,

$$\Delta_n \preccurlyeq \phi_n^m(t)a'c. \tag{4.23}$$

Now comes the heart of the argument. Set $x := \phi_n^m(t)a'$. Lemma 4.2.9 implies $H(xc) = H(xH(c))$, or here, $H(xc) = H(xs)$, thus, as Δ_n is simple, $\Delta_n \preccurlyeq xc$ implies $\Delta_n \preccurlyeq xs$. Here, (4.23) implies

$$\Delta_n \preccurlyeq \phi_n^m(t)a's. \tag{4.24}$$

It only remains to revert. Multiplying (4.24) on the left by Δ_n^m, we first find $\Delta_n^{m+1} \preccurlyeq \Delta_n^m \phi_n^m(t)a's$, or $\Delta_n^{m+1} \preccurlyeq t\Delta_n^m a's$ by pushing Δ_n^m to the right, thus, in B_n,

$$\Delta_n^{m+1} \preccurlyeq tg's. \tag{4.25}$$

However, $ts = \Delta_n$ implies $t = \Delta_n s^{-1}$, and (4.25) thus implies $\Delta_n^m \preccurlyeq s^{-1}g's$, or $\mathsf{inf}(s^{-1}g's) \geqslant m$, as desired. □

From this, it becomes easy to show that when two elements of an interval are conjugate, they are via a sequence of elements all belonging to the same interval, and two by two conjugate by a simple braid.

Lemma 4.4.9 *If two braids g and g' of B_n are conjugate and belong to the interval $[p,q]$, then there exists in $[p,q]$ a finite sequence $(g_0,...,g_d)$ satisfying $g_0 := g$ and $g_d := g'$, and such that, for every k, there exists simple s satisfying $g_{k+1} = s^{-1}g_k s$.*

Proof By Lemma 4.4.4, there exists a braid c in B_n^+ satisfying $g' = cgc^{-1}$. Let $(s_1,...,s_d)$ be the normal form of c. Set $g_0 := g$ and, for $k = 1,...,d$,

$$c_k := s_{d-k+1}s_{d-k+2}\cdots s_d \quad \text{and} \quad g_k := c_k\, g\, c_k^{-1}. \tag{4.26}$$

Then, for any k, the braids g and g_k are conjugate and, by definition, $c_d = s_1\cdots s_d = c$, so $g_d = g'$. Moreover, (4.26) implies for any k the equality $g_{k+1} = sg_ks^{-1}$ for $s = s_{d-k}$, which is simple by hypothesis.

We now show the relation $\mathsf{inf}(g_k) \geqslant p$ by reverse induction on k decreasing from d to 1. For $k = d$, $g_d = g'$, and $\mathsf{inf}(g') \geqslant p$ is correct by hypothesis. Suppose $k \leqslant d-1$. By hypothesis, $\mathsf{inf}(g) \geqslant p$ and, by the induction hypothesis, $\mathsf{inf}(g_{k+1}) \geqslant p$. However $g_{k+1} = c_{k+1}\, g\, c_{k+1}^{-1}$ and, by construction, s_{d-k} is the head of c_{k+1}. By Lemma 4.4.8, we deduce $\mathsf{inf}(s_{d-k}^{-1}g_{k+1}s_{d-k}) \geqslant p$. However, $c_{k+1} = s_{d-k}c_k$, so $g_k = s_{d-k}^{-1}g_{k+1}s_{d-k}$, and from this, $\mathsf{inf}(g_k) \geqslant p$.

Now set $h := g^{-1}$, $h' := g'^{-1}$, and $h_k := g_k^{-1}$ for every k. By reversing the relations linking the elements g_k, we obtain $h_k = c_k\, h\, c_k^{-1}$, and $h_{k+1} = sh_ks^{-1}$ for $s = s_{d-k}$: in other words, the elements h_k are linked by the same conjugation relations as the elements g_k. However, by Proposition 4.3.10, $\mathsf{inf}(h) = -\mathsf{sup}(g) \geqslant -q$ and $\mathsf{inf}(h') = -\mathsf{sup}(g') \geqslant -q$. Then the argument above implies $\mathsf{inf}(h_k) \geqslant -q$ for every k, hence, again by Proposition 4.3.10, $\mathsf{sup}(g_k) = -\mathsf{inf}(h_k) \leqslant q$. Finally, all the elements g_k belong to the interval $[p,q]$. □

As with the solution of Proposition 3.1.10 for the word problem of the monoid B_n^+, we construct a first solution for the conjugation problem of the group B_n^+ whose importance is more theoretical than practical.

Proposition 4.4.10 (Decidability) *For every n, the conjugation problem of the group B_n is decidable.*

Proof Let g, g' be arbitrary in B_n. Let $p := \min(\mathsf{inf}(g), \mathsf{inf}(g'))$ and $q := \max(\mathsf{sup}(g), \mathsf{sup}(g'))$. Then the braids g and g' are conjugate if and only if g' belongs to

$$\mathsf{Conj}_{p,q}(g) := \{h \in B_n \mid h \in [p, q] \text{ and } h \simeq g\}.$$

Any element of $[p, q]$ can be written $\Delta_n^p b$ with b in B_n^+ satisfying $b \preccurlyeq \Delta_n^{q-p}$. However, the relation $b \preccurlyeq \Delta_n^{q-p}$ implies $|b| \leqslant |\Delta_n^{q-p}|$. Consequently, the set $[p, q]$ is finite, of cardinality at most N with $N := (n-1)^{|\Delta_n^{q-p}|} = (n-1)^{n(n-1)(q-p)/2}$.

Set $X_0 := \{g\}$, and then, inductively,

$$X_{k+1} := X_k \cup \{shs^{-1} \mid h \in X_k,\ s \text{ simple, and } shs^{-1} \in [p, q]\}.$$

As each of the sets X_k is contained in $[p, q]$, and as $X_{k+1} = X_k$ implies $X_{k'} = X_k$ for every $k' \geqslant k$, there certainly exists an integer $m \leqslant N$ satisfying $X_{m+1} = X_m$. By construction, every simple conjugate of an element of X_m of $[p, q]$ is in X_m. Then, by Lemma 4.4.9, $\mathsf{Conj}_{p,q}(g) = X_m$.

Consequently, to test $g' \simeq g$, it suffices to determine the sets X_k up until finding m satisfying $X_{m+1} = X_m$, which must happen for some $m \leqslant N$, and then test $g' \in X_m$. □

Example 4.4.11 (Conjugation) For our favourite braid $g := \sigma_2^{-2}\sigma_1^{-2}\sigma_2^2\sigma_1^2$, the smallest interval containing g is $[-3, 3]$. The size of this interval is 234 in B_3^+, a moderate value albeit difficult to tackle by hand. For greater values of n, the size of the interval $[-3, 3]_n$ quickly becomes enormous: 45,252 for $n = 4$, 29,375,460 for $n = 5$, and more than $49 \cdot 10^9$ for $n = 6$. Implementing the method of Proposition 4.4.10 thus appears quite problematic.

4.4.3 Cycling and Decycling

The method of 4.4.10 is certainly a solution to the conjugation problem in the group B_n, but it is calamitous from the point of view of complexity: even for braids specified by modestly sized words, the sets $\mathsf{Conj}_{[p,q]}(g)$ defined above are enormous and it is hopeless to expect using them in a practical method. We will now describe an improvement that is much more efficient, even if remaining of exponential complexity with respect to the length of the initial braid words.

The principle is to restrict ourselves to a smaller subset of $\mathsf{Conj}_{[p,q]}(g)$, obtained by replacing the initial interval $[p,q]$ by an interval minimal in a certain sense.

Definition 4.4.12 (Cycling, decycling) For g in B_n with Δ_n-normal form $(\Delta_n^m \mid s_1, ..., s_d)$, the *cycling* and *decycling* of g are:

$$\mathsf{cycl}(g) := \Delta_n^m s_2 \cdots s_d \phi_n^m(s_1), \tag{4.27}$$

$$\mathsf{decycl}(g) := \Delta_n^m \phi_n^m(s_d) s_1 \cdots s_{d-1}. \tag{4.28}$$

For $d = 0$,[18] we set $\mathsf{cycl}(g) = \mathsf{decycl}(g) := g$.

The braids $\mathsf{cycl}(g)$ and $\mathsf{decycl}(g)$ are conjugates of g, corresponding to shifting, in one sense or the other, the simple braids of the portion 'non-Δ': with the notations of (4.27) and (4.28), we have

$$\phi_n^m(s_1)\,\mathsf{cycl}(g) = g\,\phi_n^m(s_1) \quad \text{and} \quad \mathsf{decycl}(g)\,s_d = s_d\,g. \tag{4.29}$$

The two operations are duals one of the other.

Lemma 4.4.13 *For any braid g of B_n,*

$$\mathsf{decycl}(g)^{-1} = \phi_n(\mathsf{cycl}(g^{-1})). \tag{4.30}$$

The proof follows from Proposition 4.3.10, and is left as an exercise.

It is also easy to see that cycling and decycling do not augment the complexity of a braid.

Lemma 4.4.14 *For any braid g in B_n, the braids* $\mathsf{cycl}(g)$ *and* $\mathsf{decycl}(g)$ *belong to the interval* $[\mathsf{inf}(g), \mathsf{sup}(g)]_n$.

Proof With the notations of Definition 4.4.12, the definition of $\mathsf{cycl}(g)$ directly implies $\mathsf{inf}(\mathsf{cycl}(g)) \geqslant \mathsf{inf}(g)$, whereas $\mathsf{sup}(\mathsf{cycl}(g)) \leqslant \mathsf{sup}(g)$ results from the general fact that every positive braid product of d simple braids divides Δ_n^d (on the left). The argument is similar for $\mathsf{decycl}(g)$. □

We now arrive to the key result. A priori, the cycling and decycling are very particular conjugations. In fact, the following result shows that, as for augmenting $\mathsf{inf}(g)$ or diminishing $\mathsf{sup}(g)$, these operations are in a way exhaustive.

Lemma 4.4.15 *Suppose g is a braid in B_n and there exists at least one conjugate g' of g satisfying* $\mathsf{inf}(g') > \mathsf{inf}(g)$ *(resp.,* $\mathsf{sup}(g') < \mathsf{sup}(g)$*). Then there exists p satisfying* $\mathsf{inf}(\mathsf{cycl}^p(g)) > \mathsf{inf}(g)$ *(resp.,* $\mathsf{sup}(\mathsf{decycl}^p(g)) < \mathsf{sup}(g)$*).*

[18] i.e. if g is a power of Δ_n.

Proof For ℓ an integer and g, g' in B_n, declare $g \simeq_\ell g'$ true if there exists c in B_n^+ satisfying $g' = cgc^{-1}$ with $|c| \leqslant \ell$. By Lemma 4.4.4, two conjugate braids are positively conjugate, hence the conjugation relation $\simeq$ is the union of the relations $\simeq_\ell$. We introduce two properties of a braid g:

$\mathcal{P}_\ell(g)$: there exists g' satisfying $g' \simeq_\ell g$ and $\mathrm{inf}(g') > \mathrm{inf}(g)$,

$\mathcal{Q}_\ell(g)$: there exists $p \leqslant \ell$ satisfying $\mathrm{inf}(\mathrm{cycl}^p(g)) > \mathrm{inf}(g)$.

We show by induction on $\ell \geqslant 0$ the implication

$$\forall g \in B_n (\mathcal{P}_\ell(g) \Rightarrow \mathcal{Q}_\ell(g)). \tag{4.31}$$

For $\ell = 0$, the property $\mathcal{P}_\ell(g)$ is always false, as the equality $g = g'$ excludes $\mathrm{inf}(g') > \mathrm{inf}(g)$, hence the implication (4.31) is true.

Consider now $\ell > 0$, and suppose $\mathcal{P}_\ell(g)$ is satisfied: then there exists c in B_n^+ with $|c| \leqslant \ell$ satisfying $\mathrm{inf}(cgc^{-1}) > \mathrm{inf}(g)$. Set $g' := cgc^{-1}$, and let $g_1 := \mathrm{cycl}(g)$. By Lemma 4.4.14, $\mathrm{inf}(g_1) \geqslant \mathrm{inf}(g)$, and two cases are possible.

case 1: $\mathrm{inf}(g_1) > \mathrm{inf}(g)$. Then, by definition, the properties $\mathcal{Q}_1(g)$, and thus, a fortiori, $\mathcal{Q}_\ell(g)$, are satisfied, hence the implication (4.31) is true.

case 2: $\mathrm{inf}(g_1) = \mathrm{inf}(g)$. Set $g'_1 := \phi_n(g')$. The argument, more delicate, consists in establishing that g'_1 is a conjugate of g_1 via a positive conjugant of length $\ell_1 < \ell$ and satisfying $\mathrm{inf}(g'_1) > \mathrm{inf}(g_1)$. In this way, g'_1 witnesses the property $\mathcal{P}_{\ell_1}(g_1)$, and we can apply the induction hypothesis to g_1.

Certain points are immediate. First, since ϕ_n is an automorphism of B_n^+, $\mathrm{inf}(g'_1) = \mathrm{inf}(g')$, so $\mathrm{inf}(g'_1) > \mathrm{inf}(g_1)$ as, by hypothesis, $\mathrm{inf}(g') > \mathrm{inf}(g)$ and $\mathrm{inf}(g_1) = \mathrm{inf}(g)$. Then, by construction, g'_1 is conjugate to g' since ϕ_n is the conjugation by Δ_n, hence to g, and finally to g_1 which, by construction, is conjugate to g. The remaining point to show is that g'_1 is conjugate to g_1 via an element c_1 of B_n^+ satisfying $|c_1| < |c|$. For this we compute such an element c_1 explicitly.

Let $m := \mathrm{inf}(g)$. Write $g = \Delta_n^m a$ and $g' = \Delta_n^m a'$ with a, a' in B_n^+. By hypothesis, $\Delta_n \not\leqslant a$ and $\Delta_n \leqslant a'$. Then $g' = cgc^{-1}$ implies $g'c = cg$, or $\Delta_n^m a'c = c\Delta_n^m a$, so $\Delta_n^m a'c = \Delta_n^m \phi_n^m(c)a$, and then

$$a'c = \phi_n^m(c)a. \tag{4.32}$$

The hypothesis $\Delta_n \leqslant a'$ implies $\Delta_n \leqslant a'c$, or, by (4.32), $\Delta_n \leqslant \phi_n^m(c)a$. Let s_1 be the head of a. By Lemma 4.2.9, $\Delta_n \leqslant \phi_n^m(c)a$ implies $\Delta_n \leqslant \phi_n^m(c)s_1$. Applying ϕ_n^m and taking into account that ϕ_n^{2m} is the identity and ϕ_n^m leaves Δ_n invariant, we deduce $\Delta_n \leqslant c\phi_n^m(s_1)$, or $\Delta_n \leqslant cs$ after introducing $s := \phi_n^m(s_1)$. Hence there exists in B_n^+ an element c_1 satisfying

$$\Delta_n c_1 = cs. \tag{4.33}$$

However, the hypothesis $\Delta_n \not\preccurlyeq a$ implies $s_1 \neq \Delta_n$, then $s \neq \Delta_n$, so $|s| < |\Delta_n|$. From this, (4.33) implies $|c_1| < |c|$. We find:

$$\begin{aligned} \Delta_n g'_1 c_1 &= g' \Delta_n c_1 && \text{by definition of } g'_1 \text{ from } g' \\ &= g' cs && \text{by (4.33)} \\ &= cgs && \text{since } g' \text{ is } cgc^{-1} \\ &= csg_1 && \text{by definition of } g_1 \text{ from } g \\ &= \Delta_n c_1 g_1 && \text{again by (4.33),} \end{aligned}$$

hence $g'_1 c_1 = c_1 g_1$. As desired, g'_1 thus witnesses that $\mathcal{P}_{\ell_1}(g_1)$ is satisfied. Then, by the induction hypothesis, the implication (4.31) is valid for ℓ_1, and we conclude that $\mathcal{Q}_{\ell_1}(g_1)$ is true, that is, that there exists $p_1 \leqslant \ell_1$ satisfying $\mathsf{inf}(\mathsf{cycl}^{p_1}(g_1)) > \mathsf{inf}(g_1)$. Let $p := p_1 + 1$. Then $p \leqslant \ell_1 + 1 \leqslant \ell$, and $\mathsf{cycl}^p(g) = \mathsf{cycl}^{p_1}(g_1)$. Hence p witnesses that the property $\mathcal{Q}_\ell(g)$ is satisfied, so the implication (4.31) is valid in this case.

The result for $\mathsf{sup}(g)$ follows from the one for $\mathsf{inf}(g)$ by Lemma 4.4.13. □

To recognize the braids conjugate to a braid g belonging to an interval $[p, q]$, instead of blindly using all the conjugates of g belonging to $[p, q]$ as in Proposition 4.4.10, the preceding result allows us to restrict ourselves to a minimal interval.

Definition 4.4.16 (Super summit set) For $g \in B_n^+$, set

$$\mathsf{inf}^{\simeq}(g) := \max\{\mathsf{inf}(g') \mid g' \simeq g\}, \mathsf{sup}^{\simeq}(g) := \min\{\mathsf{sup}(g') \mid g' \simeq g\}.$$

The *super summit set* of g is the set $\mathsf{SSS}(g)$ of the conjugates of g belonging to the interval $[\mathsf{inf}^{\simeq}(g), \mathsf{sup}^{\simeq}(g)]$.

An immediate application of 4.4.15 is that, for any braid g, iterating the cycling leads in a finite number of steps to a braid g_1 satisfying $\mathsf{inf}(g_1) = \mathsf{inf}^{\simeq}(g)$, then iterating the decycling starting with g_1 leads in a finite number of steps to a braid g_2 satisfying $\mathsf{inf}(g_2) \geqslant \mathsf{inf}(g_1)$, hence $\mathsf{inf}(g_2) = \mathsf{inf}^{\simeq}(g)$, and $\mathsf{sup}(g_2) = \mathsf{sup}^{\simeq}(g_1)$, thus $\mathsf{sup}(g_2) = \mathsf{sup}^{\simeq}(g)$. Consequently, we have found a conjugate g_2 of g belonging to $\mathsf{SSS}(g)$. Then, by (4.4.9), we can exhaustively enumerate the set $\mathsf{SSS}(g)$ by saturating$\{g_2\}$ with respect to conjugation by the simple braids. This done, to decide if $g \simeq g'$, it suffices, starting from g', to find an element g'_2 of $\mathsf{SSS}(g')$, and to test $g'_2 \in \mathsf{SSS}(g)$.

The algorithm sketched above is much better in practice than that of Proposition 4.4.10, but its complexity remains exponential: there exist sequences of braid words (w_N) such that the cardinality of $\mathsf{SSS}([w_N])$ depends exponentially on $|w_N|$. Even more efficient methods in practice are known,[19] but all remain of exponential complexity in the worst case, and the existence of a solution of polynomial complexity for the conjugation problem of the group B_n remains open to this day for $n \geqslant 5$.

[19] The most efficient method known today, called the 'sliding circuits' method, replaces the cycling and decycling by a unique operation, the conjugation by the gcd of the first and last terms of the Δ-normal form (Gebhardt and González-Meneses, 1969).

5
The Artin Representation

After the algebraic approach of the preceding chapters, we return to a topological approach. Developed by Emil Artin as early as the 1920s, it is based on the action of braids on the fundamental group of a punctured disk and even to a faithful representation of the group B_n in the group of automorphisms of a free group of rank n. We derive at the same time a new solution (not very efficient) of the braid isotopy problem, and a proof of the equivalence of diverse versions of isotopy for geometric braids, left in suspense since Chapter 1. It must be pointed out that certain results, in particular those mentioned above, will only be established here modulo a supplementary property of braids ('comparison property') whose proof will only be given in the next chapter.[1]

5.1 The Action of Braids on Loops

5.1.1 Braid Groups and Mapping Class Groups

In what follows, we restrict ourselves to the geometric braids traced in the cylinder $[0, 1]\times\mathbb{D}$ where $\mathbb{D}$ is the disk of radius $n/2$ centred at $((n+1)/2, 0)$. The method developed in Proposition 1.1.20 shows that the family of braids obtained is not diminished, as every braid in $[0, 1]\times\mathbb{R}^2$ is isotopic to a braid in $[0, 1]\times\mathbb{D}$.

[1] We must of course assure the absence of vicious circles: the proof in Chapter 6 relies on certain results of the current chapter; however, it does not involve any of the results for which this property is a prerequisite condition.

Notation 5.1.1 ($\mathbb{D}_n$ and $\mathbb{D}_n^-$)
In $\mathbb{R}^2$, denote $\mathbb{D}_n$ the disk $\mathbb{D}$ with n marked points $P_1, ..., P_n$, where P_k is $(k, 0)$. Denote $\mathbb{D}_n^-$ the complement of $\{P_1, ..., P_n\}$ in $\mathbb{D}_n$. Then $\mathbb{D}_n^-$ is a punctured disk with n holes.

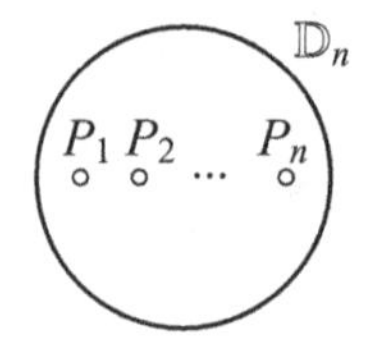

An n-strand geometric braid can be seen as the movie of a dance without collisions of n points of $\mathbb{D}$ between the instants $t = 0$ and $t = 1$, where the planar section $\{t\}\times\mathbb{D}$ is the image at the instant t.

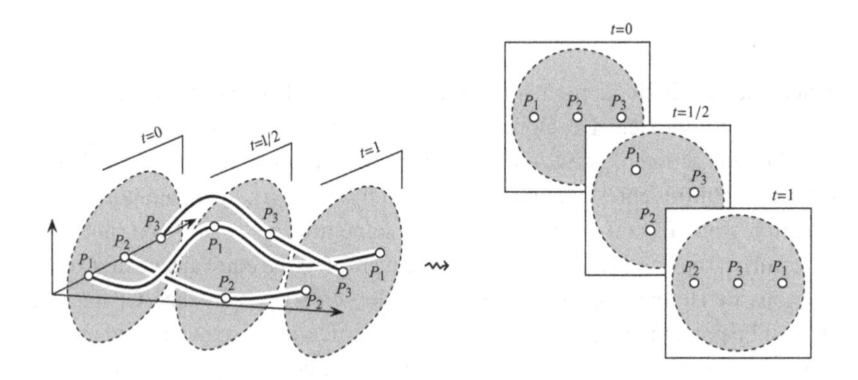

Figure 5.1 An n-strand geometric braid in the cylinder $[0, 1]\times\mathbb{D}$ seen as the dance of n points in $\mathbb{D}$ (here $n = 3$): the intersection of the strands with the plane $\{t\}\times\mathbb{R}^2$ is the family of n points at the instant t.

Specifying the dance of n points of $\mathbb{D}$ does not specify the movement of the other points of the disk. However, we shall see that such an extension is possible, providing an injective correspondence between the isotopy classes of n-strand geometric braids with certain isotopy classes of homeomorphisms of the disk with n marked points $\mathbb{D}_n$.

This correspondence is expressed by introducing the *mapping class group* of the marked surface $\mathbb{D}_n$.

Definition 5.1.2 (Mapping class group) Let Σ be a compact oriented surface (with or without boundary), and let $\{P_1, ..., P_n\}$ be a finite family of points in the interior of Σ. The *mapping class group* $\mathcal{MCG}_+(\Sigma, \{P_1, ..., P_n\})$ of Σ relative to $\{P_1, ..., P_n\}$ is the group of isotopy classes of homeomorphisms of Σ preserving the orientation of Σ, fixing the boundary $\partial\Sigma$ point by point, and leaving the set $\{P_1, ..., P_n\}$ globally invariant.

The notion of isotopy in question here is that of Definition 1.1.7: two homeomorphisms ϕ, ϕ' of Σ are isotopic if there exists a path linking ϕ to ϕ' in

Homeo(Σ), that is, a continuous mapping $F: \Sigma\times[0,1]\to\Sigma$ satisfying $F(\text{-},0)=\phi$ and $F(\text{-},1)=\phi'$, and such that $F(\text{-},s)$ is a homeomorphism of Σ for every s.[2] The mapping class groups are important tools for the study of 3-dimensional surfaces and manifolds.

Exercise 5.1.3 (Permutation) Show that isotopic homeomorphisms of $\mathbb{D}_n$ necessarily induce the same permutation of the marked points. [Hint: the set of permutations of n points is discrete.]

Returning to braids and the disk $\mathbb{D}_n$, we can now describe exactly the correspondence induced by the vision of a braid as a movie of the dance of n points in $\mathbb{D}$.

Proposition 5.1.4 (Mapping class group) *The braid group B_n is isomorphic to the mapping class group $\mathcal{MCG}_+(\mathbb{D}_n)$.*

Sketch of the proof Let β be an n-strand geometric braid, that we can suppose included in the cylinder $[0,1]\times\mathbb{D}$. By definition, the starting points of the n strands are the marked points of $\{0\}\times\mathbb{D}_n$, and the arrival points are the marked points of $\{1\}\times\mathbb{D}_n$. As illustrated in Figure 5.1, β is the graph of the movement of n points of $\mathbb{D}$ as the time progresses from 0 to 1. We could then extend this movement to a family of homeomorphisms of $\mathbb{D}$ beginning with the identity, for $t=0$, and ending, for $t=1$, by a homeomorphism preserving $\{P_1,...,P_n\}$: for this, we could imagine the disk filled with a viscous jelly communicating the movements to neighbouring points.[3] The crucial point is that, if we require the movement to be null in the neighbourhood of the boundary of $\mathbb{D}$, then the extension is not unique; however, it is unique up to an isotopy. In this way, we obtain a unique isotopy class of homeomorphisms of $\mathbb{D}$. We then associate with β the element of $\mathcal{MCG}_+(\mathbb{D}_n)$ which is the final homeomorphism class: since this one preserves $\{P_1,...,P_n\}$ globally, its class belongs to $\mathcal{MCG}_+(\mathbb{D}_n)$.

Conversely, any homeomorphism φ of $\mathbb{D}$ into itself representing an element of $\mathcal{MCG}_+(\mathbb{D}_n)$, so leaving the boundary $\partial\mathbb{D}$ fixed point by point, is necessarily isotopic to the identity (the 'Alexander trick'). To prove this, we zoom out. To simplify the notation suppose $\mathbb{D}$ is the unit disk centred at 0, and, for P a point of $\mathbb{D}$ and $s\in[0,1]$, set $\Phi(P,s):=P$ for $||P||\geqslant s$, and $\Phi(P,s):=s\varphi(P/s)$ for $||P||<s$. Then Φ realizes an isotopy between $\mathrm{id}_{\mathbb{D}}$ and φ. We then associate φ with a geometric braid by considering the graph of the restriction of the isotopy

2 As in the case of the homotopies and isotopies of geometric braids, this third condition on the intermediate values can be removed.

3 More formally, we can use the techniques developed in Chapter 1 to extend the homotopy to an isotopy.

to the set of marked points $\{P_1, ..., P_n\}$: the geometric braid obtained depends on the choice of the isotopy, but its class does not.

The two mappings specified above, the first from B_n to $\mathcal{MCG}_+(\mathbb{D}_n)$, the second from $\mathcal{MCG}_+(\mathbb{D}_n)$ to B_n, are inverses of each other. □

Thus we can from now on see an n-strand braid as an isotopy class of homeomorphisms of $\mathbb{D}_n$ preserving the orientation, leaving $\partial\mathbb{D}$ point by point invariant and the set of n marked points globally invariant.

5.1.2 The Artin Representation

The above approach leads naturally to the actions of the group B_n on the fundamental group of the punctured disk $\mathbb{D}_n^-$.

Notation 5.1.5 (Homeomorphisms) The group of homeomorphisms of $\mathbb{D}$ leaving $\partial\mathbb{D}$ fixed point by point is denoted $\mathsf{Homeo}(\mathbb{D}, \partial\mathbb{D})$, whereas the group of isotopy classes of $\mathsf{Homeo}(\mathbb{D}, \partial\mathbb{D})$ is denoted $\mathsf{Homeo}(\mathbb{D}, \partial\mathbb{D})/{\approx}$.

The group $\mathsf{Homeo}(\mathbb{D}, \partial\mathbb{D})$ includes as a strict subset the set of homeomorphisms brought into play in the mapping class group of $\mathbb{D}_n$: these are in addition obliged to preserve the orientation and leave the marked points globally fixed.

We begin with the observation that there exists an action of the homeomorphisms of $\mathbb{D}$ on the loops in $\mathbb{D}$ – see Appendix D for a reminder of the definitions.

Lemma 5.1.6 *Let P_* be fixed on $\partial\mathbb{D}$. For any φ in* $\mathsf{Homeo}(\mathbb{D}, \partial\mathbb{D})$ *and for any loop γ in $\mathbb{D}$ with basepoint P_*, setting*

$$(\gamma{\scriptstyle\bullet}\varphi)(t) := \varphi(\gamma(t)) \tag{5.1}$$

induces a well-defined mapping $\widehat{\varphi}$ sending the isotopy class of γ to that of $\gamma{\scriptstyle\bullet}\varphi$, which is an automorphism of $\Omega_1(\mathbb{D}, P_*)/{\sim}$. The mapping $\rho\colon \varphi \mapsto \widehat{\varphi}$ is a homomorphism of $\mathsf{Homeo}(\mathbb{D})/{\approx}$ *into $\Omega_1(\mathbb{D}, P_*)/{\sim}$.*

Proof In what follows, a loop γ is identified with a parametrization, that is, a continuous mapping of the interval $[0, 1]$ into $\mathbb{D}$.

First, the class of $\gamma{\scriptstyle\bullet}\varphi$ depends only on the class of γ and of the isotopy class of φ. Indeed, if ϕ is a homotopy from γ to γ', hence a continuous mapping of $[0, 1] \times [0, 1]$ into $\mathbb{D}$, then, since φ is a homeomorphism, the mapping $(t, t') \mapsto \varphi(\phi(t, t'))$ is a homotopy from $\gamma{\scriptstyle\bullet}\varphi$ to $\gamma'{\scriptstyle\bullet}\varphi$. Similarly, if Φ is an isotopy from φ to φ', hence a continuous mapping of $\mathbb{D} \times [0, 1]$ into $\mathbb{D}$, then the mapping $(t, t') \mapsto \Phi(\gamma(t), t')$ is a homotopy from $\gamma{\scriptstyle\bullet}\varphi$ to $\gamma{\scriptstyle\bullet}\varphi'$.

Thus, for any homeomorphism φ of $\mathbb{D}$, we obtain a well-defined action $\widehat{\varphi}$ of the loops with basepoints on $\partial\mathbb{D}$ onto themselves. Then, by construction, if γ_1, γ_2 are two loops in $\mathbb{D}$, we have $(\gamma_1\gamma_2){\bullet}\varphi = (\gamma_1{\bullet}\varphi)(\gamma_2{\bullet}\varphi)$, hence $\widehat{\varphi}$ is an endomorphism. Moreover, we obtain $(\gamma{\bullet}\varphi){\bullet}\varphi^{-1} = \gamma = (\gamma{\bullet}\varphi^{-1}){\bullet}\varphi$, thus $\widehat{\varphi^{-1}}$ is the inverse of $\widehat{\varphi}$. Consequently, $\widehat{\varphi}$ is an automorphism.

The mapping $\rho : \varphi \mapsto \widehat{\varphi}$ thus goes from the group $\mathsf{Homeo}(\mathbb{D})/\approx$ towards the group of classes of loops $\Omega_1(\mathbb{D}, P_*)/\sim$. Finally, ρ is itself a homomorphism, as (5.1) directly implies $\widehat{\varphi_1\varphi_2} = \widehat{\varphi_1}\,\widehat{\varphi_2}$. □

When we restrict the family of homeomorphisms $\mathbb{D}$ to those globally preserving the marked points, considering their restriction to the complement $\mathbb{D}_n^-$ of the family of marked points makes sense, and we obtain an action on the loops in $\mathbb{D}_n^-$.

Proposition 5.1.7 (Action) *The action of* (5.1) *induces a homomorphism of* $\mathcal{MCG}_+(\mathbb{D}_n)$ *into* $\pi_1(\mathbb{D}_n^-)$.

Proof Let ϕ be a homeomorphism of $\mathbb{D}_n$ globally preserving the marked points $P_1, ..., P_n$. Then ϕ sends $\mathbb{D}_n^-$ into itself, and its restriction to $\mathbb{D}_n^-$ is a homeomorphism of $\mathbb{D}_n^-$. Moreover, isotopic homeomorphisms have isotopic restrictions. Consequently, the mapping $\phi \mapsto \phi{\restriction}_{\mathbb{D}_n^-}$ induces an injective homomorphism of $\mathcal{MCG}_+(\mathbb{D}_n)$ into $\mathsf{Homeo}(\mathbb{D}_n^-)/\approx$. Composing with the homomorphism ρ thus gives a homomorphism of $\mathcal{MCG}_+(\mathbb{D}_n)$ into $\Omega_1(\mathbb{D}_n^-, P_*)/\sim$, that is, into the fundamental group $\pi_1(\mathbb{D}_n^-)$. □

We thus obtain a homomorphism of $\mathcal{MCG}_+(\mathbb{D}_n)$ in the group of automorphisms $\mathsf{Aut}(\pi_1(\mathbb{D}_n^-))$. We saw in Proposition 5.1.4 that the mapping class group $\mathcal{MCG}_+(\mathbb{D}_n)$ is isomorphic to the braid group B_n, and it is a classic result that the fundamental group of a disk with n holes is a free group of rank n – see Appendix D. We thus obtain in this way a representation of the group B_n in the automorphisms of the group F_n.

Definition 5.1.8 (Artin representation) The *Artin representation*, following Proposition 5.1.7, is the homomorphism ρ_{A} of the group B_n into $\mathsf{Aut}(F_n)$ induced by the action (5.1).

As will be recalled in Proposition 5.3.11, the classes of loops $\gamma_1, ..., \gamma_n$ such that γ_k makes a single turn around P_k, for example in the counterclockwise direction (Figure 5.3), form a basis of $\pi_1(\mathbb{D}_n^-)$. Figure 5.2 shows the action of the braid σ_1 on the 3 fundamental loops of $\mathbb{D}_3^-$.

It then suffices to read the action of σ_i on γ_k to obtain explicit formulas for the Artin representation.

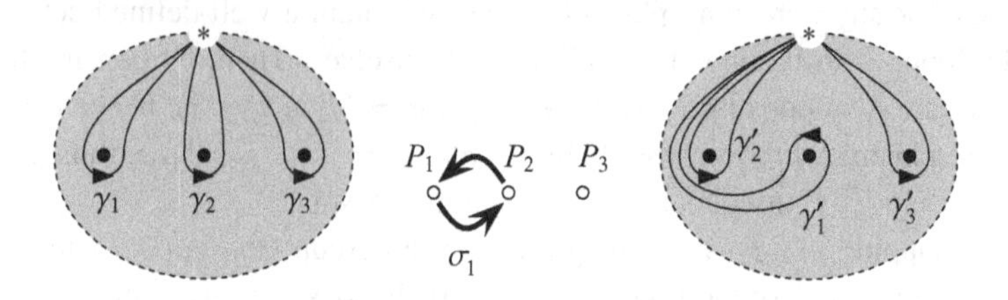

Figure 5.2 Action of the braid σ_i, seen as a half-turn exchanging P_i and P_{i+1} in $\mathbb{D}_n$, on the loops γ_k constituting the standard basis of $\pi_1(\mathbb{D}_n^-)$, here for $i = 1$ and $n = 3$: the image of γ_k is γ_k'.

Proposition 5.1.9 (Artin representation) *Let x_k be the isotopy class of the loop γ_k for $1 \leqslant k \leqslant n$. For every i, the action of the braid σ_i on the basis $(x_1, ..., x_n)$ of $\pi_1(\mathbb{D}_n^-)$ is given by*

$$\rho_A(\sigma_i)(x_k) = x_k \bullet \sigma_i = \begin{cases} x_i x_{i+1} x_i^{-1} & \textit{for } k = i, \\ x_i & \textit{for } k = i+1, \\ x_k & \textit{for } k \neq i, i+1. \end{cases} \tag{5.2}$$

Proof The values can be read in Figure 5.2. The only delicate case is the loop γ_1', which we must express as isotopic to a product of loops $\gamma_k^{\pm 1}$. However, the sequence of diagrams

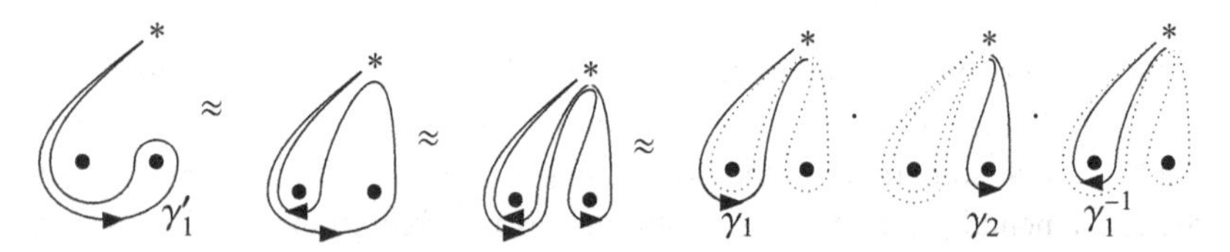

shows that γ_1' is isotopic to the concatenation of the loops γ_1, γ_2, and the inverse of γ_1, hence, by generalizing, the formulas of (5.2). □

Note that once given the formulas of (5.2), we can forget their topological origin and, by a direct verification, show that they provide a representation of the group B_n. For this, we can extend by composition the values of (5.2) to arbitrary braid words by setting

$$\rho_A(\sigma_{i_1}^{e_1} \cdots \sigma_{i_\ell}^{e_\ell}) := \rho_A(\sigma_{i_1})^{e_1} \circ \cdots \circ \rho_A(\sigma_{i_\ell})^{e_\ell},$$

and then verifying the compatibility with the braid relations, that is, verifying the equalities $\rho_A(\sigma_i\sigma_j) = \rho_A(\sigma_j\sigma_i)$ for $|i - j| \geqslant 2$ and $\rho_A(\sigma_i\sigma_j\sigma_i) = \rho_A(\sigma_j\sigma_i\sigma_j)$ for $|i - j| = 1$. This is easy.

5.2 The Injectivity of the Artin Representation

A natural question for the Artin representation is whether it is injective,[4] in other words, knowing if it provides a complete invariant with respect to braid isotopy. As was proved by E. Artin, the response is positive, and in fact we will provide a proof. Incidentally, the method used here, different from Artin's proof, uses two ingredients, one ('the acyclicity property') to be shown in this chapter, the other ('the comparison property') in the next chapter.

5.2.1 σ-Positive Braids

The starting point is a particular family of braids, defined by the existence of certain types of expressions.

Definition 5.2.1 (σ-positive braid) A braid word is said to be *σ_i-positive* if it contains at least one letter σ_i but not σ_i^{-1} nor $\sigma_j^{\pm 1}$ with $j < i$. A braid is called *σ_i-positive* if it admits at least one σ_i-positive expression. A braid is said to be *σ-positive* if it is σ_i-positive for at least one value of i.

Example 5.2.2 (σ-positive braid) The word $\sigma_2^{-1}\sigma_1\sigma_2$ is σ_1-positive, but $\sigma_1\sigma_2\sigma_1^{-1}$ is not, as it contains both σ_1 and σ_1^{-1}. Similarly, $\sigma_1\sigma_2$ is σ_1-positive, but not σ_2-positive since it contains a letter σ_i with $i < 2$. Note that, in general, a word can be σ_i-positive for at most one value of i (see Exercise 5.2.5(iii)).

The braid $\sigma_2^{-1}\sigma_1\sigma_2$ is σ_1-positive, as it is expressed by the word $\sigma_2^{-1}\sigma_1\sigma_2$, σ_1-positive. However so is the braid $\sigma_1\sigma_2\sigma_1^{-1}$... since it is the same braid as the former: by definition, a braid is σ_i-positive if, among its diverse expressions, there exists *at least one* that is σ_i-positive. We do not require this of *all* the expressions.

To complement the σ_i-positive words and braids, we symmetrically introduce the σ_i-negative words and braids.

Definition 5.2.3 (σ-negative and σ_i-neutral braids) A braid word w is said to be *σ_i-negative* (*resp.*, *σ_i-neutral*) if $\overline{w}$ is σ_i-positive (*resp.*, if w does not contain any letter $\sigma_i^{\pm 1}$ with $j \leqslant i$). A braid is called *σ_i-negative* (*resp.*, *σ_i-neutral*) if it admits at least one σ_i-negative (*resp.*, σ_i-neutral) expression. A braid is said to be *σ-negative* if it is σ_i-negative for at least one value of i.

By construction, a braid g is σ_i-negative if and only if g^{-1} is σ_i-positive, and similarly, a braid g is σ-negative if and only if g^{-1} is σ-positive.

[4] We also speak of 'fidelity'.

We introduce two properties of σ-positive braids. This will be proved later, but for the moment we introduce them as hypotheses to which we attribute a name.

Definition 5.2.4
(i) (Acyclicity property) A σ-positive braid is not trivial.
(ii) (Comparison property) If g is a non-trivial braid, then at least one of the braids g or g^{-1} is σ-positive.

Exercise 5.2.5 (Trichotomy)
(i) Admitting the acyclicity property, show that, for any braid g, the properties 'g is σ_i-positive', 'g is σ_i-neutral', and 'g is σ_i-negative' are mutually exclusive.
(ii) Admitting the acyclicity property, show that, for any braid g, the properties 'g is σ-positive', 'g is trivial', and 'g is σ-negative' are mutually exclusive.
(iii) Show that a braid can be σ_i-positive for at most one value of i.

5.2.2 A Proof of the Acyclicity Property

We now use the Artin representation to establish the acyclicity property. In what follows, it is convenient to denote $\widehat{\sigma_i}$ the automorphism $\rho_{\mathsf{A}}(\sigma_i)$. Moreover, we have seen in Section 2.3.4 that the free group F_n can be identified with the set of reduced words on the alphabet $\{x_1^{\pm 1}, ..., x_n^{\pm 1}\}$, that is, not including any factor $x_k x_k^{-1}$ or $x_k^{-1} x_k$. Recall that, for any word w on the preceding alphabet, there exists a unique reduced word denoted $\mathsf{red}(w)$ representing the same element of F_n as w.

We are going to study specifically the image by the automorphisms $\widehat{\sigma_i}^{\pm 1}$ of the reduced words terminating with the letter x_1^{-1}.

Lemma 5.2.6 *The automorphisms $\widehat{\sigma_i}^{\pm 1}$ with $i \geqslant 2$ send every reduced word terminating with x_1^{-1} to a reduced word terminating with x_1^{-1}.*

Proof Suppose $i \geqslant 2$ and consider a reduced word terminating with x_1^{-1}, say vx_1^{-1}. By construction, $\widehat{\sigma_i}(vx_1^{-1})$ is the product in F_n of $\widehat{\sigma_i}(v)$ and $\widehat{\sigma_i}(x_1^{-1})$, which is x_1^{-1}, so, by definition of this product,

$$\widehat{\sigma_i}(vx_1^{-1}) = \mathsf{red}(\widehat{\sigma_i}(v)\, x_1^{-1}).$$

Suppose $\widehat{\sigma_i}(vx_1^{-1})$ does not end with x_1^{-1}. This means the final letter x_1^{-1} disappears during the reduction, annihilated by a letter x_1 of $\widehat{\sigma_i}(v)$. However, by construction, a letter x_1 of $\widehat{\sigma_i}(v)$ can only come from a letter x_1 of v. There thus exists a (reduced) decomposition $v = v_1 x_1 v_2$ satisfying

$$\mathsf{red}(\widehat{\sigma_i}(v)\, x_1^{-1}) = \mathsf{red}(\widehat{\sigma_i}(v_1)\, x_1\, \underline{\widehat{\sigma_i}(v_2)\, x_i^{-1}}),$$

where the underlined factor reduces to the empty word (unit of F_n). As $\widehat{\sigma_i}$ is injective, $\widehat{\sigma_i}(v_2) = 1$ implies $v_2 = 1$. Hence the word vx_1^{-1} is $v_1x_1x_1^{-1}$, which is not reduced. This contradicts the hypothesis that $\widehat{\sigma_1}(vx_1^{-1})$ does not end with x_1^{-1}.

The argument is the same for $\widehat{\sigma_i}^{-1}$. □

We treat the case of $\widehat{\sigma_1}$ in the same manner.

Lemma 5.2.7 *The automorphism $\widehat{\sigma_1}$ sends every reduced word terminating in x_1^{-1} to a reduced word terminating in x_1^{-1}.*

Proof Consider again a reduced word terminating in x_1^{-1}, say vx_1^{-1}. As above, we find

$$\widehat{\sigma_1}(vx_1^{-1}) = \mathsf{red}(\widehat{\sigma_1}(v)x_1x_2^{-1}x_1^{-1}).$$

Suppose $\widehat{\sigma_1}(vx_1^{-1})$ does not finish with x_1^{-1}. This means that, in the above reduction, the final letter x_1^{-1} disappears by reduction with a letter x_1 which, as it cannot be the one preceding x_2^{-1}, necessarily comes from $\widehat{\sigma_1}(v)$. By definition of $\widehat{\sigma_1}$, a letter x_1 in $\widehat{\sigma_1}(v)$ comes either from a letter $x_1^{\pm 1}$ in v, or from a letter x_2.

Suppose the letter x_1 in play comes from a letter x_1^e, $e = \pm 1$, in v, so $v = v_1x_1^e v_2$. We thus find

$$\widehat{\sigma_1}(vx_1^{-1}) = \mathsf{red}(\widehat{\sigma_1}(v_1)\, x_1\, \underline{x_2^e x_1^{-1}\widehat{\sigma_1}(v_2)x_1x_2^{-1}}\, x_1^{-1}),$$

where the underlined factor reduces to the empty word. Thus $\widehat{\sigma_1}(v_2) = x_1x_2^{1-e}x_1^{-1}$, which, by the definition of $\widehat{\sigma_1}$, requires $v_2 = x_1^2$ for $e = -1$, and $v_2 = 1$ for $e = 1$. Then the word vx_1^{-1} is either $v_1x_1^{-1}x_1^2x_1^{-1}$, or $v_1x_1x_1^{-1}$, and neither is reduced. This case is thus impossible.

Suppose now that the letter x_1 in play comes from a letter x_2 in v, so $v = v_1x_2v_2$. We then find

$$\widehat{\sigma_1}(vx_1^{-1}) = \mathsf{red}(\widehat{\sigma_1}(v_1)\, x_1\, \underline{\widehat{\sigma_1}(v_2)x_1x_2^{-1}}\, x_1^{-1}),$$

where the underlined factor reduces to the empty word. Hence $\widehat{\sigma_1}(v_2) = x_2x_1^{-1}$, implying $v_2 = x_2^{-1}x_1$. Thus the word vx_1^{-1} is $v_1x_2x_2^{-1}x_1x_1^{-1}$, which is not reduced. This case is equally impossible, so the hypothesis that $\widehat{\sigma_1}(vx_1^{-1})$ does not finish in x_1^{-1} is a contradiction. □

Exercise 5.2.8 (Image) Show that $\widehat{\sigma_1}$ sends any reduced word terminating in x_1 to a reduced word terminating in x_1^{-1}.

It is now easy to assemble the pieces of the puzzle.

Lemma 5.2.9 *If g is a σ_i-positive braid, the word $\rho_{\mathsf{A}}(g)(x_i)$ finishes with the letter x_i^{-1}.*

Proof First suppose g is σ_1-positive. By definition, g admits an expression w that is a σ_1-positive word, hence can be written $w_0\,\sigma_1\,w_1\,\sigma_1 \cdots \sigma_1\,w_p$, where p is at least 1 and where the words w_k contain only the letters $\sigma_i^{\pm 1}$ with $i \geqslant 2$. Consider the word $\rho_{\mathsf{A}}(g)(x_1)$. It is convenient here (as in (5.2)) to write $x_1 \bullet w$ for $\rho_{\mathsf{A}}(w)(x_1)$. By definition, we have

$$\rho_{\mathsf{A}}(g)(x_1) = x_1 \bullet w = ((x_1 \bullet w_0) \bullet \sigma_1) \bullet (w_1 \sigma_1 \cdots \sigma_1 w_p).$$

We thus find $x_1 \bullet w_0 = x_1$, then $(x_1 \bullet w_0) \bullet \sigma_1 = x_1 x_2 x_1^{-1}$, a reduced word terminating in x_1^{-1}. From this, repeated applications of Lemmas 5.2.6 and 5.2.7 show that the word $\rho_{\mathsf{A}}(g)(x_1)$, image of the preceding one under the action of $w_1 \sigma_1 \cdots \sigma_1 w_p$, also ends with x_1^{-1}.

Now suppose g is σ_i-positive with $i \geqslant 2$. We can then write $g = \mathsf{dec}^{i-1}(g')$ where g' is σ_1-positive, and dec is the operation of shifting by one all the indices of the generators σ_i. If we extend the notation to words in $x_k^{\pm 1}$, then, by construction, the Artin representation ρ_{A} commutes with the shift of the indices, hence the equality $\rho_{\mathsf{A}}(g)(x_i) = \mathsf{dec}^{i-1}(\rho_{\mathsf{A}}(g')(x_1))$. By the results above, the word $\rho_{\mathsf{A}}(g')(x_1)$ finishes with the letter x_1^{-1}, and $\mathsf{dec}^{i-1}(\rho_{\mathsf{A}}(g')(x_1))$ thus finishes with x_i^{-1}. □

The precise results of Lemma 5.2.9 imply:

Proposition 5.2.10 (Representation) *If g is a σ-positive braid, the automorphism $\rho_{\mathsf{A}}(g)$ is not the identity.*

Proof By Lemma 5.2.9, if g is σ_i-positive, the image of x_i by $\rho_{\mathsf{A}}(g)$ is a reduced word terminated by x_i^{-1}, hence certainly not equal to x_i. Hence $\rho_{\mathsf{A}}(g)$ is not the identity, which sends x_i to x_i. □

This in turn implies the desired result.

Corollary 5.2.11 (Acyclicity property) *A σ-positive braid is not trivial.*

Proof By Proposition 5.2.10, the image of a σ-positive braid g by the Artin representation is not the identity. This implies in particular that g itself is not the trivial braid. □

5.2.3 Injectivity of the Artin Representation

As announced, we establish here the conditional results based on a hypothesis that will be proved in the next chapter, that is, the comparison property of Definition 5.2.4.

Proposition 5.2.12 (Faithfulness) *If the comparison property holds, the Artin representation is faithful.*

Proof Let g be a non-trivial braid. By the comparison property, at least one of the braids g or g^{-1} is σ-positive. In the first case, Proposition 5.2.10 states that $\rho_{\mathsf{A}}(g)$ is not the identity. In the second case, Proposition 5.2.10 says that $\rho_{\mathsf{A}}(g^{-1})$ is not the identity: however $\rho_{\mathsf{A}}(g)$ is $\rho_{\mathsf{A}}(g^{-1})^{-1}$, so it is not the identity either. Hence, in every case, $g \neq 1$ implies $\rho_{\mathsf{A}}(g) \neq \mathsf{id}$. □

Under the same hypothesis, we obtain a new solution to the braid isotopy problem.

Proposition 5.2.13 (New solution of the isotopy problem) *If the comparison property is true, then an n-strand braid word represents* 1 *in the group* B_n *if and only if* $\rho_{\mathsf{A}}(w)(x_k) = x_k$ *for* $k = 1, ..., n$.

Proof By Proposition 5.2.12, w represents 1 in B_n if and only if the automorphism $\rho_{\mathsf{A}}(w)$ is the identity. As F_n is a free group with basis $\{x_1, ..., x_n\}$, a homomorphism ϕ of F_n is the identity if and only if ϕ sends each of the generators x_k onto itself. □

It is easy to transform the criterion of Proposition 5.2.13 into an algorithm (do it!). From a practical point of view, the algorithm thus obtained is not very efficient: given the definition of the Artin representation, the only clear upper bound[5] for the length of the words $\rho_{\mathsf{A}}(w)(x_k)$ is in $O(3^{|w|})$, because when we go from w to $w\sigma_i$, each letter x_i is replaced (before reduction) by $x_i x_{i+1} x_i^{-1}$, which can a priori multiply the length of the image by 3.

Example 5.2.14 ($\sigma_2^{-2}\sigma_1^{-2}\sigma_2^2\sigma_1^2$) Consider once again the 'difficult' word of Example 1.3.11, that is, $w := \sigma_2^{-2}\sigma_1^{-2}\sigma_2^2\sigma_1^2$. We find here

$$\rho_{\mathsf{A}}(w)(x_1) = x_1x_2x_1^{-1}x_3x_1x_2^{-1}x_1^{-1}x_3^{-1}x_1x_3x_1x_2x_1^{-1}x_3^{-1}x_1x_2^{-1}x_1^{-1},$$

which is not x_1: this is enough to conclude that $\rho_{\mathsf{A}}(w)$ is not the identity, hence that w does not represent 1 in B_3. Note that this result is independent of the comparison property: in every case, if the Artin representation is not the identity, w does not represent 1. The comparison property is only necessary for the converse.

5.2.4 The Equivalence of the Diverse Notions of Isotopy

Thanks to the results on the Artin representation, we can now return to the question, left open in Chapter 1, of the equivalence between the diverse forms of isotopy.

[5] In fact, numerous factors are repeated in the words in play, so that these can be compressed with suitable encodings (Schleimer, 2008).

Lemma 5.2.15 *If β, β' are two geometric braids isotopic in the unrestrained sense, the automorphisms $\rho_{\mathsf{A}}(\beta)$ and $\rho_{\mathsf{A}}(\beta')$ coincide.*

Proof Suppose Φ is an isotopy in the unrestrained sense linking β to β'. By definition, Φ is a continuous path in $\mathsf{Homeo}([0,1]{\times}\mathbb{D})$ linking the identity to a homeomorphism ϕ satisfying $\beta' = \beta \circ \phi$. Denote Φ_s the homeomorphism $\Phi(s)$: by hypothesis, Φ_s is a homeomorphism of the cylinder $[0,1]{\times}\mathbb{D}$ leaving the two ends invariant point by point, but not necessarily (contrary to the case of the special homeomorphisms considered in the strong version of isotopy) preserving each of the vertical disks $\{t\}{\times}\mathbb{D}$. Moreover, Φ_0 is the identity, whereas Φ_1 sends β onto β'. By Proposition 5.1.4, we see the geometric braids as the homeomorphisms of $\mathbb{D}_n$ induced by the movements of the n marked points of $\mathbb{D}_n$, that is, we identify β with a family $(\phi_t)_{0\leqslant t\leqslant 1}$ of homeomorphisms of $\mathbb{D}_n$, and β' to a similar family $(\phi'_t)_{0\leqslant t\leqslant 1}$. In terms of homeomorphisms of $\mathbb{D}_n$, we thus have a family of homeomorphisms depending continuously on two indices, s and t, that we can denote $\phi_{t,s}$. Consider the images by these homeomorphisms $\phi_{t,s}$ of the loops γ_k of the standard basis of $\pi_1(\mathbb{D}_n^-)$ (cf. Figure 5.3). As above, we denote x_k the class of γ_k in $\pi_1(\mathbb{D}_n^-)$. By definition, $\phi_{0,0}(\gamma_k) = \gamma_k$, whereas $\phi_{1,0}(\gamma_k)$ is a loop whose class in $\pi_1(\mathbb{D}_n^-)$ is $\rho_{\mathsf{A}}(\beta)(x_k)$. For any s, we have $\phi(0,s)(\gamma_k) = \gamma_k$ since, by hypothesis, Φ_0 is the identity. Finally, for $s = 1$, as Φ_1 sends the family $(\phi_t)_t$ onto $(\phi'_t)_t$, we deduce that $\phi_{1,1}(\gamma_k)$ is a loop whose class in $\pi_1(\mathbb{D}_n^-)$ is $\rho_{\mathsf{A}}(\beta')(x_k)$. Note that, for $0 < s < 1$, nothing says that the image of γ_k is situated in a vertical plane; however, we can be sure that it is a continuous closed curve avoiding the images of the marked points. Moreover, for $s = 1$, the image $\phi_{1,s}(\gamma_k)$ is situated in the plane $\{1\}{\times}\mathbb{R}^2$, and it makes sense to consider its class in $\pi_1(\mathbb{D}_n^-)$. We thus obtain a continuous function from $[0,1]$ into $\pi_1(\mathbb{D}_n^-)$, taking the values $\rho_{\mathsf{A}}(\beta)(x_k)$ for $s = 0$, and $\rho_{\mathsf{A}}(\beta')(x_k)$ for $s = 1$. As the free group $\pi_1(\mathbb{D}_n^-)$ is a discrete topological space, the above function can only be constant, hence $\rho_{\mathsf{A}}(\beta)(x_k) = \rho_{\mathsf{A}}(\beta')(x_k)$, so $\rho_{\mathsf{A}}(\beta) = \rho_{\mathsf{A}}(\beta')$ by varying k. □

We can now complete the equivalence results announced in Chapter 1 and left hanging.

Proposition 5.2.16 *If the comparison property holds, then for any pair of geometric braids β, β', the following properties are equivalent: $\beta \approx^{\mathsf{h}} \beta'$ (homotopy), $\beta \approx \beta'$ ('stratified' isotopy), and $\beta \approx^{\mathsf{nr}} \beta'$ ('unrestrained' isotopy).*

Proof Given the results previously established, the only implication remaining to prove is $\beta \approx^{\mathsf{nr}} \beta' \Rightarrow \beta \approx \beta'$. However, by Lemma 5.2.15, $\beta \approx^{\mathsf{nr}} \beta'$ implies $\rho_{\mathsf{A}}(\beta) = \rho_{\mathsf{A}}(\beta')$, hence $\beta \approx \beta'$ by Proposition 5.2.12. □

Always modulo a proof of the comparison property, these results definitively conclude the search for what is both the most natural and the most robust notion for an equivalence of geometric braids, in conformance with the aspirations of Chapter 1.

5.3 Appendix D: The Fundamental Group of a Punctured Disk

We used above the fact that the fundamental group of a disk with n holes is a free group with n generators. This appendix contains a rapid introduction to the fundamental group, as well as proof of these results.

5.3.1 The Fundamental Group

Every topological space $\mathcal{X}$ is associated with various homotopy invariants, that is, diverse objects (numbers, functions, groups, ...) depending only on the type of homeomorphism of $\mathcal{X}$ and constructed by means of homotopy classes. The fundamental group is one of such homotopy invariants.

Recall that a curve γ traced in the plane, or more generally on a surface Σ, is identified with a parametrization, a continuous mapping of the interval $[0, 1]$ into the surface Σ.

Definition 5.3.1 (Loop) Let $\mathcal{X}$ be a topological space, and P_* a point of $\mathcal{X}$. A *loop* in $\mathcal{X}$ with basepoint P_* is a curve $\gamma\colon [0, 1] \to \mathcal{X}$ satisfying $\gamma(0) = \gamma(1) = P_*$. The family of all loops of $\mathcal{X}$ with basepoint P_* is denoted $\Omega_1(\mathcal{X}, P_*)$.

As in Chapter 1, we have a natural notion of homotopic loops: two loops γ, γ' with basepoint P_* are said to be *homotopic*, denoted $\gamma \sim \gamma'$, if there exists continuous $\phi\colon [0, 1] \times [0, 1] \to \mathcal{X}$ satisfying $\phi(\text{-}, 0) = \gamma$, $\phi(\text{-}, 1) = \gamma'$, and $\phi(0, t) = \phi(1, t) = P_*$ for every t. Thus, two loops are homotopic if we can continuously deform one to the other.

Notation 5.3.2 The quotient-set $\Omega_1(\mathcal{X}, P_*)/\!\sim$ is denoted $\pi_1(\mathcal{X}, P_*)$. The class of a loop γ in $\pi_1(\mathcal{X}, P_*)$ is denoted $[\gamma]$.

As with the geometric braids of Chapter 2, we can define a product on the space of loops.

Definition 5.3.3 (Product) For γ_1, γ_2 in $\Omega_1(\mathcal{X}, P_*)$, the *product* γ of γ_1 and γ_2 is defined by

$$\gamma(t) = \begin{cases} \gamma_1(2t) & \text{for } 0 \leqslant t \leqslant 1/2, \\ \gamma_2(2t-1) & \text{for } 1/2 \leqslant t \leqslant 1. \end{cases} \tag{5.3}$$

The coherence of the values at $t = 1/2$ ensures that the product is well-defined.

As with the product of geometric braids, the product of loops leads to a group structure.

Lemma 5.3.4 *The product of loops is compatible with the homotopy relation, and it induces a group structure on $\pi_1(\mathcal{X}, P_*)$.*

The verification is simple (and similar to the case of B_n).

It remains to see that, in the favourable cases, the group $\pi_1(\mathcal{X}, P_*)$ does not depend on the choice of the basepoint P_*.

Lemma 5.3.5 *If the space $\mathcal{X}$ is path-connected,*[6] *then, for any P_*, P'_* in $\mathcal{X}$, the groups $\pi_1(\mathcal{X}, P_*)$ and $\pi_1(\mathcal{X}, P'_*)$ are isomorphic.*

Proof Let π be a path from P_* to P'_* in $\mathcal{X}$. Then the mapping $\gamma \mapsto \pi^{-1}\gamma\pi$ induces a well-defined morphism of $\pi_1(\mathcal{X}, P_*)$ to $\pi_1(\mathcal{X}, P'_*)$ (check this), and $\gamma \mapsto \pi\gamma\pi^{-1}$ induces its inverse morphism.[7] □

By Lemma 5.3.5, we can ignore the basepoints.

Definition 5.3.6 (Fundamental group) If $\mathcal{X}$ is a path-connected topological space, the *fundamental group* $\pi_1(\mathcal{X})$ of $\mathcal{X}$ is (the class of isomorphisms of) $\pi_1(\mathcal{X}, P_*)$, where P_* is any point in $\mathcal{X}$.

Exercise 5.3.7 (Fundamental group)
(i) Show that the fundamental group is an invariant of homeomorphism: two homeomorphic spaces have isomorphic fundamental groups.
(ii) Show that, if $\mathcal{X}$ is a subspace of $\mathbb{R}^n$ such that, for any P in $\mathcal{X}$, the segment $[0, P]$ is included in $\mathcal{X}$ ('starred' space), then the fundamental group of $\mathcal{X}$ is trivial.

5.3.2 The Fundamental Group of a Disk with One Hole

Our aim now is to compute the fundamental group of a disk with n holes. We begin with the case $n = 1$.

[6] i.e. if for any points P and P' of $\mathcal{X}$, there exists at least one path (continuous curve) joining P to P' in $\mathcal{X}$.

[7] As in the case of braids, we write $\gamma_1\gamma_2$ for 'γ_1 then γ_2'.

Proposition 5.3.8 (Group $\pi_1(\mathbb{S}^1)$) *The fundamental group of a circle is* $\mathbb{Z}$.

Proof Let $\mathbb{S}^1$ be the circle with equation $x^2 + y^2 = 1$ in $\mathbb{R}^2$, and P_* the point $(1, 0)$. For an integer n, let γ_n be the curve $t \mapsto (\cos(2\pi nt), \sin(2\pi nt))$. Then γ_n is a loop in $\mathbb{S}^1$ with basepoint P_*. We show that mapping $I\colon n \mapsto [\gamma_n]$, a homomorphism since the product $\gamma_n\gamma_m$ is homotopic to γ_{n+m} (check this), is an isomorphism of $\mathbb{Z}$ to $\pi_1(\mathbb{S}^1)$.

For this, denote $\mathbb{H}$ the circular helix with equation $t \mapsto (\cos t, \sin t, t)$ with $t \in \mathbb{R}$, and let pr be the restriction to $\mathbb{H}$ of the projection $(x, y, z) \mapsto (x, y)$. Then pr is a surjective continuous mapping of $\mathbb{H}$ onto $\mathbb{S}^1$. It is not globally injective, but it is locally: there exists a finite covering of $\mathbb{S}^1$ by open subsets $U_1, ..., U_p$ (for example, two arcs of length ℓ with $\pi < \ell < 2\pi$) such that, for each point M in U_i and any preimage $\widetilde{M}$ of M in $\mathbb{H}$, there exists an open neighbourhood $\widetilde{U}$ of $\widetilde{M}$ such that pr induces a homeomorphism of $\widetilde{U}$ onto U_i.

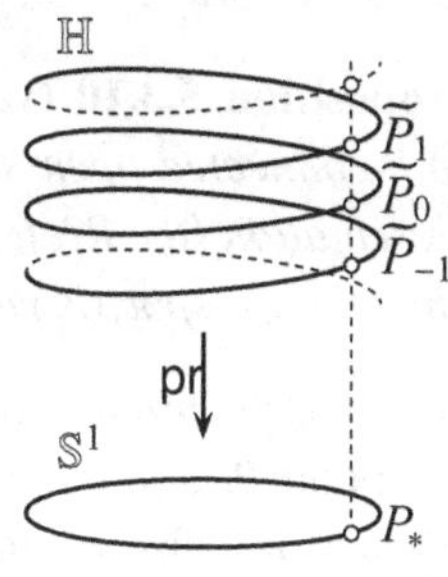

Set $\widetilde{P}_n := (1, 0, n)$. The points $\widetilde{P}_n$ are the preimages of P_* in $\mathbb{H}$. Then any loop γ with basepoint P_* in $\mathbb{S}^1$ can be lifted into a unique path $\widetilde{\gamma}$ in $\mathbb{H}$ starting at P_*, finishing on one of the points $\widetilde{P}_n$, and projecting onto γ: the unicity results from pr being a homeomorphism on each arc U_i.

Suppose ϕ is a homotopy between γ_n and $\gamma_{n'}$. Then, since pr is locally a homeomorphism, ϕ lifts uniquely to a homotopy between the paths $\widetilde{\gamma_n}$ and $\widetilde{\gamma_{n'}}$. Hence, the final points of $\widetilde{\gamma_n}$ and $\widetilde{\gamma_{n'}}$ necessarily coincide as values at 0 and 1 of a continuous function with values in the discrete space $\{\widetilde{P}_n \mid n \in \mathbb{Z}\}$. Thus the homomorphism I is injective.

Moreover, if γ is a loop in $\mathbb{S}^1$ such that $\widetilde{\gamma}$ ends at $\widetilde{P}_n$, the path $\widetilde{\gamma}^{-1}\widetilde{\gamma}_n$ is a loop in $\mathbb{H}$. However, $\mathbb{H}$ is homeomorphic to $\mathbb{R}$, hence its fundamental group is trivial (Exercise 5.3.7). Consequently, there exists a homotopy of this loop to the constant loop, hence the projection is a homotopy of $\gamma^{-1}\gamma_n$ to the constant loop of $\mathbb{S}^1$. In other words, $[\gamma] = [\gamma_n]$, and the homomorphism I is surjective. □

Recall that $\mathbb{D}_n^-$ is a disk with n holes.

Corollary 5.3.9 (Group $\pi_1(\mathbb{D}_1^-)$) *The fundamental group of* $\mathbb{D}_1^-$ *is* $\mathbb{Z}$.

Proof The topological space $\mathbb{D}_1^-$ is homeomorphic to a ring, thus to a thickened version of a circle. Every loop of the circle is a loop in $\mathbb{D}_1^-$. Conversely, by a radial projection, we associate with any loop of $\mathbb{D}_1^-$ a loop homotopic to $\mathbb{S}^1$. □

5.3.3 The Fundamental Group of a Disk with n Holes

Up to a homeomorphism, a disk with n holes can be seen as a disk with one hole glued to a disk with $n-1$ holes. Thus, to determine by induction $\pi_1(\mathbb{D}_n^-)$, it suffices to have a method to determine the fundamental group of two spaces glued together from their respective fundamental groups.

This is provided by van Kampen's theorem, here presented in a somewhat restricted form[8].

Proposition 5.3.10 (van Kampen's theorem) *Suppose $\mathcal{X}_1$ and $\mathcal{X}_2$ are two path-connected open sets covering[9] a space $\mathcal{X}$, the group $\pi_1(\mathcal{X}_i)$ admits the presentation $\langle S_i \,|\, R_i \rangle$ for $i = 1, 2$, $\mathcal{X}_1 \cap \mathcal{X}_2$ is path-connected, and π_1 is trivial. Then the group $\pi_1(\mathcal{X})$ admits the presentation $\langle S_1 \cup S_2 \,|\, R_1 \cup R_2 \rangle$.*

Sketch of the proof Since $\mathcal{X}_1$ and $\mathcal{X}_2$ are path-connected, so is $\mathcal{X}$. Choose a basepoint P_* in $\mathcal{X}_1 \cap \mathcal{X}_2$. Then any loop in $\mathcal{X}$ can be decomposed into a finite product of paths γ_k alternatively included in $\mathcal{X}_1$ and $\mathcal{X}_2$ and with extremities in $\mathcal{X}_1 \cap \mathcal{X}_2$. As $\mathcal{X}_1 \cap \mathcal{X}_2$ is path-connected, we can suppose that the extremities are the point P_*, and the hypothesis that $\pi_1(\mathcal{X}_1 \cap \mathcal{X}_2)$ is trivial implies that the choice of paths bringing the extremities to P_* is indifferent. Hence γ_k is a loop, thus a finite product of loops corresponding to elements of S_1 and S_2. Consequently, $S_1 \cup S_2$ generates $\pi_1(\mathcal{X})$.

Clearly the relations of R_1 and R_2 are satisfied in $\pi_1(\chi)$. The point is then to show that, conversely, if there exists a homotopy ϕ linking a loop γ to the constant loop γ_0, then γ can be deformed to γ_0 by using only the relations of $R_1 \cup R_2$. If all the loops in play in ϕ admitted the same decomposition into a finite sequence of loops contained in $\mathcal{X}_1$ and in $\mathcal{X}_2$, we could follow each of the fragments and use the hypothesis that a trivial loop in $\mathcal{X}_i$ can be deformed to γ_0 using the relations of R_i. However, a priori, the decomposition could change along ϕ: new alternations $\mathcal{X}_1$-$\mathcal{X}_2$ could appear. However, as $[0, 1]$ is compact, the total number of fragments appearing in ϕ is bounded by a finite number N, and it suffices to apply the relations of R_1 or of R_2 to each of these N fragments. □

We are now ready to tackle the case of a disk with n holes.

Proposition 5.3.11 (Group $\pi_1(\mathbb{D}_n^-)$) *For $k = 1, \ldots, n$, let γ_k be a loop starting from the boundary of $\mathbb{D}_n^-$ and turning counterclockwise once around the kth*

[8] The general form of van Kampen's theorem concerns the case where we do not suppose $\pi_1(\mathcal{X}_1 \cap \mathcal{X}_2)$ trivial: in order to not count twice the non-trivial loops in the interior of $\mathcal{X}_1 \cap \mathcal{X}_2$, we must pass to the quotient of the group-sum of $\pi(\mathcal{X}_1)$ and $\pi_1(\mathcal{X}_2)$ to there identify the subgroups coming from the embeddings of $\pi_1(\mathcal{X}_1 \cap \mathcal{X}_2)$ (passage to the 'amalgamated sum').

[9] i.e. $\mathcal{X}_1 \cup \mathcal{X}_2 = \mathcal{X}$.

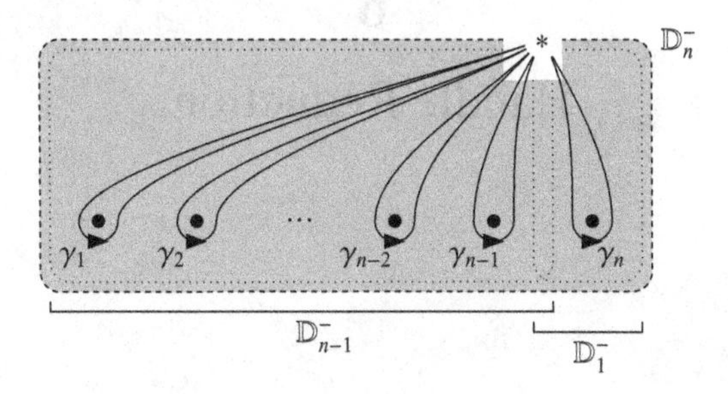

Figure 5.3 Standard basis of the fundamental group of the punctured disk $\mathbb{D}_n^-$, obtained by adjoining to the basis of the free group $\pi_1(\mathbb{D}_{n-1}^-)$ a unique element, the class of the loop γ_n in $\mathbb{D}_1^-$ turning once around the central well (here in the counterclockwise sense).

hole of $\mathbb{D}_n^-$. Then the fundamental group of $\mathbb{D}_n^-$ is a free group of rank n, whose basis is formed by the classes $x_1, ..., x_n$ of $\gamma_1, ..., \gamma_n$.

Proof We reason by induction on n. For $n = 1$, we use Corollary 5.3.9. For $n \geqslant 2$, $\mathbb{D}_n^-$ admits a covering by $\mathbb{D}_{n-1}^-$ and $\mathbb{D}_1^-$ (or by two spaces respectively homeomorphic to them), intersecting in a space homeomorphic to a non-punctured disk (Figure 5.3). The fundamental group of the latter is trivial, whereas, by the induction hypothesis, $\pi_1(\mathbb{D}_{n-1}^-)$ is $\langle [\gamma_1], ..., [\gamma_{n-1}] \,|\, \text{-} \rangle$ and $\pi_1(\mathbb{D}_1^-)$ is $\langle [\gamma_n] \,|\, \text{-} \rangle$ by Corollary 5.3.9. Van Kampen's theorem thus implies that the fundamental group of $\mathbb{D}_n^-$ is the free group $\langle [\gamma_1], ..., [\gamma_n] \,|\, \text{-} \rangle$. $\square$

6

Handle Reduction

At the end of the last chapter, several results about braids, including the faithfulness of the Artin representation and diverse applications ensuing from it, have an uncertain status, to the extent that the proofs given rely on a hypothesis not yet proved, the comparison property. The principal aim of this chapter is to provide a proof of this property, accomplished with the aid of a method known as *handle reduction*. The interest of this method extends beyond this proof in that it provides a new solution to the isotopy problem which, in practice, heuristically seems more efficient than the reduction to the normal form of Chapter 4 and, even if this remains an open question, could be the fastest solution to date.

6.1 The Handles of a Braid

6.1.1 The Notion of a Handle

In this chapter, we study diverse expressions of a braid g, that is, different braid words w satisfying $[w] = g$, thus it is important to distinguish between braid words and braids. We return to the σ-positive and σ-negative words introduced in Definition 5.2.1. A nonempty braid word w is σ-positive if the letter σ_m with the smallest index present in w – known as the *minimal* letter of w – appears only positively: σ_m is present, but not σ_m^{-1}, and in this case we say more precisely that w is σ_m-positive. Symmetrically, we speak of a σ_m*-negative* word if σ_m^{-1} appears but not σ_m.

Our goal here is to establish that every braid word is equivalent to a word either empty, σ-positive, or σ-negative, as will be shown in Proposition 6.3.1. For this, we observe that a word that is neither empty, σ-positive, nor

σ-negative contains at least one alternation of $\sigma_m \cdots \sigma_m^{-1}$ or $\sigma_m^{-1} \cdots \sigma_m$ of its minimal letter, and we give a name to this phenomenon.

Definition 6.1.1 (Handle) A positive (*resp.*, negative) *σ_m-handle* is a braid word of the form $\sigma_m u \sigma_m^{-1}$ (*resp.*, $\sigma_m^{-1} u \sigma_m$) where u does not contain any letter $\sigma_i^{\pm 1}$ with $i \leqslant m$; the handle is said to be *permitted* if at least one of the letters σ_{m+1} or σ_{m+1}^{-1} is not found in u.

Note that a σ_m-handle $\sigma_m^e u \sigma_m^{-e}$ is permitted if and only if the word u is σ_{m+1}-positive, σ_{m+1}-negative, or σ_{m+1}-neutral, in other words it does not include any σ_{m+1}-handle as a factor, see Figure 6.1.

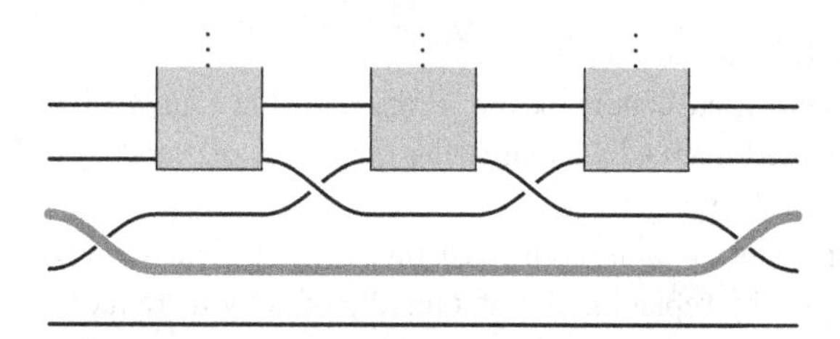

Figure 6.1 A positive σ_2-handle: a braid word starting with σ_2, ending with σ_2^{-1}, and between them only letters $\sigma_i^{\pm 1}$ with $i \geqslant 3$ appear; here the centre word is σ_3-positive, hence the handle is permitted. The terminology comes from the fact that the thickened strand in the figure looks (vaguely) like the handle of a suitcase.

Recall that a (braid) word w can be seen as a mapping with domain $\{1, ..., |w|\}$.

Notation 6.1.2 (Factor) For a braid word w and $1 \leqslant p \leqslant q \leqslant |w|$, the word obtained from w by extracting the letters between the pth and qth positions is called the *(p, q)-factor* of w, denoted $w\restriction_{(p,q)}$.

A *prefix* of w is a factor beginning with the first letter.

We distinguish one special handle among the diverse handles of a braid word.

Definition 6.1.3 (First handle) A handle v is said to be a *first handle* in a word w if there exists (p, q) satisfying $w\restriction_{(p,q)} = v$ and $w\restriction_{(p',q')}$ is not a handle for any pair (p', q') with $q' < q$.

The first handle of w is thus the one first completed when scanning w from left to right.

A first observation is that every braid word with a handle necessarily admits a permitted handle.

Lemma 6.1.4 *The first handle of a word with a handle is permitted.*

Proof Let q be minimal such that $w\lceil_{(1,q)}$ includes a handle. By hypothesis, there exist p and m such that $w\lceil_{(p,q)}$ is a σ_m-handle. By construction, this handle is the first handle of w. If it were not permitted, there would exist (p', q') with $q' \leqslant q-1$ such that $w\lceil_{(p',q')}$ is a σ_{m+1}-handle, contradicting the choice of q. □

6.1.2 Handle Reduction

Our aim here is to get rid of these handles. To this end, observe that every handle is equivalent to a word where the first and last letters $\sigma_m^{\pm 1}$ are removed.

Definition 6.1.5 (Reduction)
(i) If v is a permitted handle, say $v = \sigma_m^e u \sigma_m^{-e}$, the *reduct* of v is defined as the word obtained from v by suppressing the letters $\sigma_m^{\pm 1}$ and by replacing each letter σ_{m+1}^d by $\sigma_{m+1}^{-e}\sigma_m^d\sigma_{m+1}^e$.
(ii) If a braid word w admits at least one handle, denote $\mathsf{red_P}(w)$ the word obtained from w by replacing the first handle of w by its reduct.[1]

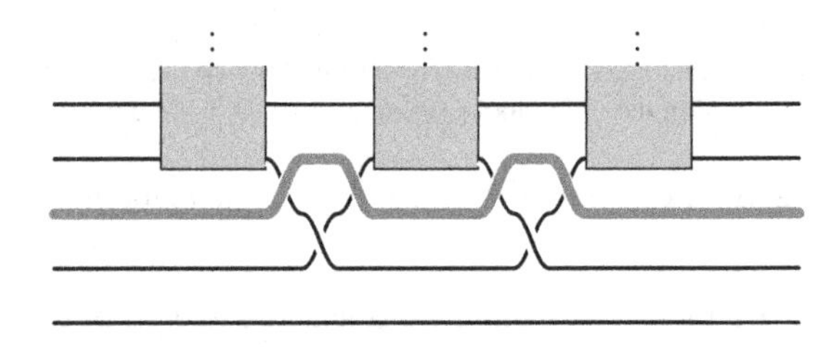

Figure 6.2 Reduct of the handle of Figure 6.1: the letters $\sigma_2^{\pm 1}$ on the ends are erased, and each letter σ_3 is replaced by $\sigma_3^{-1}\sigma_2\sigma_3$. The crossings in the interior of the three boxes are not modified.

Every braid word $\sigma_i\sigma_i^{-1}$ or $\sigma_i^{-1}\sigma_i$ is a permitted handle, and its reduct is the empty word. Thus, handle reduction extends the notion of free reduction of Definition 2.3.31. In this case, clearly the handle and its reduct represent the same element of the group B_n. In fact this can be extended to a general result.

Lemma 6.1.6 *Every permitted handle is equivalent to its reduct.*

[1] We could in the same way reduce any permitted handle of w, whether it is the first or not: the only difference is that, in doing so, the reduction is no longer a deterministic process: a word can be reduced in several distinct manners, and there is no longer unicity of the result.

Proof A comparison of Figures 6.1 and 6.2 clearly shows that the diagrams are isotopic: the thickened strand, which in the beginning passed under the two crossings at the position σ_{m+1} (in the sense of 'closest to the bottom of the figure'), now passes above these two crossings, now translated to the position σ_m. □

Exercise 6.1.7 (Reduction) Verify the result algebraically.

Passing from a braid word w containing a handle to the word $\mathsf{red}_{\mathsf{P}}(w)$ allows us to get rid of the first handle of w. However, this does not guarantee that $\mathsf{red}_{\mathsf{P}}(w)$ is without handles, since the handles of w other than the first persist, and, even worse, new handles may have been introduced. We are thus lured to iterating the process. Denote $\mathsf{red}_{\mathsf{P}}^k(w)$ the word obtained from w by reducing k times the first handle, if indeed one still remains.

Examples 6.1.8 (Handle reduction) Iteratively reducing the handles of our favourite word $\sigma_2^{-2}\sigma_1^{-2}\sigma_2^2\sigma_1^2$ gives the following steps, where we use a, b, ... for $\sigma_1, \sigma_2, \ldots$ and A, B, ... for $\sigma_1^{-1}, \sigma_2^{-1}, \ldots$:

1: BBAAbbaa
2: BBAbaBbaBa
3: BBbaBBbaBa
4: BaBBbaBa
5: BaBaBa,

and the reduction stops as the last word obtained does not include any handles (in fact it is σ_1-positive). This example can seem a bit too simple. Here is another, for a 4-strand braid word:

0: abaCBCABAcbca
1: abCBAbCBAcbca
2: aCBcAbCBAcbca
3: CBAbcbCBAcbca
4: CBAbcCBcAcbca
5: CBAbBcAcbca
6: CBAcAcbca
7: CBAccbaBc
8: CBccbaBBc
9: CcbCcbCaBBc
10: bCcbCaBBc
11: bbCaBBc,

and, once again, the reduction stops, with a final word σ_1-positive.

Two examples do not make a proof, but we shall see that the behaviour observed above is indeed general and handle reduction always converges.

Proposition 6.1.9 (Convergence) *For any braid word w, there exists an integer N such that the word $\mathsf{red}_{\mathsf{P}}^N(w)$ does not contain any handles.*

The proof of this result, which is not very easy but brings into play several interesting techniques, is the subject of the next section.

6.2 Convergence of the Reduction

6.2.1 Drawn Words in a Subset

Examples 6.1.8 show that, in contrast with the special case of free reduction, handle reduction does not necessarily diminish the length of the words to which it is applied and, indeed, it could a priori be the case that, starting from a word w, we obtain an infinite sequence of words $\mathsf{red}_{\mathsf{P}}^{k}(w)$ of increasing length. While we fail to obtain an upper bound on the length of the words $\mathsf{red}_{\mathsf{P}}^{k}(w)$, we can show that these remain drawn – in the sense of Definition 6.2.1 – in a certain finite subset of B_n^+, an indication that they can not meander about arbitrarily.

Recall that a *prefix* of a (braid) word w is a subword of w beginning with its first letter, that is, a word u such that there exists v satisfying $w = uv$.

Definition 6.2.1 (Drawn braid words) For $X \subseteq B_n$ and $a \in X$, a braid word w is said to be *drawn from a in X* if, for any prefix u of w, the braid $a[u]$ belongs to X.

For $X = B_n$, every word is drawn from any arbitrary braid, and the notion is not interesting. However, for X finite, the property is non-trivial. Note that arbitrarily long words can be drawn in a finite set: for example, every word $(\sigma_1\sigma_1^{-1})^k$ is drawn from 1 in $\{1, \sigma_1\}$.

The above terminology is natural when interpreted in terms of the Cayley graph of the group B_n.

Definition 6.2.2 (Cayley graph) The *Cayley graph* of B_n relative to $\sigma_1, ..., \sigma_{n-1}$ is the labelled graph $\Gamma(B_n)$ whose set of vertices is B_n and such that, for a, b in B_n, there exists an edge labelled σ_i from a to b if and only if $b = a\sigma_i$.

For any braid a and braid word w, the sequence of vertices $a[u]$ with u prefix of w defines a *path* $\gamma(a, w)$ starting from a in $\Gamma(B_n)$: the word w is the sequence of labels of the edges composing $\gamma(a, w)$, with the convention that an edge σ_i traversed backwards contributes σ_i^{-1}. Then, saying that a braid word w is drawn from a in X means that the path $\gamma(a, w)$ does not exit from X.

Recall that $\preccurlyeq$ designates the relation of left divisibility in a (braid) monoid: $a \preccurlyeq b$ holds if there exists x in the monoid in question satisfying $ax = b$.

Notation 6.2.3 (Set $\mathsf{Div}(b)$) If b is a positive n-strand braid, denote

$$\mathsf{Div}(b) := \{x \in B_n^+ \mid x \preccurlyeq b\}.$$

By definition, $a \preccurlyeq b$ implies $|a| \leqslant |b|$, hence every set $\mathsf{Div}(b)$ in B_n^+ is finite, of cardinality bounded above by $(n-1)^{|b|}$.

The goal is to show that, for any n-strand braid word w, every word of the form $\mathsf{red}_{\mathsf{P}}^{k}(w)$ is drawn in a certain finite set depending only on w. The starting point is the following.

Lemma 6.2.4 *For any braid word w, there exist two positive braids a, b such that w is drawn from a in* $\mathsf{Div}(b)$.

Proof Suppose $|w| = \ell$ and let $n-1$ be the maximal index of the letters $\sigma_i^{\pm 1}$ present in w. For $p \leqslant \ell$, let w_p be the prefix of w of length p. For any p, the braid $[w_p]$ belongs to B_n, hence by Proposition 4.3.1, there exists a non-negative integer d_p such that $\Delta_n^{d_p}[w_p]$ belongs to B_n^+, then, by Lemma 3.1.30, there exists a non-negative integer e_p satisfying $\Delta_n^{d_p}[w_p] \preccurlyeq \Delta_n^{d_p+e_p}$. Set $d := \max\{d_1, ..., d_\ell\}$ and $e := \max\{e_1, ..., e_\ell\}$. Then, by construction, $\Delta_n^d[w_p] \preccurlyeq \Delta_n^{d+e}$ for any p, in other words, w is drawn from Δ_n^d in $\mathsf{Div}(\Delta_n^{d+e})$. □

The next point is to show that, if w is drawn from a in $\mathsf{Div}(b)$, then so is $\mathsf{red_P}(w)$. For this, we decompose the passage from w to $\mathsf{red_P}(w)$ into four types of elementary transformations, and show that each of these remains within the set $\mathsf{Div}(b)$.

Definition 6.2.5 (Special rule) For w, w' braid words, w is said to pass to w' by a *special rule* if w' is obtained by transforming a factor of w as follows:
- type 1: $\sigma_i\sigma_j \mapsto \sigma_j\sigma_i$ with $|i - j| \geqslant 2$;
- type 2: $\sigma_i^{-1}\sigma_j^{-1} \mapsto \sigma_j^{-1}\sigma_i^{-1}$ with $|i - j| \geqslant 2$;
- type 3: $\sigma_i^{-1}\sigma_j \mapsto \sigma_j\sigma_i^{-1}$ with $|i - j| \geqslant 2$,
  or $\sigma_i^{-1}\sigma_j \mapsto \sigma_j\sigma_i\sigma_j^{-1}\sigma_i^{-1}$ with $|i - j| = 1$,
  or $\sigma_i^{-1}\sigma_i \mapsto \varepsilon$;
- type 4: $\sigma_i\sigma_j^{-1} \mapsto \sigma_j^{-1}\sigma_i$ with $|i - j| \geqslant 2$,
  or $\sigma_i\sigma_j^{-1} \mapsto \sigma_j^{-1}\sigma_i^{-1}\sigma_j\sigma_i$ with $|i - j| = 1$,
  or $\sigma_i\sigma_i^{-1} \mapsto \varepsilon$.

The rules of type 1 and 2 correspond to relations of positive braids and their inverses, those of type 3 are reversings of factors as in Definition 3.3.13, and those of type 4 are their symmetric versions. At each time, we pass from a braid word to an equivalent word.

As announced, the handle reduction can be decomposed.

Lemma 6.2.6 *For any braid word w such that $\mathsf{red_P}(w)$ is defined, w can be transformed to $\mathsf{red_P}(w)$ by a finite sequence of special rules.*

Proof We must show that, if v is a permitted handle and if v' is its reduct, we can get from v to v' by the special rules. By definition, there exist exponents d and e equal to ± 1 such that v has the form

$$v = \sigma_i^e \quad u_0 \quad \sigma_{i+1}^d \quad u_1 \quad \cdots \quad u_{r-1} \quad \sigma_{i+1}^d \quad u_r \quad \sigma_i^{-e}, \tag{6.1}$$

where $u_0, ..., u_r$ contains only letters $\sigma_j^{\pm 1}$ with $j \geqslant i + 2$, and then

$$v' = u_0 \quad \sigma_{i+1}^{-e}\sigma_i^d\sigma_{i+1}^e \quad u_1 \quad \cdots \quad u_{r-1} \quad \sigma_{i+1}^{-e}\sigma_i^d\sigma_{i+1}^e \quad u_r. \tag{6.2}$$

First suppose $d = 1$ and $e = -1$. The words in play are

$$\begin{array}{lllllllll} v = & \sigma_i^{-1} & u_0 & \sigma_{i+1} & u_1 & \cdots & u_{r-1} & \sigma_{i+1} & u_r \ \ \sigma_i \,, \\ v' = & & u_0 & \sigma_{i+1}\sigma_i\sigma_{i+1}^{-1} & u_1 & \cdots & u_{r-1} & \sigma_{i+1}\sigma_i\sigma_{i+1}^{-1} & u_r \,. \end{array}$$

Starting from v, we use rules of type 2 and 3 to push the initial letter σ_i^{-1} to the right until it meets the final letter σ_i. First, σ_i^{-1} traverses u_0 using rules of type 3 for the positive letters, and of type 2 for the negative letters. In this way, we arrive to the word

$$u_0 \quad \sigma_i^{-1}\sigma_{i+1} \quad u_1 \quad \cdots \quad u_{r-1} \quad \sigma_{i+1} \quad u_r \quad \sigma_i.$$

An additional type 3 rule pushes σ_i^{-1} across σ_{i+1}, giving

$$u_0 \quad \sigma_{i+1}\sigma_i\sigma_{i+1}^{-1} \quad \sigma_i^{-1}u_1 \quad \cdots \quad u_{r-1} \quad \sigma_{i+1} \quad u_r \quad \sigma_i.$$

We can continue by pushing σ_i^{-1} across u_1, then across the next letter σ_{i+1}, and so on. After r steps, we arrive to the word

$$u_0 \quad \sigma_{i+1}\sigma_i\sigma_{i+1}^{-1} \quad u_1 \quad \cdots \quad u_{r-1} \quad \sigma_{i+1}\sigma_i\sigma_{i+1}^{-1} \quad u_r \quad \sigma_i^{-1}\sigma_i \,,$$

and, with an additional type 3 rule, we obtain the desired word v'.

The argument for the case $d = -1$, $e = +1$ is similar, using rules of type 1 and 4 instead of 2 and 3. For the case $d = 1$, $e = 1$, the argument is symmetric: starting from v, we push the final letter σ_i to the left until it meets the initial letter σ_i^{-1} with the aid of rules of type 2 and 4. Finally, the case $d = e = -1$ is similar, with rules of type 1 and 3 instead of 2 and 4. □

We are getting there... It only remains to study the impact of the special rules on the words drawn in a domain $\mathsf{Div}(b)$.

Lemma 6.2.7 *Suppose a braid word w is drawn from a in* $\mathsf{Div}(b)$ *and w is transformed into w' by the special rules. Then w' is drawn from a in* $\mathsf{Div}(b)$.

Proof It suffices to consider the case where w is transformed to w' by applying a single rule; we consider each rule in turn.

We begin with the type 1. By definition, there exist w_1, w_2 and i, j with $|i - j| \geqslant 2$ satisfying

$$w = w_1 \ \sigma_i\sigma_j \ w_2 \quad \text{and} \quad w' = w_1 \ \sigma_j\sigma_i \ w_2.$$

We must show that, for every prefix u of w', the braid $a[u]$ belongs to $\mathsf{Div}(b)$. By construction, the only prefix of w' not a prefix of w is $u_1 = w_1\sigma_j$, and so we need to show that $a[u_1]$ is in $\mathsf{Div}(b)$, that is, $1 \preccurlyeq a[u_1] \preccurlyeq b$. Let $c := a[w_1]$ and $d := a[w_1]\sigma_i\sigma_j$. By definition, $c \preccurlyeq a[u_1] \preccurlyeq d$, and it suffices to establish $1 \preccurlyeq c$ and $d \preccurlyeq b$. However, these two relations follow from the hypothesis that

w is drawn from a in $\mathsf{Div}(b)$, since w_1 and $w_1\sigma_i\sigma_j$ are prefixes of w. Hence w' is drawn from a in $\mathsf{Div}(b)$.

We next consider the type 2. We find

$$w = w_1\,\sigma_i^{-1}\sigma_j^{-1}\,w_2 \quad \text{and} \quad w' = w_1\,\sigma_j^{-1}\sigma_i^{-1}\,w_2,$$

as always with $|i - j| \geqslant 2$. The only prefix of w' not a prefix of w is now $u_1 = w_1\sigma_j^{-1}$. Set $c := a[w_1]\sigma_i^{-1}\sigma_j^{-1}$ and $d := a[w_1]$. By construction, $c \preccurlyeq a[u_1] \preccurlyeq d$, and again, it suffices to show $1 \preccurlyeq c$ and $d \preccurlyeq b$. However, these inequalities follow from the hypothesis that w is drawn from a in $\mathsf{Div}(b)$, since $w_1\sigma_i^{-1}\sigma_j^{-1}$ and w_1 are prefixes of w. Thus w' is drawn from a in $\mathsf{Div}(b)$.

Coming to the type 3, we consider the case

$$w = w_1\,\sigma_i^{-1}\sigma_j\,w_2 \quad \text{and} \quad w' = w_1\,\sigma_j\sigma_i\sigma_j^{-1}\sigma_i^{-1}\,w_2$$

with $|i - j| = 1$. The other two cases, where $|i - j| \geqslant 2$ or $i = j$, are similar but simpler. Here, three prefixes of w' are not prefixes of w, namely $u_1 = w_1\sigma_j$, $u_2 = w_1\sigma_j\sigma_i$, and $u_3 = w_1\sigma_j\sigma_i\sigma_j^{-1}$. Let $c := a[w_1]\sigma_i^{-1}$ and $d := a[w_1]\sigma_j\sigma_i$. By construction, $c \preccurlyeq a[u_k] \preccurlyeq d$ for $k = 1, 2, 3$, and, once again, it suffices to establish that $1 \preccurlyeq c$ and $d \preccurlyeq b$. However $1 \preccurlyeq c$ results from the hypothesis that w is drawn from a in $\mathsf{Div}(b)$. Moreover, by hypothesis, we have at the same time $c\sigma_i \preccurlyeq b$ and $c\sigma_j \preccurlyeq b$, hence $\mathsf{lcm}(c\sigma_i, c\sigma_j) \preccurlyeq b$. However, the lcm in question is d. Thus w' is drawn from a in $\mathsf{Div}(b)$.

Finally, we consider the type 4, with $|i - j| = 1$, or

$$w = w_1\,\sigma_i\sigma_j^{-1}\,w_2 \quad \text{and} \quad w' = w_1\,\sigma_j^{-1}\sigma_i^{-1}\sigma_j\sigma_i\,w_2.$$

Three prefixes of w' are not prefixes of w, namely $u_1 = w_1\sigma_j^{-1}$, $u_2 = w_1\sigma_j^{-1}\sigma_i^{-1}$, and $u_3 = w_1\sigma_j^{-1}\sigma_i^{-1}\sigma_j$. Set $c := a[w_1]\sigma_j^{-1}\sigma_i^{-1}$ and $d := a[w_1]\sigma_i$. Then $c \preccurlyeq a[u_k] \preccurlyeq d$ for $k = 1, 2, 3$ and, once more, we must establish that $1 \preccurlyeq c$ and $d \prec b$. The second relation follows directly from the hypothesis that w is drawn from a in $\mathsf{Div}(b)$. For the first, w_1 and $w_1\sigma_i\sigma_j^{-1}$ are prefixes of w, hence the hypothesis that w is drawn from a in $\mathsf{Div}(b)$ implies $1 \preccurlyeq d\sigma_i^{-1}$ and $1 \preccurlyeq d\sigma_j^{-1}$, hence $1 \preccurlyeq \mathsf{gcd}(d\sigma_i^{-1}, d\sigma_j^{-1})$. However, the gcd in question is c. Thus w' is drawn from a in $\mathsf{Div}(b)$. □

Assembling the pieces, we obtain the desired result.

Proposition 6.2.8 (Bounds) *For any braid word w, there exist positive braids a and b such that w and every word obtained from w by iterated handle reduction is drawn from a in* $\mathsf{Div}(b)$.

Proof By Lemma 6.2.4, there exist a and b such that w is drawn from a in $\mathsf{Div}(b)$. By Lemma 6.2.7, any word obtained by the special rules from w is also drawn from a in $\mathsf{Div}(b)$. Finally, by Lemma 6.2.7, any word obtained by reduction of handles from w is obtained from w by the special rules, hence justiciable by the preceding result. □

6.2.2 The Critical Prefix

The result of the preceding section is still insufficient to establish the convergence of the reduction of handles, in particular because no orientation is privileged in the transformation leading from a word w to the word $\mathsf{red}_\mathsf{P}(w)$: a priori, there could exist loops in the reduction process, that is, equalities $\mathsf{red}_\mathsf{P}^{k+p}(w) = \mathsf{red}_\mathsf{P}^{k}(w)$ with $p > 0$. To exclude them, we must analyse more finely the reductions of handles and, in particular, discern quantities that, in a certain manner, always evolve in the same sense.

To this end, we introduce two new parameters.

Definition 6.2.9 (σ_1-profile, sign) For a braid word w:
(i) the *σ_1-profile* of w, denoted $\mathsf{Pr}(w)$, is the sequence of lengths of the blocks of letters σ_1 and σ_1^{-1} when all the letters $\neq \sigma_1^{\pm 1}$ are erased;
(ii) if $\sigma_1^{\pm 1}$ is present in w, the *sign* of w, denoted $\mathsf{sg}(w)$, is $+1$ (*resp.*, -1) if the first letter $\sigma_1^{\pm 1}$ in w is σ_1 (*resp.*, σ_1^{-1}).

Example 6.2.10 (σ_1-profile) For $w := \sigma_1\sigma_2\sigma_1\sigma_3^{-1}\sigma_1^{-1}\sigma_1\sigma_1\sigma_1\sigma_2$, the word remaining when the letters other than $\sigma_1^{\pm 1}$ are removed is $\sigma_1\sigma_1\sigma_1^{-1}\sigma_1\sigma_1\sigma_1$, and the σ_1-profile of w is thus the sequence $(2, 1, 3)$, of length 3. Here $\mathsf{sg}(()w) = +1$ since the first letter $\sigma_1^{\pm 1}$ of w is σ_1.

By definition, the number of σ_1-handles in a word w is $|\mathsf{Pr}(w)| - 1$ for $\mathsf{Pr}(w) \neq ()$, and 0 otherwise. In particular, w is σ_1-positive or σ_1-negative if and only if $\mathsf{Pr}(w)$ is of length 1.

The first important observation is that, during handle reduction, the number of σ_1-handles cannot increase.

Lemma 6.2.11 *Let w be a braid word admitting at least one handle. Set $w' := \mathsf{red}_\mathsf{P}(w)$. Then three cases are possible:*
(i) $|\mathsf{Pr}(w)| \leqslant 1$ *and* $\mathsf{Pr}(w') = \mathsf{Pr}(w)$,
(ii) $|\mathsf{Pr}(w)| \geqslant 2$ *and* $|\mathsf{Pr}(w')| = |\mathsf{Pr}(w)|$ *and* $\mathsf{sg}(w') = \mathsf{sg}(w)$,
(iii) $|\mathsf{Pr}(w)| \geqslant 2$ *and* $|\mathsf{Pr}(w')| < |\mathsf{Pr}(w)|$.

Proof If w does not contain a σ_1-handle, that is, if $\mathsf{Pr}(w)$ is of length 0 or 1, we pass from w to w' by reducing a σ_i-handle with $i \geqslant 2$, and by construction, w' has the same letters $\sigma_1^{\pm 1}$ as w. Thus $\mathsf{Pr}(w') = \mathsf{Pr}(w)$, and we are in case (i).

Now suppose w has at least one σ_1-handle, so $|\mathsf{Pr}(w)| \geqslant 2$. Two cases are possible. If the first handle of w is a σ_i-handle with $i \geqslant 2$, then as above, w' has the same letters $\sigma_1^{\pm 1}$ as w. Thus $\mathsf{Pr}(w') = \mathsf{Pr}(w)$, and we are in case (ii).

Suppose the first handle of w is a σ_1-handle. Let $(p, q, ...)$ be the beginning of $\mathsf{Pr}(w)$, and set $e := \mathsf{sg}(w)$. The generic form of w is

$$w = v_0\, \sigma_1^e\, v_1\, \sigma_1^e \cdots v_{p-2}\, \sigma_1^e\, v_{p-1}\, \underline{\sigma_1^e\, v_p\, \sigma_1^{-e}}\, v_{p+1}\, \sigma_1^{-e} \cdots, \tag{6.3}$$

where the words v_k do not contain $\sigma_1^{\pm 1}$ and where the underlined factor is the first handle of w. By Lemma 6.1.4, this one is permitted, and hence there exists $d = \pm 1$ such that the letter σ_2^{-d} is absent from v_p. Then

$$v_p = u_0\, \sigma_2^d\, u_1\, \sigma_2^d \cdots u_{r-1}\, \sigma_2^d\, u_r$$

with $r \geqslant 0$ and the words u_k not containing $\sigma_1^{\pm 1}$ nor $\sigma_2^{\pm 1}$, giving

$$\begin{aligned} w' = v_0\, \sigma_1^e\, v_1\, \sigma_1^e \cdots v_{p-2}\, \sigma_1^e\, v_{p-1}\, u_0\, \sigma_2^{-e} \sigma_1^d \sigma_2^e \\ \sigma_2^{-e} \sigma_1^d \sigma_2^e \cdots u_{r-1}\, \sigma_2^{-e} \sigma_1^d \sigma_2^e\, u_r\, v_{p+1}\, \sigma_1^{-e} \cdots \end{aligned} \tag{6.4}$$

This value shows that the blocks of $\sigma_1^{\pm 1}$ in w' are obtained from those of w by replacing $(p, q, ...)$ by $(p', q', ...)$ with $p' = p - 1 + r$ and $q' = q - 1$ for $d = e$, and $p' = p - 1, q' = q - 1 + r$ for $d = -e$: the reduction removes a letter σ_1^e and a letter σ_1^{-e}, and introduces r new letters σ_1 or σ_1^{-1}. Two cases arise: for $p'q' \neq 0$, the profile of w' is the indicated sequence: it has the same length as $\mathsf{Pr}(w)$, and by construction, the sign of w' is that of w, thus we are in the case (ii). On the other hand, if one of the integers p' or q' is zero, the profile of w' is obtained from the indicated sequence by suppressing the null term and, if applicable, regrouping the adjacent terms separated by this null term. In this case, the length of w' is that of $\mathsf{Pr}(w)$ diminished by one or two, and we are in the case (iii). □

Going even further requires an additional notion.

Definition 6.2.12 (Critical prefix) The *critical prefix* $\mathsf{pc}(w)$ of a braid word w admitting at least one σ_1-handle is the prefix of w ending on the first letter of the first σ_1-handle of w.

For example, the critical prefix of the word considered in Definition 6.2.9 is the word $\sigma_1\sigma_2\sigma_1$ (of length 3) whereas for the word of (6.3), it is $v_0\, \sigma_1^e v_1\, \sigma_1^e \cdots v_{p-2} \sigma_1^e v_{p-1} \sigma_1^e$.

We finally arrive at the key result.

Lemma 6.2.13 *Suppose w is a braid word drawn from a in $\mathsf{Div}(b)$, admitting at least one handle, and that case (ii) of Lemma 6.2.11 holds. Set $w' := \mathsf{red}_{\mathsf{P}}(w)$*

and $e := \mathsf{sg}(w)$, *and suppose the first handle of* w *is a* σ_i*-handle. Then there exists a word* $\mathsf{trans}(w)$ *satisfying* $\mathsf{pc}(w') \equiv \mathsf{pc}(w)\,\mathsf{trans}(w)$ *and such that*

(i) $\mathsf{trans}(w)$ *contains a letter* σ_1^{-e} *but not a letter* σ_1^{e} *in the case* $i = 1$, *and is empty in the case* $i \geqslant 2$,

(ii) $\mathsf{trans}(w)$ *is drawn from* $a[\mathsf{pc}(w)]$ *in* $\mathsf{Div}(b)$.

Proof We retain the notation of the proof of Lemma 6.2.11. In the case $i \geqslant 2$, that is, if we pass from w to w' by reducing a handle that is not a σ_1-handle, the letters $\sigma_1^{\pm 1}$ of w are not altered and the reduction takes place in the interior of one of the words v_k with $k \leqslant p$. Thus, whether $k < p$ or $k = p$, the critical prefixes $\mathsf{pc}(w)$ and $\mathsf{pc}(w')$ are equivalent, and choosing $\mathsf{trans}(w) := \varepsilon$ gives the stated results.

Suppose now $i = 1$, so that we pass from the form (6.3) to (6.4): we distinguish three cases according to the values of r and d. First suppose $r = 0$, meaning $\sigma_2^{\pm 1}$ is absent from v_p, which becomes u_0. In this case, we have

$$w = v_0\,\sigma_1^e\,v_1\,\sigma_1^e\,\cdots\,v_{p-2}\,\sigma_1^e\,v_{p-1}\,\underline{\sigma_1^e\,u_0\,\sigma_1^{-e}}\,v_{p+1}\,\sigma_1^{-e}\,\cdots,$$

$$w' = v_0\,\sigma_1^e\,v_1\,\sigma_1^e\,\cdots\,v_{p-2}\,\underline{\sigma_1^e\,v_{p-1}\quad u_0\quad v_{p+1}\,\sigma_1^{-e}}\,\cdots,$$

with the first σ_1-handles underlined. We read $\mathsf{pc}(w) = \mathsf{pc}(w')\,v_{p-1}\sigma_1^e$, hence $\mathsf{pc}(w') \equiv \mathsf{pc}(w)\sigma_1^{-e}v_{p-1}^{-1}$, and we obtain the desired results by setting $\mathsf{trans}(w) := \sigma_1^{-e}v_{p-1}^{-1}$: indeed, $\mathsf{trans}(w)$ is drawn in $\mathsf{Div}(b)$ from $a[\mathsf{pc}(w)]$ since $v_{p-1}\sigma_1^e$ is a suffix of $\mathsf{pc}(w)$ which, by hypothesis, is drawn from a in $\mathsf{Div}(b)$.

Now suppose $r \geqslant 1$ and $d = -e$, meaning σ_2^{-e} is present in v_p. Again underlining the first σ_1-handles, we find

$$w = v_0\,\sigma_1^e \cdots \sigma_1^e\,v_{p-1}\,\underline{\sigma_1^e\,u_0\quad \sigma_2^{-e}\quad u_1 \cdots u_{r-1}\quad \sigma_2^{-e}\quad u_r\,\sigma_1^{-e}}\,v_{p+1}\sigma_1^{-e}\cdots,$$

$$w' = v_0\,\sigma_1^e \cdots \underline{\sigma_1^e\,v_{p-1}\quad u_0\,\sigma_2^{-e}\sigma_1^{-e}}\sigma_2^e\,u_1 \cdots u_{r-1}\,\sigma_2^{-e}\sigma_1^{-e}\sigma_2^e\,u_r\quad v_{p+1}\sigma_1^{-e}\cdots.$$

As in the preceding case, we find $\mathsf{pc}(w) = \mathsf{pc}(w')\,v_{p-1}\sigma_1^e$, and, by the same argument, obtain the result for $\mathsf{trans}(w) := \sigma_1^{-e}v_{p-1}^{-1}$.

Finally suppose $r \geqslant 1$ and $d = e$, that is, σ_2^e is present in v_p. In this case, as always underlining the first σ_1-handles, we find

$$w = v_0\,\sigma_1^e \cdots \sigma_1^e\,v_{p-1}\,\underline{\sigma_1^e\,u_0\quad \sigma_2^{e}\quad u_1 \cdots u_{r-1}\quad \sigma_2^{e}\quad u_r\,\sigma_1^{-e}}\,v_{p+1}\sigma_1^{-e}\cdots,$$

$$w' = v_0\,\sigma_1^e \cdots \sigma_1^e\,v_{p-1}\quad u_0\,\sigma_2^{-e}\sigma_1^{e}\sigma_2^e\,u_1 \cdots u_{r-1}\,\sigma_2^{-e}\underline{\sigma_1^{e}\sigma_2^e\,u_r\quad v_{p+1}\sigma_1^{-e}}\cdots.$$

With our notation, we read this time the relation $\mathsf{pc}(w)v_p\sigma_1^{-e} \equiv \mathsf{pc}(w')\sigma_2^e u_r$, and deduce $\mathsf{pc}(w') \equiv \mathsf{pc}(w)\mathsf{trans}(w)$ with $\mathsf{trans}(w) := v_p\sigma_1^{-e}u_r^{-1}\sigma_2^{-e}$. It remains to see that $\mathsf{trans}(w)$ is drawn from $a[\mathsf{pc}(w)]$ in $\mathsf{Div}(b)$. However, by hypothesis, w and, a fortiori, its prefix $\mathsf{pc}(w)v_p\sigma_1^{-e}$, are drawn from a in

$\mathsf{Div}(b)$, hence $v_p\sigma_1^{-e}$ is drawn from $a[\mathsf{pc}(w)]$ in $\mathsf{Div}(b)$. On the other hand, by Proposition 6.2.8, w' and, a fortiori, its prefix $\mathsf{pc}(w')\sigma_2^e u_r$, are drawn from a in $\mathsf{Div}(b)$. Consequently, $\sigma_2^e u_r$ is drawn from $a[\mathsf{pc}(w')]$ in $\mathsf{Div}(b)$. As the braids $a[\mathsf{pc}(w)v_p\sigma_1^{-e}]$ and $a[\mathsf{pc}(w')\sigma_2^e u_r]$ coincide, we conclude that $\mathsf{trans}(w)$ is drawn from a in $\mathsf{Div}(b)$, thus completing the proof. □

6.2.3 Conclusion of the Proof of Proposition 6.1.9

We are now almost ready to assemble the pieces and establish the convergence of handle reduction as announced in Proposition 6.1.9. The last piece of the puzzle is a consequence of the acyclicity property shown in Chapter 5.

Lemma 6.2.14 *If b is a positive braid and w a σ_1-positive (resp., σ_1-negative) word drawn in* $\mathsf{Div}(b)$, *then the number of letters σ_1 (resp., σ_1^{-1}) in w is at most the cardinality of* $\mathsf{Div}(b)$.

Proof Suppose w is σ_1-positive and drawn from a in $\mathsf{Div}(b)$. Let r be the number of letters σ_1 in w. For $k = 1, ..., r$, let u_k be the prefix of w finishing just before the kth letter σ_1. By hypothesis, all the braids $a[u_k]$ belong to $\mathsf{Div}(b)$. Moreover, $k < k'$ implies $a[u_k] \neq a[u_{k'}]$: indeed, by construction, $u_{k'} = u_k v$, where v contains at least one letter σ_1 but no letters σ_1^{-1}. By the acyclicity property of Proposition 5.2.11, the braid $[v]$ is not trivial. The r braids $a[u_k]$ are thus pairwise distinct elements of $\mathsf{Div}(b)$, hence $r \leqslant \#(\mathsf{Div}(b))$.

If w is σ_1-negative and drawn from a in $\mathsf{Div}(b)$, the word w^{-1} is σ_1-positive and drawn from $a[w]$ in $\mathsf{Div}(b)$, and we apply the preceding result. □

We can (finally!) conclude.

Proof of Proposition 6.1.9 We show by induction on $n \geqslant 2$ that, for any braid word w in $\mathcal{BW}_n$, there exists N such that $\mathsf{red}_{\mathsf{P}}^N(w)$ does not contain a handle, and thus $\mathsf{red}_{\mathsf{P}}^{N+1}(w)$ does not exist.

For $n = 2$, the word w contains only σ_1 and σ_1^{-1}, and the reduction of handles coincides with the free reduction of Definition 2.3.32. The result is thus clear, and $N \leqslant |w|/2$.

We now consider the case $n \geqslant 3$. First observe that if w is a word of $\mathcal{BW}_n$ not containing $\sigma_1^{\pm 1}$, there exists a word w' in $\mathcal{BW}_{n-1}$ such that w is $\mathsf{dec}(w')$, and the induction hypothesis implies the existence of N such that $\mathsf{red}_{\mathsf{P}}^N(w')$ is without handles: then the compatibility of reduction with a shift of the indices implies that $\mathsf{red}_{\mathsf{P}}^N(w)$ is equally without handles.

In search of a contradiction, we now suppose w is a word in $\mathcal{BW}_n$ such that $\mathsf{red}_{\mathsf{P}}^k(w)$, denoted w_k, is defined for every k. By Lemma 6.2.11, the lengths of the profiles of the words w_k form a non-increasing sequence, hence finally constant. Up to eventually suppressing a finite number of initial terms, we can suppose that there exists ℓ such that $|\mathsf{Pr}(w_k)| = \ell$ is satisfied for every k, and the case $\ell = 0$ is excluded by the above remark.

Let E be the set of integers k such that the first handle of w_k is a σ_1-handle. Let k be an arbitrary integer in E. We can write

$$w_k = v_0\, \sigma_1^e\, v_1\, \sigma_1^e\, v_2 \cdots v_{r-1}\, \sigma_1^e\, v_r\, v$$

where v is either empty (case $\ell = 1$), or beginning with σ_1^{-e} (case $\ell \geqslant 2$) and where the words v_j do not contain $\sigma_1^{\pm 1}$. By the remark above, there exist integers $N_0, ..., N_r$ such that $\mathsf{red}_\mathsf{P}^{N_j}(v_j)$ has no handles. Thus, by construction, we find for $k' := k + N_0 + \cdots + N_r$

$$w_{k'} = \mathsf{red}_\mathsf{P}^{N_0}(v_0)\, \sigma_1^e\, \mathsf{red}_\mathsf{P}^{N_1}(v_1)\, \sigma_1^e\, \mathsf{red}_\mathsf{P}^{N_2}(v_2) \cdots \mathsf{red}_\mathsf{P}^{N_{r-1}}(v_{r-1})\, \sigma_1^e\, \mathsf{red}_\mathsf{P}^{N_r}(v_r)\, v.$$

If v were empty, the word $w_{k'}$ would be without handles, contradicting the hypothesis that $w_{k'+1}$ exists. Hence v begins with σ_1^{-e}, and the first handle of $w_{k'}$ is a σ_1-handle. Consequently, k' belongs to E, and E is thus infinite.

Moreover, by Proposition 6.2.8, there exist two positive braids a and b such that all the words w_k are drawn from a in $\mathsf{Div}(b)$. Let M be the cardinality of $\mathsf{Div}(b)$. As the cases (i) and (iii) are excluded here, each word w_k falls under (ii) of Lemma 6.2.11, and hence, by Lemma 6.2.13, the word $\mathsf{trans}(w_k)$ is defined. For any m, set $u_m := \mathsf{trans}(w_0)\mathsf{trans}(w_1) \cdots \mathsf{trans}(w_{m-1})$. By construction, all the words u_m are drawn from $a[\mathsf{pc}(w)]$ in $\mathsf{Div}(b)$ and are either σ_1-positive (case $e = -1$), or σ_1-negative (case $e = 1$). Thus, by Lemma 6.2.14, no word u_m can contain more that M letters σ_1^e. However, by construction, the word u_m contains as many letters σ_1^e as there are reductions of σ_1-handles between w_0 and w_m. Hence, if E is infinite,[2] the word u_m contains more than M letters σ_1^e for m large enough. This contradiction shows that the hypothesis of the existence of $\mathsf{red}_\mathsf{P}^k(w)$ for every k must be rejected. □

The preceding result is qualitative. With a bit more effort, we could extract an explicit upper bound on the number of reductions possible to apply to a word w as a function of its length. Nevertheless, as the cardinality of $\mathsf{Div}(b)$ depends exponentially on $|b|$, the bound obtained is in $O(C^{|w|})$, whereas practical experiments suggest a bound in $O(|w|^2)$.

[2] Or simply if E has more than M elements.

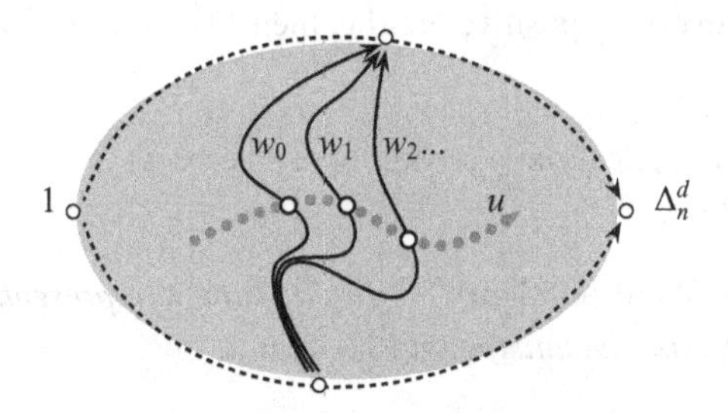

Figure 6.3 The key point: the number N of reductions of the first σ_1-handle must be finite. This is because there exists a transversal witness word u, drawn in the finite portion of the Cayley graph determined by 1 and Δ_n^d, such that u does not contain the letter σ_1^{-1}, and contains N letters σ_1; however, a path without σ_1^{-1} cannot pass twice along a same edge σ_1: as the number of edges σ_1 in a finite portion of a Cayley graph is finite, N is finite.

6.3 Applications

6.3.1 A Proof of the Comparison Property

The first direct application of the convergence of handle reduction is the property introduced in Chapter 5.

Proposition 6.3.1 (Comparison property) *If g is a non-trivial braid, then at least one of the braids g or g^{-1} is σ-positive.*

Proof Let w be an arbitrary braid word representing the braid g. By Proposition 6.1.9, there exists N such that the word $w' := \mathsf{red}_{\mathsf{P}}^N(w)$, also representing g, includes no handles. The case $w' = \varepsilon$ is excluded since g is supposed non-trivial. The only possibilities are that either w' is σ-positive, and hence g is σ-positive, or σ-negative, and hence g^{-1} is σ-positive. □

We can thus conclude that the results established in Chapter 5 under the hypothesis of the validity of the comparison property are all unconditionally valid. In particular, as announced in Proposition 5.2.12, the Artin representation is faithful and provides a solution to the braid isotopy problem. In general, the comparison property furnishes a criterion of injectivity.

Corollary 6.3.2 (Injectivity) *Suppose F is a function on B_n satisfying $F(g) \neq F(1)$ as soon as g is σ-positive or σ-negative. Then $F(g) \neq F(1)$ for every non-trivial braid.*

Exercise 6.3.3 (Injectivity) Suppose F is a function on B_n such that $F(g) \neq F(1)$ implies $F(\mathsf{dec}(g)) \neq F(1)$ and $F(g^{-1}) \neq F(1)$. Show that, if

$F(g) \neq F(1)$ for every σ_1-positive braid g, then $F(g) \neq F(1)$ for any non-trivial braid.

Another direct application is a solution to the braid isotopy problem based on handle reduction.

Corollary 6.3.4 (Word problem) *A braid word w represents the braid* 1 *if and only if there exists N satisfying* $\mathsf{red}_{\mathsf{P}}^{N}(w) = \varepsilon$.

Proof Let w be an arbitrary braid word. By Proposition 6.1.9, there exists N such that $\mathsf{red}_{\mathsf{P}}^{N}(w)$ is without handles. If $\mathsf{red}_{\mathsf{P}}^{N}(w)$ is the empty word, then w, which is equivalent to $\mathsf{red}_{\mathsf{P}}^{N}(w)$, represents the braid 1. If not, $\mathsf{red}_{\mathsf{P}}^{N}(w)$ is σ-positive or σ-negative and then, by the acyclicity property, neither $\mathsf{red}_{\mathsf{P}}^{N}(w)$, nor w, equivalent to it, represent the braid 1. □

The implementation of the preceding solution in the form of an algorithm is immediate.

Algorithm 6.3.5 (Word problem of B_∞ by handle reduction)
Input: a braid word w in $\mathcal{BW}_\infty$
Output: **yes** if w represents 1 in B_∞,
no otherwise

```
1: while w includes a handle do
2:     REDUCE(w)
3: end while
4: if w = ε then
5:     return yes
6: else
7:     return no
8: end if
```

```
1: function REDUCE(w: signed braid word)
2:     v ← leftmost factor σ_i^e ··· σ_i^{-e} in w      ▹ v is modified in place within w
3:     REMOVE σ_i^{±1} from v
4:     REPLACE σ_{i+1}^d by σ_{i+1}^{-e} σ_i^d σ_{i+1}^e in v
5: end function
```

Two examples were given in Examples 6.1.8. In both cases, the word without handles obtained at the end is nonempty, hence the braids represented are non-trivial.

Numerous variants exist. In particular, it is not necessary to continue all the way to a word without handles: as soon as a σ-positive or σ-negative word is obtained, we can conclude. We can also improve the efficiency with a strategy of the type 'divide and conquer': decompose w into w_1w_2 with $|w_1| \simeq |w_2| \simeq |w|/2$ and reduce w_1 and w_2 separately, allowing us to conclude for w except when the words w'_1 and w'_2 obtained are the one σ-positive and the other σ-negative, providing on average a gain.[3] In practice, we can thereby treat in less than a second braid words with several thousand letters (and an arbitrary number of strands).[4]

We mention here without the details the existence of alternative algorithms which, from a braid word w in $\mathcal{BW}_n$, determine an equivalent word which is either empty, σ-positive, or σ-negative. In particular, the 'rotating normal form' developed by J. Fromentin (Fromentin, 2011; Fromentin and Paris, 2012) provides a complexity in time $O(|w|^2)$.

6.3.2 Ordering Braids

Another application of the acyclicity and comparison properties is the existence of a remarkable total order on the groups B_n.

Proposition 6.3.6 (Order) *For any braids a, b, declare $a < b$ if the braid $a^{-1}b$ is σ-positive. Then the relation $<$ is a total order, compatible with the left product: $a < b$ implies $ca < cb$ for every braid c.*

Proof First observe that the product of two σ-positive words is σ-positive: if u is σ_i-positive and v is σ_j-positive, then uv is $\sigma_{\min(i,j)}$-positive.

Write $a \leqslant b$ for '$a<b$ or $a = b$'. By definition, $\leqslant$ is reflexive. Suppose we have at the same time $a \leqslant b$, $b \leqslant a$, and $a \neq b$. Then $a<b$ and $b<a$, thus $a^{-1}b$ and $b^{-1}a$ are σ-positive. By the above remark, we deduce that 1, the product of $a^{-1}b$ and $b^{-1}a$, is σ-positive, contradicting the acyclicity property. The conjunction of $a \leqslant b$ and $b \leqslant a$ hence implies $a = b$. Thus $\leqslant$ is antisymmetric. Finally, suppose $a \leqslant b$ and $b \leqslant c$: then $a^{-1}b$ and $b^{-1}c$ are σ-positive. Their product $a^{-1}c$ is also σ-positive, hence $a \leqslant c$. Thus $\leqslant$ is transitive, and is an order relation, where $<$ is the strict version.

This order is total. Indeed, let a, b be two distinct braids. By the comparison property, one of the braids $a^{-1}b$ or its inverse $b^{-1}a$ is σ-positive, signifying that $a < b$ or $b < a$ is satisfied.

Finally, we always have $(ca)^{-1}(cb) = a^{-1}b$, hence $a < b$ implies $ca < cb$. □

For any n, the relation $<$ provides a total order on the group B_n. Note that it is not necessary to specify the reference index n: the notion of σ-positivity is

[3] Especially if the decomposition is repeated recursively until reaching short words.

[4] On a theoretical plane, the reduction to the normal form of Chapter 4 is better, as it is proved to be of quadratic complexity, but, in practice, handle reduction seems to be (much!) faster: a better analysis of its complexity would be appreciated...

intrinsic, and the relation $<$ of B_n is the restriction to B_n of that of B_{n+1}. The existence of a total order compatible with the left product is usually expressed as follows.

Corollary 6.3.7 (Orderability) *For every n, the group B_n is an orderable group.*

Examples 6.3.8 (Order) The order on B_2 is (isomorphic to) that of $\mathbb{Z}$: for p, q in $\mathbb{Z}$, the relation $\sigma_1^p < \sigma_1^q$ is equivalent to $p < q$, since $\sigma_1^{-p}\sigma_1^q$ is σ_1^{q-p}.

For $n \geqslant 3$, the order on B_n gives priority to generators with small indices: in particular, if a, b, c are expressions without $\sigma_1^{\pm 1}$, then $a < b\sigma_1 c$, since the quotient is $a^{-1}b\sigma_1 c$, explicitly σ_1-positive, hence σ-positive. For example, $\sigma_2^p < \sigma_1$ is true for every p.

Other cases are less evident. For example, $\sigma_1\sigma_2 < \sigma_2\sigma_1$: the quotient is $\sigma_2^{-1}\sigma_1^{-1}\sigma_2\sigma_1$, or $\sigma_2^{-1}\sigma_2\sigma_1\sigma_2^{-1}$, which is σ_1-positive.

Except for the case of B_2, the order $<$ is not compatible with the right product: we have $\sigma_2 < \sigma_1$, so $\Delta_3\sigma_2 < \Delta_3\sigma_1$, however $\Delta_3\sigma_2\Delta_3^{-1} = \sigma_1$ and $\Delta_3\sigma_1\Delta_3^{-1} = \sigma_2$, so $\Delta_3\sigma_2\Delta_3^{-1} \not< \Delta_3\sigma_1\Delta_3^{-1}$.

Exercise 6.3.9 (Bi-orderability) Suppose $n \geqslant 3$. Show that no total order on B_n can be compatible with both the left and right products.

Exercise 6.3.10 (Right orderability) Show that the relation $a^{-1} < b^{-1}$ is a total order on the braids compatible with the right product.

For $n \geqslant 3$, the order on B_n is a complicated object, many of whose properties are not yet well understood. We mention here that this order admits numerous equivalent characterizations, algebraic or topological (Dehornoy et al., 2008). A fundamental property, established by R. Laver (Laver, 1996), is that, for every n, the restriction of the order to B_n^+ is a *well-ordering*[5] whose order type is the ordinal $\omega^{\omega^{n-2}}$, meaning that the increasing enumeration of the braids of B_n^+ is indexed by the ordinals less than $\omega^{\omega^{n-2}}$.

Exercise 6.3.11 (Non-Conradian order) Let $a := \sigma_2^{-1}\sigma_1$ and $b := \sigma_2^{-2}\sigma_1$. Show that $a > 1$ and nonetheless $ba^p < b$ for every p ('non-Conradian order').

Exercise 6.3.12 (Flipped order)
(i) Show that the order $<$ is compatible with the shift on B_∞.
(ii) For a, b in B_n, declare $a \mathrel{\tilde{<}}_n b$ when $\phi_n(a) < \phi_n(b)$, where ϕ_n is the automorphism exchanging σ_i and σ_{n-i} for every i. Show that $\tilde{<}_n$ is a total order on B_n, compatible with the left product, and that the order $\tilde{<}_n$ on B_n is the restriction of the order $\tilde{<}_{n+1}$ on B_{n+1}. What is the smallest element of B_n^+ according to $\tilde{<}$?

[5] i.e. every nonempty set has a smallest element.

7
The Dynnikov Coordinates

In Chapter 5, we used the topological aspects of braids to define an action of the group B_n on the marked disk $\mathbb{D}_n$, and then on the automorphisms of a free group. Considering the disk $\mathbb{D}$ as embedded in the sphere $\mathbb{S}$, we are here going to action the braids on families of curves traced in $\mathbb{S}$. By counting the intersections between these curves and the edges of a fixed triangulation of $\mathbb{S}$, we obtain an action on sequences of integers known as the Dynnikov coordinates (Dynnikov, 2002). This approach leads to surprising formulas bringing into play the operations[1] max and +, and a new solution, extremely efficient, for the braid isotopy problem.

7.1 The Action of Braids on Laminations

In this section, we explain how the action of braids on the laminations of the sphere $\mathbb{S}$ leads to attributing to every braid in B_n a sequence of $2n$ integers (the 'Dynnikov coordinates'), and provide explicit formulas for the computation of these coordinates. As the goal is to derive these formulas, which can then be used independently of their origin, we often content ourselves here to somewhat vague explanations and sketches of proofs.

7.1.1 Laminations

As in Chapter 5, we start from the isomorphism of the group B_n with the modular group of a disk $\mathbb{D}_n$ with n marked points: we consider a braid as an isotopy class of homeomorphisms of $\mathbb{D}$ preserving the boundary point by point and globally preserving a set $\{P_1, ..., P_n\}$ of marked points.

[1] We speak here of 'tropical formulas'.

Instead of considering the actions of braids on the loops of $\mathbb{D}_n$, we examine their actions on finite families of n topologically concentric circles known as *laminations*. Here we suppose $\mathbb{D}_n$ embedded in a 2-dimensional sphere $\mathbb{S}$, and consider the laminations included in $\mathbb{S}$, but not (necessarily) in $\mathbb{D}_n$.

Notation 7.1.1 (Lamination L_*) Denote L_* the lamination composed of n disjoint circles $C_1, ..., C_n$ where C_k surrounds the points $P_1, ..., P_k$, plus an additional point P_0 external to $\mathbb{D}_n$, see Figure 7.1.

If L is a lamination and β a geometric braid seen as a homeomorphism of $\mathbb{D}$ and extended trivially to $\mathbb{S}$ outside of $\mathbb{D}$, denote $L.\beta$ the lamination obtained by applying β to L.

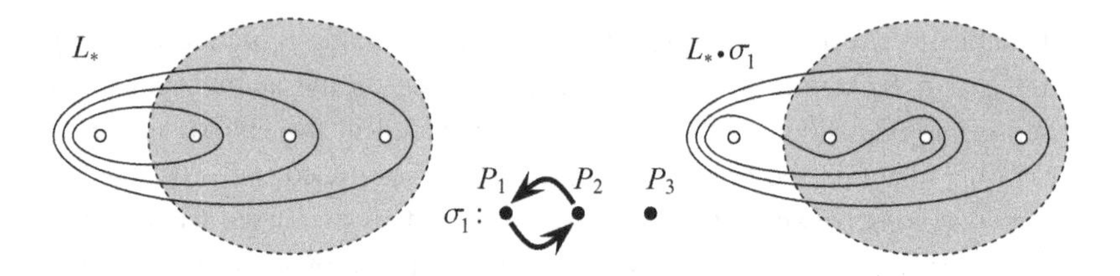

Figure 7.1 On the left, the lamination L_*: a collection of n circles in $\mathbb{S}$ surrounding the marked points of $\mathbb{D}_n$ and a supplementary point P_0 (here with $n = 3$). To the right, the image of L_* by the action of σ_1: certain curves of L_* are deformed, and certain positions with respect to the points P_k are modified.

7.1.2 Triangulations

A *triangulation* is a finite family of adjacent triangles covering the surface under consideration, here $\mathbb{S}$, and such that the intersection of two triangles is either empty or consists of a common edge. The sphere $\mathbb{S}$ is here identified with a plane completed with a (unique) point at infinity P_∞.

Notation 7.1.2 (Triangulation T_*) Denote T_* the triangulation of $\mathbb{S}$ represented in Figure 7.2.

The natural idea is to analyse a lamination L by counting its intersections with the edges of a triangulation T of the sphere $\mathbb{S}$ where $\mathbb{D}_n$ is embedded.

Definition 7.1.3 (T-coordinates) If T is a triangulation of $\mathbb{S}$ and $e_1, ..., e_N$ a fixed enumeration of the edges of T, the T-*coordinates* of a lamination L is defined as the sequence $(z_1, ..., z_N)$ where z_k is the cardinality of the intersection of e_k and the curves forming L.

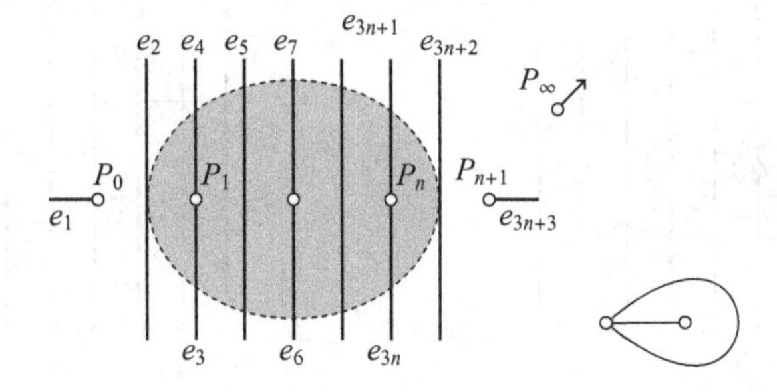

Figure 7.2 The triangulation T_* admits $2n+2$ triangles, $3n+3$ edges, and $n+3$ vertices, which are the n marked points of $\mathbb{D}_n$, two points P_0 and P_{n+1} outside of $\mathbb{D}$, and the point at infinity P_∞; in addition to a vertex at infinity, giving edges that are lines and half-lines, the triangles are degenerate with two coincident vertices, on the model of the diagram on the right.

For the T-coordinates to be useful, we must proscribe certain pathologies. A lamination avoiding the two types of intersections ('contact' and 'digon') shown on the right is said to be *in normal position* with respect to T. This gives the desired result.

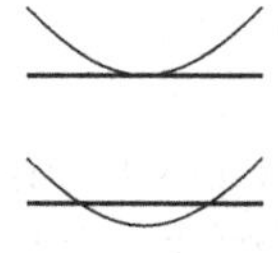

Lemma 7.1.4 *If a lamination L is in normal position with respect to T_*, its T_*-coordinates characterize its homotopy type.*

7.1.3 The Dynnikov Coordinates

The principle of the Dynnikov coordinates of a geometric braid β is to compare the T_*-coordinates of the lamination $L_* \bullet \beta$ to those of L_*. As illustrated in Figure 7.3, the action of β deforms the curves of the lamination and modifies its intersections with the edges of T_*.

Rather than considering the $3n + 3$ coordinates, we reduce to a sequence of $2n$ integers by extracting the (half)-differences between adjacent terms. In what follows, the coordinates refer to the enumeration of the edges of T_* in Figure 7.2.

Definition 7.1.5 (Dynnikov coordinates) Let g be a braid, β a representation of g such that $L_* \bullet \beta$ is in normal position for T_*, and $(z_1, ..., z_{3n+3})$ the T_*-coordinates of $L_* \bullet \beta$. The *Dynnikov coordinates* $\rho_D(g)$ of g are $(x_1, y_1, ..., x_n, y_n)$, with $x_i := (z_{3i} - z_{3i+1})/2$ and $y_i := (z_{3i-1} - z_{3i+2})/2$.

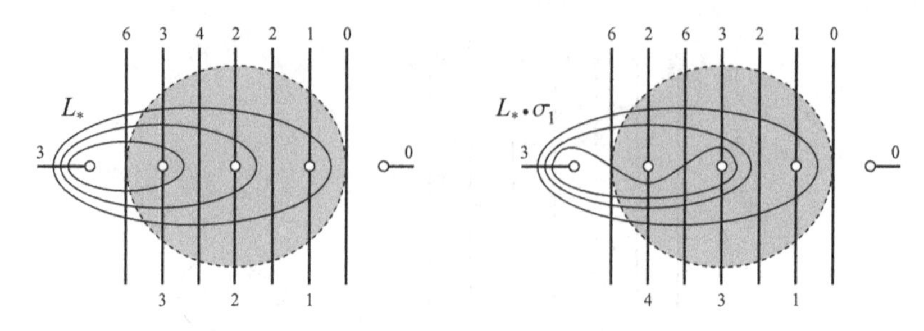

Figure 7.3 The T_*-coordinates of the laminations L_* and $L_*\bullet\sigma_1$.

Examples 7.1.6 (Dynnikov coordinates) On the left diagram of Figure 7.3, the coordinates of the unit braid of B_3 are seen to be $(0, 1, 0, 1, 0, 1)$, and on the right, those of σ_1 are $(1, 0, 0, 2, 0, 1)$. Note that by Lemma 7.1.3 the coordinates of a braid do not depend on the choice of the representation.

The problem is to show that we have not lost information about the braid g by passing from the complete sequence of T_*-coordinates to the sequence of Dynnikov coordinates. We especially wish to calculate the latter, that is, to obtain explicit formulas as a function of the generators $\sigma_i^{\pm 1}$, expressing the coordinates of $g\sigma_i^{\pm 1}$ as a function of those of g and $\sigma_i^{\pm 1}$. Here is the place for a remark, very simple but quite astute.

Lemma 7.1.7 *For any lamination L (in normal position with respect to T_*) and any geometric braid β, the T_*-coordinates of $L{\bullet}\beta$ are the $(T_*{\bullet}\beta^{-1})$-coordinates of L.*

Proof As the braid β is (seen as) a homeomorphism of $\mathbb{S}$, it is bijective and hence, for each curve γ traced in $\mathbb{S}$, the number of intersections of γ and an edge e of T_* is equal to the number of intersections of the images $\gamma{\bullet}\beta$ and $e{\bullet}\beta$ of γ and e by the action of β. □

Consequently, instead of relating the T_*-coordinates of L and $L{\bullet}\sigma_i^{\pm 1}$, it suffices to relate the T_*-coordinates and the $(T_*{\bullet}\sigma_i^{\mp 1})$-coordinates of L. The benefit is that we no longer need to consider a priori arbitrary triangulations, but only to compare the 'fixed' triangulations T_* and $T_*{\bullet}\sigma_i^{\mp 1}$, which is much simpler. The key is that two triangulations of the same surface are always related by a finite sequence of transformations called *flips*.

Definition 7.1.8 (Flip)
If T, T' are two triangulations of a surface Σ and if e is a common edge to two triangles of T, T' is said to be derived from T by *flipping* e if T' is obtained by replacing e by the other diagonal e' of the quadrilateral formed by the triangles.

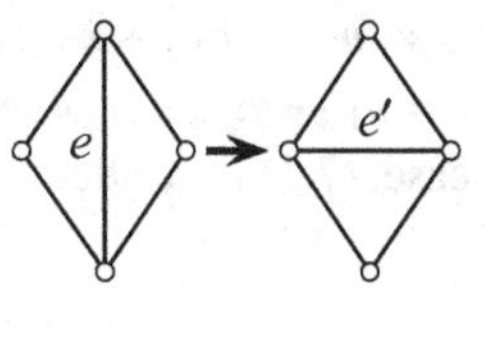

Hence it is certainly possible to go from the (singular) triangulation T_* to the triangulation $T_*{\boldsymbol\cdot}\sigma_i^{\mp 1}$ by a finite sequence of flips. Indeed, for σ_i^{-1}, we find the following solution.

Lemma 7.1.9 *T_* goes to $T_*{\boldsymbol\cdot}\sigma_i^{-1}$ by composing the four flips of Figure 7.4.*

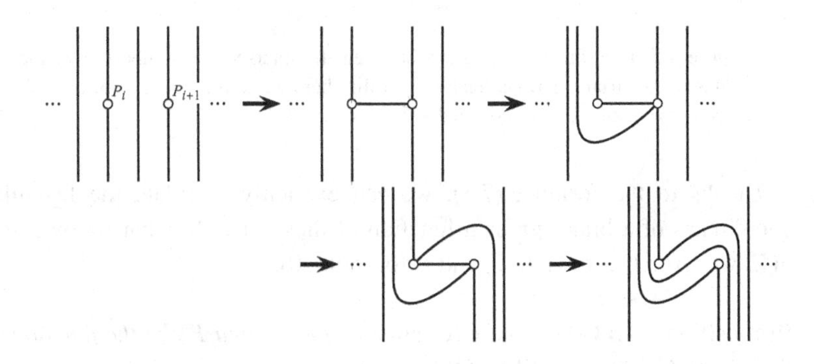

Figure 7.4 Decomposition of the action of σ_i^{-1} on T_* (a counterclockwise half turn exchanging P_i and P_{i+1}) in a sequence of four flips (convince yourself that these are indeed flips!); only the edges in a neighbourhood of the vertices P_i and P_{i+1} are shown.

It only remains to analyse the influence of a flip on the coordinates, that is, on their intersection numbers. Here we see appearing a simple but somewhat strange formula, a sort of 'tropical' version of Ptolemy's theorem.

Lemma 7.1.10 *If L is a lamination in normal position with respect to two triangulations, images of each other by the flip of an edge e, the numbers $z_1, \ldots, z_4, z, z'$ counting the intersections of L with the edges $e_1, \ldots, e_4, e, e'$ of Figure 7.5 are related by*

$$z + z' = \max(z_1 + z_3, z_2 + z_4). \tag{7.1}$$

Proof Let $n_{i,j}$ be the number of curves cutting both e_i and e_j. Then $z_i = \sum_{j \neq i} n_{i,j}$. The crucial point is that since, by hypothesis, the curves are disjoint, we cannot simultaneously have $n_{1,3} \neq 0$ and $n_{2,4} \neq 0$. However, in the case $n_{2,4} = 0$ (as in Figure 7.5), we find $z = n_{1,2} + n_{1,3} + n_{3,4}$ and $z' = n_{1,4} + n_{1,3} + n_{2,3}$, hence $z + z' = z_1 + z_3$. Similarly, for $n_{1,3} = 0$, we find $z = n_{1,2} + n_{2,4} + n_{3,4}$ and $z' = n_{1,4} + n_{2,4} + n_{2,3}$, hence $z + z' = z_2 + z_4$. In every case, (7.1) is verified. □

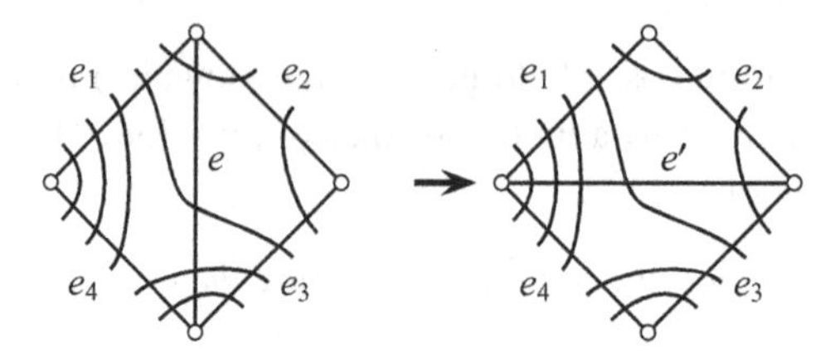

Figure 7.5 The 'Ptolemy' relation between the intersection numbers with the edges of two triangulations related by a flip; here we find $z_1 = z_4 = 5$, $z_2 = 2$, $z_3 = 4$, $z = 4$, and $z' = 5$: indeed, $4 + 5 = \max(5 + 4, 2 + 5)$.

Thanks to the formula (7.1), we can explicitly calculate the Dynnikov coordinates of a braid $g\sigma_i$ as a function of those of g. In what follows, for t in $\mathbb{Z}$, we write t^+ for $\max(t, 0)$ and t^- for $\min(t, 0)$.

Proposition 7.1.11 (Dynnikov formulas) *Let F^+ and F^- be the functions of $\mathbb{Z}^4$ into itself defined by $F^+ := (F_1^+, ..., F_4^+)$ and $F^- := (F_1^-, ..., F_4^-)$ with*

$$F_1^+(x_1,y_1,x_2,y_2):=x_1+y_1^++(y_2^+-t_1)^+, \quad F_2^+(x_1,y_1,x_2,y_2):=y_2-t_1^+,$$
$$F_3^+(x_1,y_1,x_2,y_2):=x_2+y_2^-+(y_1^-+t_1)^-, \quad F_4^+(x_1,y_1,x_2,y_2):=y_1+t_1^+,$$
$$F_1^-(x_1,y_1,x_2,y_2):=x_1-y_1^+-(y_2^++t_2)^+, \quad F_2^-(x_1,y_1,x_2,y_2):=y_2+t_2^-,$$
$$F_3^-(x_1,y_1,x_2,y_2):=x_2-y_2^--(y_1^--t_2)^-, \quad F_4^-(x_1,y_1,x_2,y_2):=y_1-t_2^-,$$

where $t_1 := x_1 - y_1^- - x_2 + y_2^+$ and $t_2 := x_1 + y_1^- - x_2 - y_2^+$. Then, if $(x_1, y_1, ..., x_n, y_n)$ are the Dynnikov coordinates of a braid g, those of the braid $g\sigma_i^e$, $e = \pm 1$, are $(x'_1, y'_1, ..., x'_n, y'_n)$ with $x'_k = x_k$ and $y'_k = y_k$ for $k \neq i, i+1$, and

$$(x'_i, y'_i, x'_{i+1}, y'_{i+1}) = \begin{cases} F^+(x_i, y_i, x_{i+1}, y_{i+1}) & \text{for } e = +1, \\ F^-(x_i, y_i, x_{i+1}, y_{i+1}) & \text{for } e = -1. \end{cases}$$

Proof Suppose $(z_1, ..., z_{3n+3})$ are the T_*-coordinates of a lamination L. Applying (7.1) to the four flips of Figure 7.4, we find that the $(T_*\bullet\sigma_i^{-1})$-coordinates $(z'_1, ..., z'_{3n+3})$ of L are given by

$$\begin{aligned}
z'_k &= z_k \text{ for } k \leqslant 3i-1 \text{ and } k \geqslant 3i+5,\\
z'_{3i} &= z_{3i+3},\\
z'_{3i+1} &= \max(z_{3i+1}+z_{3i+3}, z_{3i-1}+t) - z_{3i},\\
z'_{3i+2} &= \max(z_{3i+1}+z_{3i+3}, z'_{3i+1}+z'_{3i+3}) - t,\\
z'_{3i+3} &= \max(z_{3i+1}+z_{3i+3}, z_{3i+5}+t) - z_{3i+4},\\
z'_{3i+4} &= z_{3i+1}
\end{aligned}$$

with $t = \max(z_{3i}+z_{3i+4}, z_{3i+1}+z_{3i+3}) - z_{3i+2}$. By Lemma 7.1.7, the same formulas relate the T_*-coordinates of a lamination L with those of the lamination $L{\cdot}\sigma_i$. By extracting the Dynnikov coordinates as half-differences between the integers z_k and z'_k, we finally obtain the announced formulas in terms of the functions F_i^+ and F_i^-. The case of σ_i^{-1} is handled symmetrically. □

With the preceding analysis, we have associated every braid in B_n with a sequence of $2n$ integers, called the Dynnikov coordinates. By construction, the coordinates of the unit braid are $(0, 1, 0, 1, ..., 0, 1)$.

Exercise 7.1.12 (Dynnikov formulas) Starting from $(0, 1, 0, 1, ..., 0, 1)$, verify that the formulas of Proposition 7.1.11 give the values $(1, 0, 0, 2, 0, 1, ..., 0, 1)$ for σ_1, read from Figure 7.3. What are the values for σ_1^{-1}?

7.2 Making Use of These Coordinates

In this second part of the chapter, we take for given the Dynnikov formulas, and first show how to directly justify their existence without involving topology (Section 7.2.1), and then how the coordinates characterize braids and their position with respect to the trivial braid in the order of Chapter 6 (Section 7.2.2). Finally, we briefly discuss the solution they provide for the isotopy problem (Section 7.2.3).

7.2.1 Direct Construction

The formulas of Proposition 7.1.11 appear as the result of the analysis of Section 7.1, which at the same time furnishes their wording and their validity. Forgetting this origin, an alternative approach consists of supposing the formulas given ex nihilo, and action the braid words of $\mathcal{BW}_n$ 'abstractly' on $\mathbb{Z}^{2n}$ by

$$(x_1, y_1, , ..., x_n, y_n){\cdot}\sigma_i^e = (x'_1, y'_1, , ..., x'_n, y'_n), \tag{7.2}$$

with $x'_k = x_k$ and $y'_k = y_k$ for $k \neq i, i{+}1$, and

$$(x'_i, y'_i, x'_{i+1}, y'_{i+1}) = F^e(x_i, y_i, x_{i+1}, y_{i+1}),$$

where F^+ and F^- are the functions of Proposition 7.1.11.

A direct verification shows that this action induces a well-defined function on the braids, allowing us to continue as above.

Lemma 7.2.1 *If w, w' are equivalent words, then $(0, 1, , ..., 0, 1)\centerdot w$ and $(0, 1, , ..., 0, 1)\centerdot w'$ coincide.*

Principle of the proof As the action is not associated with a homomorphism, it does *not* suffice to consider the case of 'naked' braid relations. However, as the action is to the right and the result is an equality, it suffices to show that for any free group or braid relation $w = w'$ and for any sequence $(x_1, y_1, ..., x_n, y_n)$, we have $(x_1, y_1, ..., x_n, y_n)\centerdot w = (x_1, y_1, ..., x_n, y_n)\centerdot w'$. The definition renders the case of $\sigma_i\sigma_j = \sigma_j\sigma_i$ trivial. From this, and given the compatibility of the formulas with a shift of the indices, it suffices to consider the cases of $\sigma_i\sigma_i^{-1} = \varepsilon$ and $\sigma_i^{-1}\sigma_i = \varepsilon$ for $i = 1, 2$ and of $\sigma_1\sigma_2\sigma_1 = \sigma_2\sigma_1\sigma_2$, acting on a quadruple (x_1, y_1, x_2, y_2). These five verifications, that the quasi-unfamiliarity of computations with the operations max and + seem to render apocalyptic (especially for the last), can be accomplished by separately examining all combinations of signs imaginable... □

7.2.2 Injectivity of the Coordinates

In the case of the Artin representation ρ_A of Chapter 5, we used the topology of braids and their action on a marked disk to associate with each braid an object, an automorphism of a free group. Here we associate with each braid a sequence of numbers ρ_D. The question in both cases is to know whether or not we have lost information, that is, whether or not the automorphism $\rho_A(g)$ or the coordinates $\rho_D(g)$ characterize the braid g without ambiguity.

The response, positive, follows exactly the same plan as that used for the Artin representation of Chapter 5.

Lemma 7.2.2 *The first non-zero odd coordinate of a σ-positive (resp., σ-negative) braid is a strictly positive (resp., strictly negative) integer.*

Proof Let w be a σ_1-positive word, $w = w_0\sigma_1w_1\sigma_1 \cdots \sigma_1w_p$ where $\sigma_1^{\pm 1}$ is absent from the words w_k. We follow the first two coordinates of the sequence $(0, 1, ..., 0, 1)\centerdot w$ as the successive letters are added, starting from the beginning. We thus initially have $(0, 1)$. As long as σ_1 has not been encountered, that is, as long as we are still in w_0, we remain with $(0, 1)$. As soon as the first letter σ_1 is met, the first coordinate becomes $F_1^+(0, 1, x_2, y_2)$, or, by definition, $1 + 0^+ + z^+$ for a certain integer z (in fact x_2, but this is of no importance), hence a strictly positive integer. Next, the traversal of the words w_k does not change the first coordinate, whereas that of any (eventual) letters σ_1 can only augment it, since, by definition, we add some non-negative integers. The result is thus established for a σ_1-positive braid.

Now let w be a σ_i-positive braid word with $i \geqslant 2$. Then w is $\mathsf{dec}^{i-1}(w')$ where w' is σ_1-positive. By construction, the sequence of coordinates of w is that of w' preceded by $(0, 1, ..., 0, 1)$ with $2i - 2$ terms. The $i - 1$ first odd coordinates of w are thus zero, whereas the ith is the first of w', strictly positive as seen above.

The case of σ-negative braids is left as an exercise. □

Exercise 7.2.3 (σ-negative) Show that the first non-zero odd coordinate of a σ-negative braid is a strictly negative integer.

From the above results, we can derive a new proof[2] of the acyclicity problem: since the first coordinate of a σ-positive braid is not that of the trivial braid, such a braid cannot be trivial. Introducing again the comparison property established in Chapter 6, we can conclude for the injectivity.

Proposition 7.2.4 (Injectivity) *Two distinct braids have distinct Dynnikov coordinates.*

Proof Let g and g' be two braids in B_n having for coordinates the sequence $(x_1, y_1, ..., x_n, y_n)$. Then by hypothesis

$$(0, 1, ..., 0, 1) \bullet g = (0, 1, ..., 0, 1) \bullet g' = (x_1, y_1, ..., x_n, y_n). \tag{7.3}$$

However, since the braid gg^{-1} is trivial, its coordinates, which by definition are $(x_1, y_1, ..., x_n, y_n) \bullet g^{-1}$, are $(0, 1, ..., 0, 1)$, hence, by (7.3),

$$(0, 1, ..., 0, 1) \bullet g' g^{-1} = (0, 1, ..., 0, 1). \tag{7.4}$$

If g and g' are not equal, the quotient $g'g^{-1}$ is not trivial, and the comparison property implies that it is σ-positive or σ-negative (cf. Proposition 6.3.1). In either case, (7.4) contradicts the result of Lemma 7.2.2. Thus the only possibility is the equality of g and g'. □

In other words, the Dynnikov coordinates provide a complete invariant for braid isotopy.

The above results also provide 'for free' a characterization in terms of the Dynnikov coordinates of the braid order introduced in Section 6.3.2.

[2] Incidentally shorter than that of Chapter 5 – but the argument leading to the Dynnikov coordinates is more delicate than that leading to the Artin representation.

Proposition 7.2.5 (Order) *If g, g' are two braids in B_n, the relation $g < g'$ is verified if and only if the first non-zero odd coordinate of $g^{-1}g'$ is a strictly positive integer.*

Proof By definition, if $g<g'$ is verified, the braid $g^{-1}g'$ is σ-positive, hence by Lemma 7.2.2, its first non-zero odd coordinate is strictly positive. Conversely, if $g<g'$ does not hold, then $g^{-1}g'$ is either trivial, with coordinates $(0, 1, ..., 0, 1)$, or σ-negative, in which case, by Lemma 7.2.2, its first non-zero odd coordinate is strictly negative. In both cases, the first non-zero odd coordinate of $g^{-1}g'$ is not strictly positive. □

7.2.3 Implementation

According to Proposition 7.2.4, the Dynnikov coordinates provide a new solution to the braid isotopy problem. Presenting it in the form of an algorithm is very simple. Recall that if w is a word, $w(k)$ is the kth letter of w.

Algorithm 7.2.6 (Word problem of B_n by Dynnikov coordinates)
Input: a braid word w in $\mathcal{BW}_n$
Output: **yes** if w represents 1 in B_n
no otherwise

```
1: S ← (0, 1, ..., 0, 1)
2: for k := 1 to |w| do
3:     S ← S•w(k)          ▷ where • refers to the action of Equation (7.2)
4: end for
5: if S = (0, 1, ..., 0, 1) then
6:     return yes
7: else
8:     return no
9: end if
```

We return one last time to the 'difficult' braid of Chapter 1.

Example 7.2.7 (Dynnikov coordinates) Starting from $w := \sigma_2^{-2}\sigma_1^{-2}\sigma_2^{2}\sigma_1^{2}$, we find for $(0, 1, ..., 0, 1)\bullet w$ the sequence $(1, -19, -12, 9, 0, 13, 0, 1)$ (check this!). This sequence is not $(0, 1, ..., 0, 1)$, and we conclude once again that w does not represent 1 in B_3^+ – and, moreover, that the braid $[w]$ satisfies $[w] > 1$ for the order of Section 6.3.2.

One of the most remarkable aspects of the 'tropical' Dynnikov formulas concerns the complexity of the associated algorithm. Experiments show that the integers appearing as coordinates can be large, but there exists an explicit upper bound.

Proposition 7.2.8 (Complexity) *The algorithm 7.2.6 has complexity linear in memory and quadratic in time with respect to the length of the input word.*

Proof The augmentation of the maximal size of the binary representation of the coordinates induced by the addition of a letter $\sigma_i^{\pm 1}$ to a word cannot exceed three, as the Dynnikov formulas only bring in to play the operations 'max', which cannot increase the maximal size of the integer, and three additions, which, in the worst case, augment the size by three bits (in the case of a carry). In total, the size of the coordinates of a word of length ℓ is thus $O(\ell)$, and, since the 'max' and the sum of integers of size ℓ can be calculated in time $O(\ell)$, the complete computation of the coordinates of a word of length ℓ requires $O(\ell^2)$ steps. □

If the index of a braid n is left variable, the length of the sequence of coordinates is not fixed a priori, but the computation remains possible: when applicable, we add terms when a letter $\sigma_i^{\pm 1}$ appears with i larger than the previous bound. With careful coding, where in particular the generator $\sigma_i^{\pm 1}$ is stored using $\log i$ bits, we obtain a solution to the word problem still of complexity $O(\ell^2)$. In contrast, the solutions based on the normal form of Chapter 4 have a complexity of $O(\ell^2)$ for n fixed; however, when n is free, the global complexity becomes $O(\ell^2 n \log n)$. This is because the normalization of words of length 2, which is equivalent to a sort procedure, cannot be exhaustively precalculated and must be determined dynamically. The solution based on the Dynnikov coordinates is thus more efficient.

8
A Few Avenues of Investigation

We hope that the results of the preceding chapters have convinced the readers that the braids host a rich mathematical theory. However, the subject is far from being exhausted, and countless results go beyond what has been established here. The goal of this final chapter is to open a few further avenues of investigation, and to tackle a few points not developed due to the lack of space. We will move in two directions: first towards the groups B_n themselves (Section 8.1), with a few words on dual monoids (Section 8.1.1), on linear representations (Section 8.1.2), and on potential applications to cryptography (Section 8.1.3), then towards the numerous extensions and generalizations of braid groups (Section 8.2), with a few examples of surface braid groups (Section 8.2.1), and the Artin–Tits groups (Section 8.2.2). To conclude, we pose an open conjecture associated with an enticing reward (Section 8.2.3) ...

8.1 More on B_n

8.1.1 The Dual Braid Monoid

The family of Artin generators of B_n is remarkable: as seen in Chapters 3 and 4, the monoid B_n^+ that it generates has rich properties, in particular B_n is the fraction group of B_n^+ and there exists a greedy normal form associated with the family of divisors of Δ_n. It is natural to question whether these properties are also found for other families of generators. A complete response remains unknown, but we explain here the existence of at least one solution.

Exercise 8.1.1 (Generating family)
(i) Show that $\{\sigma_1, ..., \sigma_{n-1}\}$ is a minimal generating family of B_n, in the sense that no proper subfamily is generating.
(ii) Show that, for any $n \geqslant 3$, the family $\{\sigma_1, \cdots \sigma_{n-1}\}$ generates B_n.

We introduce a new generating family for B_n.

Definition 8.1.2 (Birman–Ko–Lee generators) For $1 \leqslant i < j < n$, set

$$a_{i,j} = \sigma_i \cdots \sigma_{j-2}\, \sigma_{j-1}\, \sigma_{j-2}^{-1} \cdots \sigma_i^{-1} \tag{8.1}$$

(see Figure 8.1).

By definition, $\sigma_i = a_{i,i+1}$ for any i, hence the Artin generators are also Birman–Ko–Lee generators. Consequently, for every n, the latter form a generating family of B_n. For $n \geqslant 3$, the inclusion is strict: $a_{1,3}$ is $\sigma_1\sigma_2\sigma_1^{-1}$, which is neither an Artin generator, nor even a positive braid.

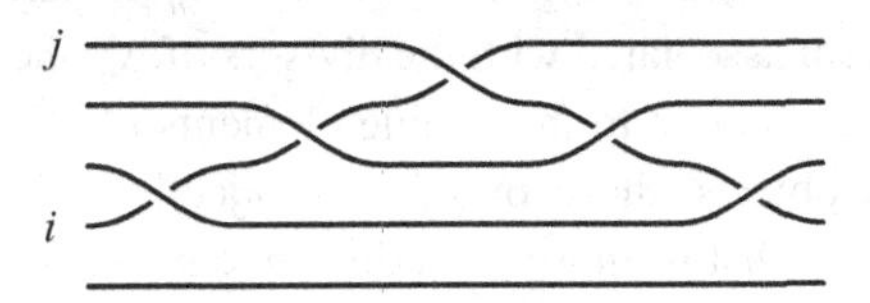

Figure 8.1 The Birman–Ko–Lee generators are particular conjugates of the Artin generators σ_i: the braid $a_{i,j}$, here $a_{2,5}$, corresponds to the crossing of the strands i and j passing behind all the intermediate strands.

Speaking of new generators means speaking of a new presentation.[1]

Lemma 8.1.3 *In terms of the generators $a_{i,j}$, the group B_n is presented by the relations $a_{i,j}a_{i',j'} = a_{i',j'}a_{i,j}$ if the intervals $[i, j]$ and $[i', j']$ are disjoint or nested, and $a_{i,j}a_{j,k} = a_{j,k}a_{i,k} = a_{i,k}a_{i,j}$ for $1 \leqslant i < j < k \leqslant n$.*

Exercise 8.1.4 Show the result.

We have seen in Chapter 3 that the monoid B_n^+ can be identified with the submonoid of B_n generated by the Artin generators σ_i. We can do the same with the Birman–Ko–Lee generators.

Definition 8.1.5 (Dual braid monoid) Denote B_n^{+*} the monoid generated by the elements $a_{i,j}$, $1 \leqslant i < j \leqslant n$, subject to the relations of Lemma 8.1.3.

We can then develop for the monoid B_n^{+*} an algebraic study in every way similar to that of B_n^+.

Proposition 8.1.6 (Dual monoid) *The monoid B_n^{+*} is simplifiable, and B_n is a left and right fraction group for B_n^{+*}.*

[1] Since the Artin generators belong to these new generators, a trivial presentation in terms of these consists in translating in terms of the $a_{i,i+1}$ the Artin relations, as well as the definitions of the supplementary generators. But this presentation is not terribly interesting...

The proof once again uses Ore's theorem. The existence of common multiples in B_n^{+*} is established by introducing

$$\Delta_n^* := a_{1,2}\, a_{2,3} \ldots a_{n-1,n}, \tag{8.2}$$

and by showing that any product of ℓ generators left-divides $\Delta_n^{*\ell}$.

It follows that, as in the case of B_n^+, the monoid B_n^{+*} can be identified as the submonoid of B_n generated by the elements $a_{i,j}$. Given these identifications, and since σ_i is $a_{i,i+1}$, the monoid B_n^+ is contained in B_n^{+*}, a strict inclusion for $n \geqslant 3$ as, by definition, $a_{1,3}$ is in B_3^{+*}, but not in B_3^+.

The properties of B_n^{+*} are analogous to those of B_n^+. In particular, there exists a greedy normal form associated with the divisors of Δ_n^*, the '*simple elements*' of B_n^{+*}. However, in contrast to the simple elements of B_n^+, bijective with the permutations of n objects, those of B_n^{+*} are bijective with the *noncrossing partitions* of $\{1, ..., n\}$, that is, those excluding $p < p' < q < q'$ with p, q in one block and p', q' in another. It is well known that the number of noncrossing partitions of $\{1, ..., n\}$ is the Catalan number $\frac{1}{n+1}\binom{2n}{n}$, and hence is the number of simple elements of B_n^{+*}.

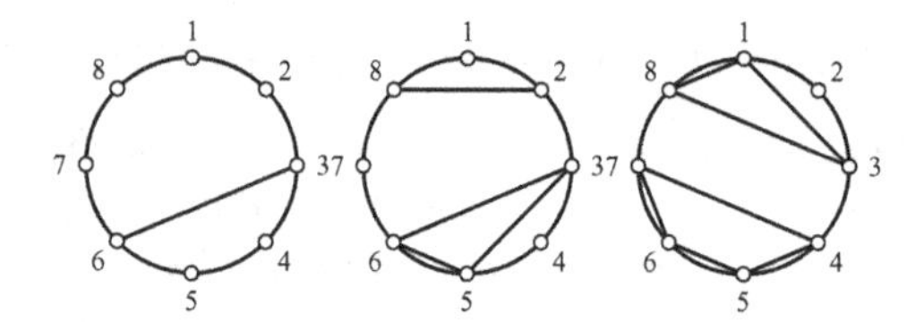

Figure 8.2 Representation of $a_{i,j}$ (here $a_{3,6}$) as a chord joining the vertices i and j on a circle with n marked vertices (on the left). The relations of Lemma 8.1.3 correspond to the fact that, when the juxtaposition is read as a product, the disjoint chords commute, and in a triangle, the clockwise product of two adjacent edges does not depend on the starting vertex: here, for example, $a_{2,8}a_{3,6} = a_{3,6}a_{2,8}$ and $a_{3,5}a_{5,6} = a_{5,6}a_{3,6} = a_{3,6}a_{3,5}$ (in the middle). The simple elements of B_n^{+*} correspond to unions of disjoint polygons and, from this, to noncrossing partitions: here the simple element $a_{1,3}a_{3,8}a_{4,5}a_{5,6}a_{6,7}$, associated with the partition $\{\{1, 3, 8\}, \{2\}, \{4, 5, 6, 7\}\}$ (on the right).

Exercise 8.1.7 (Conjugation by Δ_n^*) Show that the conjugation by Δ_n^* is the automorphism of B_n^{+*} sending $a_{i,j}$ to $a_{i+1,j+1}$ for $j < n$, and $a_{i,n}$ to $a_{1,i+1}$. What is its order? What is the associated geometric transformation to the correspondence of Figure 8.2?

Over and above the intrinsic interest of this new structure, the introduction of the dual monoid allows us to improve certain algorithms, notably in relation to the results of Chapter 6 (Fromentin and Paris, 2012).

8.1.2 Linear Representations

For any group G, it is common to study the linear representations of G, that is, the homomorphisms of G towards groups of matrices. In particular, when such a representation is faithful,[2] we obtain a realization of G as a group of matrices and hence, ipso facto, means to perform computations on G.

In the case of the group B_n, we constructed in Chapter 5 a faithful representation ρ_{A} towards the group $\mathsf{Aut}(F_n)$. This group is not a group of matrices, but it is easy to derive from ρ_{A} a linear representation that is a sort of linearized version.

Lemma 8.1.8 *For $1 \leqslant i < n$, let $\Sigma_{i,n}$ be the $n \times n$ matrix equal to the identity matrix, except for the 2×2 squares at the intersection of the rows and columns i and $i+1$, which are $\begin{pmatrix} 1-t & t \\ 1 & 0 \end{pmatrix}$. Then $\Sigma_{i,n}$ is invertible in $\mathsf{Mat}_n(\mathbb{Z}[t,t^{-1}])$, and the mapping ρ_{B}, sending σ_i to $\Sigma_{i,n}$ for every i, induces a linear representation of B_n.*

Exercise 8.1.9 Prove the result.

We thus obtain a representation of B_n in the group $\mathsf{GL}_n(\mathbb{Z}[t,t^{-1}])$ of $n \times n$ invertible matrices whose coefficients are Laurent polynomials in t and t^{-1}.

Definition 8.1.10 (Burau representation) The representation ρ_{B} is called the *Burau representation* of B_n.

This immediately leads to the question of the faithfulness of the Burau representation, long left open.

Proposition 8.1.11 (Non-faithfulness) *The Burau representation is faithful for $n = 3$, but not for $n \geqslant 5$.*

Sketch of the proof The case $n = 3$ is quite special, because of the link of B_3 with $\mathsf{SL}_2(\mathbb{Z})$. We can first extract from ρ_{B} a representation ρ'_{B} of dimension 2, in which σ_1 and σ_2 are sent to $\left(\begin{smallmatrix} -t & 0 \\ 1 & 1 \end{smallmatrix}\right)$ and $\left(\begin{smallmatrix} 1 & t \\ 0 & -t \end{smallmatrix}\right)$. For $t = -1$, this reduces to $\sigma_1 \mapsto \mathsf{a} := \left(\begin{smallmatrix} 1 & 0 \\ 1 & 1 \end{smallmatrix}\right)$ and $\sigma_2 \mapsto \mathsf{b} := \left(\begin{smallmatrix} 1 & -1 \\ 0 & 1 \end{smallmatrix}\right)$. However $\mathsf{SL}_2(\mathbb{Z})$ is generated by a and b subject to the relations $\mathsf{aba} = \mathsf{bab}$ and $(\mathsf{aba})^4 = 1$, hence a braid g

[2] Injective.

in B_3 satisfying $\rho_{\mathsf{B}}(g) = 1$ must be a power of $(\sigma_1\sigma_2\sigma_1)^4$. However, ρ'_{B} sends $(\sigma_1\sigma_2\sigma_1)^4$ to $\left(\begin{smallmatrix} t^6 & 0 \\ 0 & t^6 \end{smallmatrix}\right)$, which is not the identity.

For $n \geqslant 5$, the analysis of ρ_{B} in terms of the action of braids on the curves of a disk with n marked points leads to an explicit braid of B_5 whose image by ρ_{B} is the identity matrix (Moody, 1991). □

The faithfulness of ρ_{B} in the case of B_4 remains open.

Other linear representations of B_n have been discovered. The major result is the construction of a representation of dimension $n(n-1)/2$, based on the action of braids on pairs of points of a disk, and then the proof of its faithfulness (Bigelow, 2001; Krammer, 2000).

Proposition 8.1.12 (Lawrence–Krammer representation) *For every n, there exists a faithful representation of B_n in the group $\mathsf{GL}_{n(n-1)/2}(\mathbb{Z}[t, t^{-1}, q, q^{-1}])$.*

The construction is explicit, but the formulas are complicated and not presented here.

8.1.3 Applications to Cryptography

As the braid groups admit efficient computation procedures, we could imagine using them as basic structures for cryptographic schemes, in place of the more usual structures such as the groups $\mathbb{Z}/n\mathbb{Z}$ or the groups associated with elliptic curves. The specificity of B_n is to be non-Abelian (for $n \geqslant 3$), opening new possibilities. Typically, we exploit here the absence of an efficient solution for the conjugation problem of the groups B_n.

We present here one such scheme, in this case a 'Diffie–Hellman' type authentication scheme, proposed in (Ko et al., 2000). The general problem is as follows: Alice (the prover) wants to prove her identity to Bob (the verifier) by showing that she knows a certain *private key*, but without allowing an intruder who intercepts her communication with Bob (or even Bob!) to deduce anything at all about this secret key. Nevertheless, we want to divulge the information needed by Bob for his verification: this is what is known as the *public key*.

The keys and the data exchanged are here braids of a group B_{2n}. Denote B_n^{low} the subgroup of B_{2n} generated by $\sigma_1, ..., \sigma_{n-1}$ and B_n^{high} the subgroup generated by $\sigma_{n+1}, ..., \sigma_{2n-1}$. The public key is a pair of braids (p, p') in B_{2n} conjugate via a braid s in B_n^{low}: $p' = sps^{-1}$. Alice's private key is the braid s. The exchanges are thus as follows:

(i) Bob selects a random braid t in B_n^{high} and sends Alice the ('challenge') word $p'' := tpt^{-1}$.
(ii) Alice sends the response $p''' := sp''s^{-1}$.
(iii) Bob verifies the equality $p''' = tp't^{-1}$.

In a realistic implementation, we would use a collision-free *hash function* H in order to exchange the braids, that is, an injective coding of the braids by sequences of 0 and 1. If H is impossible to inverse in practice, it adds a supplementary measure of security.

A correct response from Alice convinces Bob that Alice knows the secret key s. Indeed, by construction, the braids s and t commute, and

$$tp't^{-1} = t(sps^{-1})t^{-1} = s(tpt^{-1})s^{-1} = sp''s^{-1}.$$

Moreover, this equality can only be produced if Alice is capable of finding the value of $sp''s^{-1}$ from the public keys p, p', and p''. The security of the protocol thus reposes on the impossibility to find, starting from the pair (p, p'), a braid s satisfying $p' = sps^{-1}$, that is, to resolve what is known as the *conjugacy search problem.*[3]

With the current state of the art, resolving the conjugacy search problem for a pair of braids of length 512 in B_{50} exceeds the capabilities of practical computation, at least for appropriately chosen initial braids.[4] For the moment, the diverse attacks proposed against braid-based protocols are quite easily foiled with a suitable choice of keys; however, formalizing proofs of security often remains a problem.[5] It must be said that after a promising start the enthusiasm for braid-based cryptography has fallen somewhat flat. See (Dehornoy, 2004) for a more complete presentation.

[3] This is different from the *conjugacy decision problem* for (p, p') (here we know that the braids p and p' are conjugate), a priori even more difficult: think of the difference between recognizing if an integer is prime, and find a factorization when it is not.

[4] In the same way, in schemes based on integers, such as RSA, the public keys cannot be arbitrary and must be carefully chosen to prevent attacks.

[5] A theoretical obstruction is that the groups B_n are not amenable so there do not exist good probability measures on the sets of braids.

8.2 Many Diverse Cousins

The braid groups can be defined using a variety of approaches: each of these leads to generalizations and diverse extensions. Due to a lack of space, we mention here only two of these.

8.2.1 The Surface Braid Groups

The approach of Chapter 1 defines a braid as an isotopy class of a family of arcs attached to n points of a plane, or, if you prefer, of a planar disk. A natural generalization consists of replacing the plane, or the disk, by an arbitrary surface.

Let Σ be a compact surface, and let $P_1, ..., P_n$ be fixed points of Σ.

Definition 8.2.1 (Surface geometric braid) An *n-strand geometric braid* on Σ is a family β of n arcs traced in $[0, 1]\times\Sigma$, starting from and arriving at $P_1, ..., P_n$, such that, for any t, the intersection of β and $\{t\} \times \Sigma$ contains exactly n points.

When Σ is a disk in the plane,[6] we find the geometric braids of Section 1.1.

As in Definition 1.1.7, we pass to the braids strictly speaking by taking the quotient with respect to a formalization of the notion of deformation.

Definition 8.2.2 (Surface braid) An *n-strand braid* on Σ is an isotopy class of geometric braids. The space of n-strand braids on Σ is denoted $B_n(\Sigma)$.

As long as Σ is path-connected, the space $B_n(\Sigma)$ obtained does not depend on the choice of base points.

The concatenation of paths induces on $B_n(\Sigma)$ a well-defined product, and $B_n(\Sigma)$ is thus equipped with a group structure: the group of n-strand braids on the surface Σ.

The properties of the groups $B_n(\Sigma)$ depend heavily on the topology of Σ, classified by its genus (number of handles), its number of boundary components, and its orientability. Note that, whatever the surface Σ, the group $B_1(\Sigma)$ is the fundamental group $\pi_1(\Sigma)$.

If Σ is of genus 0, with one boundary component, and is orientable, the surface is homeomorphic to the disk $\mathbb{D}$, and we are back to the groups B_n. We simply mention here two other examples of groups $B_n(\Sigma)$ illustrating the differences in other cases.

A natural case is that of the sphere $\mathbb{S}$: genus 0, no boundary, orientable. The embedding of the disk $\mathbb{D}$ in the sphere $\mathbb{S}$ by the adjunction of a point induces a homomorphism Φ of the group B_n (i.e. $B_n(\mathbb{D})$) into the group $B_n(\mathbb{S})$.

[6] Or the entire plane, but this is not compact.

Lemma 8.2.3 *The homomorphism Φ is surjective, and its kernel is the distinguished subgroup generated by $\tau := \sigma_1 \cdots \sigma_{n-2}\sigma_{n-1}^2\sigma_{n-2} \cdots \sigma_1$.*

The triviality of $\Phi(\tau)$ follows from the possibility to use the far side of the sphere to sneak all the way around, as suggested in Figure 8.3.

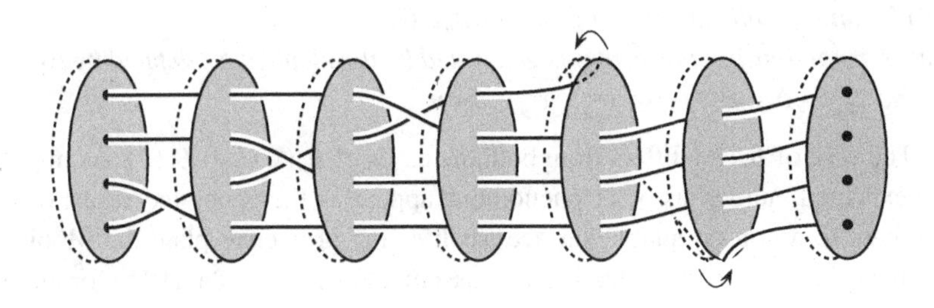

Figure 8.3 In a braid on $\mathbb{S}$, seen as a union of a visible face and a hidden face, passing by the hidden face creates additional isotopies; for example, $\sigma_1\sigma_2\sigma_3^2\sigma_2\sigma_1$ can be deformed into a trivial braid in $B_4(\mathbb{S})$: when the strand 1 has mounted to the position 4 on the visible face, we push it to the hidden face, where it can redescend without crossing anything, and reappear on the visible face in position 1; by similarly deforming the 'ascending' portion, we obtain a trivial geometric braid.

The group $B_n(\mathbb{S})$ is thus a quotient of the group B_n.

Using the same notation for the braids and their image by the homomorphism Φ, we deduce point (i) of the following result.

Proposition 8.2.4 (Group $B_n(\mathbb{S})$)
(i) *The group $B_n(\mathbb{S})$ is generated by $\sigma_1, ..., \sigma_{n-1}$ subjected to the Artin relations and the relation $\sigma_1 \cdots \sigma_{n-2}\sigma_{n-1}^2\sigma_{n-2} \cdots \sigma_1 = 1$.*
(ii) *In $B_n(\mathbb{S})$, for $n \geqslant 3$, the braids Δ_n and Δ_n^* are of order*[7] *4 and $2n$; the centre of $B_n(\mathbb{S})$ is $\{1, \Delta_n^2\}$.*

The presentation of (i) implies that $B_2(\mathbb{S})$ and $B_3(\mathbb{S})$ are finite, with 2 and 12 elements,[8] whereas $B_n(\mathbb{S})$ is infinite for $n \geqslant 4$.

To show how heavily $B_n(\Sigma)$ depends on Σ, we mention here the case of a torus $\mathbb{A}$, which can be seen as a sphere with two boundary components. We can then again define crossing braids analogous to the Artin generators σ_i. However, as $\mathbb{A}$ is not simply connected, there exist non-trivial loops in $\mathbb{A}$,

[7] Recall that an element g of a group G is said to be of *order* p if $g^p = 1$ and p is minimal with this property.

[8] In fact respectively $\mathbb{Z}/2\mathbb{Z}$ and the semidirect product $\mathbb{Z}/3\mathbb{Z} \rtimes \mathbb{Z}/4\mathbb{Z}$ which is not a direct product.

typically the loop γ that goes once around the central hole and induces an element of $B_n(\mathbb{A})$ not expressible in terms of the σ_i.

Proposition 8.2.5 (Group $B_n(\mathbb{A})$)
(i) *The group* $B_n(\mathbb{A})$ *is generated by n generators, that is,* $\sigma_1, ..., \sigma_{n-1}$, *subject to the Artin relations, plus* γ *subject to* $\sigma_1\gamma\sigma_1\gamma = \gamma\sigma_1\gamma\sigma_1$ *and* $\gamma\sigma_i = \sigma_i\gamma$ *for* $i \geqslant 2$.
(ii) *It can be realized as a subgroup of* $B_{n+1}(\mathbb{D})$.
(iii) *It is torsion-free; its centre is generated by the element* Δ_n *defined by* $\Delta_1 := \gamma$ *and* $\Delta_n := \Delta_{n-1}\sigma_{n-1}\cdots\sigma_1\gamma\sigma_1\cdots\sigma_{n-1}$ *for* $n \geqslant 2$.

The case of $B_n(\mathbb{A})$ differs from both the cases of $B_n(\mathbb{D})$ and $B_n(\mathbb{S})$. As might be expected, numerous other phenomena appear as the genus increases, or if we pass to a non-orientable surface, such as the projective plan, the Möbius strip, or the Klein bottle. Please see (Guaschi and Juan-Pineda, 2015) for more information.

8.2.2 The Artin–Tits Groups

The characterization of the braid groups and monoids by the Artin presentation invites us to consider the groups and monoids defined by syntactically similar presentations.

Definition 8.2.6 (Artin–Tits monoids and groups) A monoid (*resp.*, group) is said to be *Artin–Tits* if it admits a presentation (S, R) without any relations $x... = x...$ in R and where, for every $x \neq y$ in S, there exists at most one relation $x... = y...$, of the form $xyx... = yxy...$, with the two terms having the same length.

Examples 8.2.7 (Artin–Tits groups) The classical braid groups B_n, or $B_n(\mathbb{D})$, are Artin–Tits groups, the lengths of the relations being 2 or 3 according to the distance between the indices of the generators.

Similarly, the braid groups of the torus $B_n(\mathbb{A})$ are Artin–Tits groups, with the lengths of the relations this time 2, 3, and 4.

Other completely different examples are the free Abelian groups $\mathbb{Z}^n$ (with a relation $xy = yx$ for every pair of generators $x \neq y$), and the free groups (no relations).

We also mention[9]

$$\widehat{B}_4 := \langle \sigma_1, \sigma_2, \sigma_3 \mid \sigma_1\sigma_2\sigma_1 = \sigma_2\sigma_1\sigma_2,\ \sigma_2\sigma_3\sigma_2 = \sigma_3\sigma_2\sigma_3,\ \sigma_3\sigma_1\sigma_3 = \sigma_1\sigma_3\sigma_1 \rangle,$$

a variant of B_4 where the commutation relation between σ_1 and σ_3 is replaced by a relation of length 3 (this presentation was seen in Exercise 3.3.12).

[9] The usual name of this group is a 'type $\widetilde{\mathrm{A}}_2$ Artin–Tits group'.

A portion of the study of Chapter 3 can be extended: an Artin–Tits monoid $\langle S \mid R\rangle^+$ is cancellative and possesses conditional left and right lcms. However, $\langle S \mid R\rangle^+$ possesses common multiples only if the elements of S have one, and this is equivalent to the finiteness of the group[10] $\langle S \mid R \cup \{x^2 = 1 \mid x \in S\}\rangle$. In the case of B_n, the quotient is the symmetric group $\mathfrak{S}_n$, whereas for $\mathbb{Z}^n$, it is the group $\{0,1\}^n$, both finite. But this is neither the case for the free groups with rank $\geqslant 2$, nor for $\widehat{B}_4$, as Exercise 3.3.12 implies that $\sigma_1, \sigma_2, \sigma_3$ do not have a common multiple in $\widehat{B}_4^+$.

In the case where the elements of S have a common multiple (called 'spherical'), that is, where the associated Coxeter group is finite, Ore's theorem applies, the group $\langle S \mid R\rangle$ is a group of fractions for the monoid $\langle S \mid R\rangle^+$, and the analysis of Chapter 4 can be extended, providing a greedy normal form starting from the finite family of divisors of the lcm of the generators.

In the non-spherical case, the Artin–Tits group is not a group of fractions for the corresponding monoid,[11] and the relation between the monoid and the group is more delicate. A remarkable result (Paris, 2002) affirms nonetheless that, in every case, the monoid $\langle S \mid R\rangle^+$ can be embedded in the group $\langle S \mid R\rangle$: we can thus identify $\langle S \mid R\rangle^+$ with the submonoid of $\langle S \mid R\rangle$ generated by S.

We arrive now to the critical point. The word problem for any Artin–Tits monoid is easily resolved, either by the 'stupid' method of Proposition 3.1.10, or by the more efficient technique of Algorithm 3.3.17. However, in the non-spherical case, these solutions do not induce any solution for the word problem of the group.

Question 8.2.8 (Word problem) Is the word problem of the group $\langle S \mid R\rangle$ decidable for every Artin–Tits presentation (S, R)?

Although extremely frustrating, this question currently remains open. In addition to the spherical cases, the decidability is established for numerous special cases – including for the free groups, or for $\widehat{B}_4$ – but there exist simple examples for which we can say *nothing*, for example the variant $\widehat{B}_5$ of B_5 where the relations $\sigma_1\sigma_3 = \sigma_3\sigma_1$ and $\sigma_2\sigma_4 = \sigma_4\sigma_2$ are replaced by $\sigma_1\sigma_3\sigma_1 = \sigma_3\sigma_1\sigma_3$ and $\sigma_2\sigma_4\sigma_2 = \sigma_4\sigma_2\sigma_4$.

8.2.3 A Conjecture

We conclude this text with a conjecture, namely a 'candidate solution' to the word problem of the Artin–Tits groups, based on what is known as the

[10] This group is known as the 'Coxeter group associated with (S, R)'.

[11] Reflect on the case of a free group with rank $\geqslant 2$: a reduced word has no reason in general to be the quotient of two positive words...

reduction of multifractions. We suppose in what follows that (S, R) is a fixed Artin–Tits presentation, and we use M and G to denote $\langle S \mid R\rangle^+$ and $\langle S \mid R\rangle$.

Definition 8.2.9 (Multifraction) A *multifraction* is a finite sequence of elements of M.

We denote M^* the set of multifractions. It is in general false that every element of the group G can be written as $a_1a_2^{-1}$ with a_1, a_2 in M, but such an element can always be written as $a_1a_2^{-1}a_3a_4^{-1}...$ with the a_i in M. In other words, the mapping

$$E : (a_1, ..., a_p) \mapsto a_1a_2^{-1}a_3a_4^{-1}... \tag{8.3}$$

is a surjection of M^* onto G: the multifractions represent the elements of G. We use $\underline{a}, \underline{b}, ...$ as a generic notation for the multifractions. The length $|\underline{a}|$ of $\underline{a}$ is the number of terms in the sequence, namely p for $\underline{a} = (a_1, ..., a_p)$. In this context, the word problem of G reduces to recognizing which multifractions represent 1 in G. It is convenient, and coherent with the definition of E, to associate with a multifraction $(a_1, a_2, ...)$ the alternating path $\xrightarrow{a_1} \xleftarrow{a_2} \xrightarrow{a_3} \xleftarrow{a_4} \cdots$

We now introduce a rewriting relation on M^* which can be seen as extending the reduction of Definition 2.3.35, and is equivalent in the case of free groups. It concerns rewriting a multifraction into multifractions representing, in the sense of (8.3), the same element of G.

Recall that, for a, b in M, we denote $a\backslash b$ the (unique) element such that $a(a\backslash b)$ is the right lcm of a and b, if such exists; and denote a/b the (unique) element such that $(b/a)a$ is the left lcm of a and b, if such exists.

Definition 8.2.10 (Reduction) For $i \geqslant 1$ and s in S, the multifraction $\underline{b}$ is said to be obtained by *reducing s to the level i in the multifraction $\underline{a}$* if $|\underline{b}| = |\underline{a}|$, $b_k = a_k$ for $k \neq i{-}1, i, i{+}1$, and

$b_{i-1} = a_{i-1}(a_i\backslash s)$, $b_i = s\backslash a_i$, and $sb_{i+1} = a_{i+1}$ for i even,
$b_{i-1} = (s/a_i)a_{i-1}$, $b_i = a_i/s$, and $b_{i+1}s = a_{i+1}$ for i odd $\geqslant 3$,
$b_is = a_i$ and $b_{i+1}s = a_{i+1}$ for $i = 1$.

(See Figure 8.4.)

Notwithstanding these 'horrifying' formulas, the intuitive idea is simple: reducing s to the level i means removing s from the level $i{+}1$ and pushing it to the level $i{-}1$ passing through the level i. More precisely, we divide a_{i+1} by s, on the left or on the right according to the parity, and push s through a_i: if s also divides a_i, we take the quotient, but if, more generally, s and a_i have an lcm, then a_i is replaced by $s\backslash a_i$ or a_i/s (depending on the parity), and the 'remainder' $a_i\backslash s$ or s/a_i is integrated into a_{i-1}.

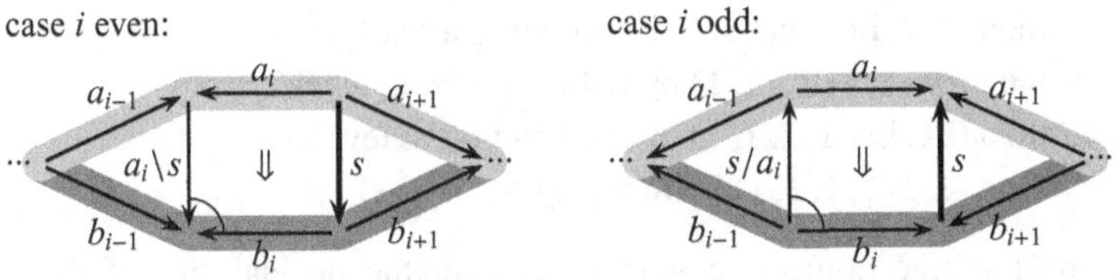

case i even: case i odd:

Figure 8.4 Reduction of multifractions: for i even, we left-divide a_i by s, push s through a_i by using the right lcm, and right-multiply a_{i-1} by the complement $a_i \backslash s$; for i odd, right and left are exchanged. Here, $\underline{a}$ corresponds to the path on top, and $\underline{b}$ to the one on the bottom. The small arcs indicate that the rectangle corresponds to the formation of an lcm.

If $\underline{a}$ reduces to $\underline{b}$, the diagrams of Figure 8.4 show that $\underline{a}$ and $\underline{b}$ represent the same element of the group G. It is not very difficult, in the spherical case in particular, to establish that the reduction is convergent: for any multifraction $\underline{a}$, there exists a unique multifraction $\mathsf{red}(\underline{a})$ such that every sequence of reductions from $\underline{a}$ leads, in a finite number of steps, to $\mathsf{red}(\underline{a})$, and that $\underline{a}$ represents 1 in the group G if and only if $\mathsf{red}(\underline{a})$ is a trivial sequence $(1, 1, ..., 1)$. In the spherical case, the multifractions $\mathsf{red}(\underline{a})$ all have the form $(a_1, a_2, 1, ..., 1)$, and we see once again that every element of G is a fraction $a_1 a_2^{-1}$.

In the general case, and notably in the case of the variants $\widehat{B_4}$ of B_4 and $\widehat{B_5}$ of B_5, the reduction is not convergent: there exist multifractions representing the same element of the group and whose reductions do not converge to a unique multifraction. Nevertheless, for the word problem of the group, we only need to know if a multifraction represents the element 1, and we leave to the sagacity of the reader[12] the following hypothesis, supported by partial results and a few billion random tests.

Conjecture 8.2.11 (Semiconvergence) *Any multifraction representing* 1 *in an Artin–Tits group reduces to a trivial multifraction* $(1, ..., 1)$.

Example 8.2.12 (Semiconvergence) Consider the group $\widehat{B_4}$, and the multifraction $\underline{a} := (1, \mathtt{c}, \mathtt{aba})$. Then $\underline{a}$ reduces to $(\mathtt{ac}, \mathtt{ca}, \mathtt{ba})$ and to $(\mathtt{bc}, \mathtt{cb}, \mathtt{ab})$, irreducible and representing the same element of $\widehat{B_4}$. It is easy to see that the quotient $(\mathtt{ac}, \mathtt{ca}, \mathtt{ba}, \mathtt{ab}, \mathtt{cb}, \mathtt{bc})$ represents 1 in $\widehat{B_4}$. The conjecture thus predicts that this multifraction must reduce to $(1, ..., 1)$ ... and this is indeed the case: the multifraction $(\mathtt{ac}, \mathtt{ca}, \mathtt{ba}, \mathtt{ab}, \mathtt{cb}, \mathtt{bc})$

gives $(\mathtt{ac}, \mathtt{cac}, \mathtt{b}, 1, \mathtt{cb}, \mathtt{bc})$ by reducing $\mathtt{ab}$ at level 3,
then $(\mathtt{ac}, \mathtt{cac}, \mathtt{bcb}, 1, 1, \mathtt{bc})$ by reducing $\mathtt{cb}$ at level 4,
then $(\mathtt{ac}, \mathtt{cac}, \mathtt{bcb}, \mathtt{bc}, 1, 1)$ by reducing $\mathtt{bc}$ at level 5,

[12] and to the available computing power: with the reversing procedure of Definition 3.3.13, it is very easy to implement multifraction reduction.

then $(1, \mathtt{c}, \mathtt{bcb}, \mathtt{bc}, 1, 1)$ by reducing $\mathtt{ac}$ at level 1,
then $(\mathtt{bc}, 1, 1, \mathtt{bc}, 1, 1)$ by reducing $\mathtt{cbc}$ at level 2,
then $(\mathtt{bc}, \mathtt{bc}, 1, 1, 1, 1)$ by reducing $\mathtt{bc}$ at level 3,
then $(1, 1, 1, 1, 1, 1)$ by reducing $\mathtt{bc}$ at level 1.

A proof of the conjecture would establish the decidability of the word problem for the Artin–Tits groups. To give credit to this approach and to motivate the ardour of the researchers, the author (and his family, now that he has sadly left us...) offers a reward of 1000 euros to the first person to either prove the conjecture, or, on the contrary, provide a counterexample, that is, to find an Artin–Tits group G and a multifraction $\underline{a}$ representing 1 in G and nevertheless not reducing to $(1, ..., 1)$.

Anyone up for the challenge???

9
Solutions to the Exercises

9.1 Exercises of Chapter 1

Exercise 1.1.22 (Shifts)
(i) Following the same schema as in Proposition 1.1.20, show that the addition of a first non-braided strand defines, for every n, an embedding shift_n (a 'shift') of the space B_n in B_{n+1}. Is this embedding surjective? Anticipating Definition 1.2.8, what is the image of σ_i by shift_n?
(ii) Show that the embeddings shift_n are compatible between themselves, and hence deduce that they induce an embedding shift of B_∞ into itself.

Solution
(i) The construction is symmetric with that of Proposition 1.1.20: instead of adding an $(n+1)$th trivial strand to the right of the n strands, we add it to the left. In this case, the technical condition to avoid any inopportune incursion of strands is to limit ourselves to braids trivial outside of the prism $[0, 1]\times]1.5, +\infty[\times\mathbb{R}$.

The embedding shift_n is not surjective: by construction, σ_1 cannot belong to the image of shift_n. The image of σ_i by shift_n is σ_{i+1}, and, symmetrically, that of $\overline{\sigma_i}$ is $\overline{\sigma_{i+1}}$.
(ii) The two operations of adding a strand on the right by e_n and adding a strand on the left by shift_n commute: for any geometric braid β satisfying ad hoc 'prism constraints', we have $\mathsf{shift}_n(e_n(\beta)) \approx e_n(\mathsf{shift}_n(\beta))$. Hence, if g belongs to B_n, there is no ambiguity in defining $\mathsf{shift}(g)$ by $\mathsf{shift}(g) := \mathsf{shift}_n(g)$. □

Exercise 1.3.2 (Closure) Show that the closures of two isotopic braid diagrams are diagrams of isotopic links, but that non-isotopic braid diagrams can have closures corresponding to isotopic link diagrams. [Hint: Compare the closures of the braids of Example 1.3.5.]

Solution An isotopy between subdiagrams can be trivially extended to an isotopy of diagrams, hence $\beta \approx \beta'$ implies $\widehat{\beta} \approx \widehat{\beta'}$.

However, the geometric braids σ_1 and $\overline{\sigma_1}$ are not isotopic as their linking numbers are distinct, respectively $1/2$ and $-1/2$, but their closures are isotopic, as suggested by the following diagrams:

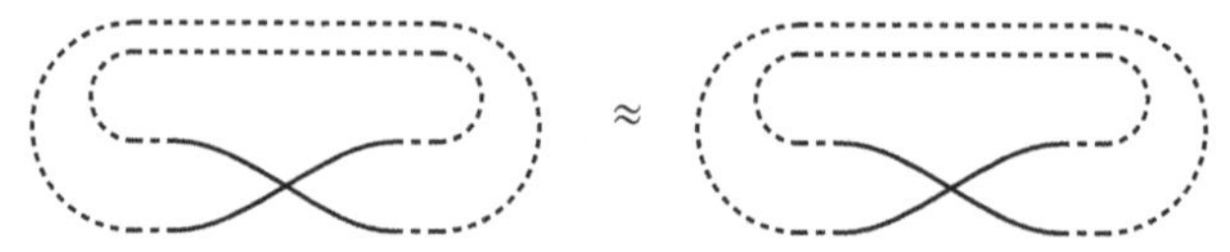

(these two diagrams are both isotopic to a simple loop, i.e. to a trivial knot). □

Exercise 1.3.8 (Symmetry of the linking number) Show that, for any geometric braid β and every i, j, $\lambda_{j,i}(\beta) = \lambda_{i,j}(\beta)$.

Solution By definition, for $i < j$, we have $\theta_{j,i}(t) = \theta_{i,j}(t) + \pi$ for any t. We thus have equality of the derivatives $\theta'_{j,i} = \theta'_{i,j}$, hence $\lambda_{j,i}(\beta) = \lambda_{i,j}(\beta)$ by integration. □

9.2 Exercises of Chapter 2

Exercise 2.1.9 (Submonoids)

(i) Show that a subset M' of a monoid M is a submonoid of M if and only if M' contains 1 and is closed under the product (the product of two elements of M' belongs to M').

(ii) Deduce that a subset X of M generates a monoid M if and only if every element of $M \setminus \{1\}$ can be written as a finite product of elements of X.

Solution

(i) Suppose M' is a submonoid of of M, that is, M' equipped with the operations induced by those of M is a monoid. Denote $\cdot'$ the restriction of $\cdot$ to M'. By hypothesis, $(M', \cdot', 1)$ is a monoid. Hence, again by hypothesis, 1 is in M'. Moreover, since $\cdot'$ is a binary operation on M, $a \cdot' b$ is in M' for any a, b in M': this also means $a \cdot b$ in M'.

Conversely, suppose 1 is in M' and M' is closed under the product. Then, by hypothesis, the restriction $\cdot'$ of $\cdot$ to M' is a binary operation defined on all of M'. Then, for any a, b, c in M', we have

$$a \cdot' (b \cdot' c) = a \cdot (b \cdot c) = (a \cdot b) \cdot c = (a \cdot' b) \cdot' c,$$

hence $\cdot'$ is associative. Moreover, for any a in M', we have

$$a \cdot' 1 = a \cdot 1 = a \quad \text{and} \quad 1 \cdot' a = 1 \cdot a = a,$$

so $(M', \cdot', 1)$ is a monoid, hence, by definition, a submonoid of M.

(ii) Suppose X generates M, and a is an element of $M \setminus \{1\}$. By hypothesis, a belongs to the smallest submonoid including X, which, by (i), is the union of $\{1\}$ and the set of finite products of elements of X. Hence a is a finite product of elements of X. Conversely, suppose every element of $M \setminus \{1\}$ is a finite product of elements of X. Then every element of $M \setminus \{1\}$ is an element of the closure of X under the product, hence, by (i), an element of the smallest submonoid of M including X: this means X generates M. □

Exercise 2.1.19 (Shift)
(i) Show that the embedding shift_n of B_n into B_{n+1} defined in Exercise 1.1.22 is a homomorphism.
(ii) Deduce that the embedding shift is a non-surjective endomorphism of B_∞ into itself, characterized by the fact that it sends σ_i to σ_{i+1} for every i.

Solution
(i) Let D_1 and D_2 be n-strand braid diagrams. The figure below shows that shift_n (here for $n = 3$) is a homomorphism of B_n into B_{n+1}:

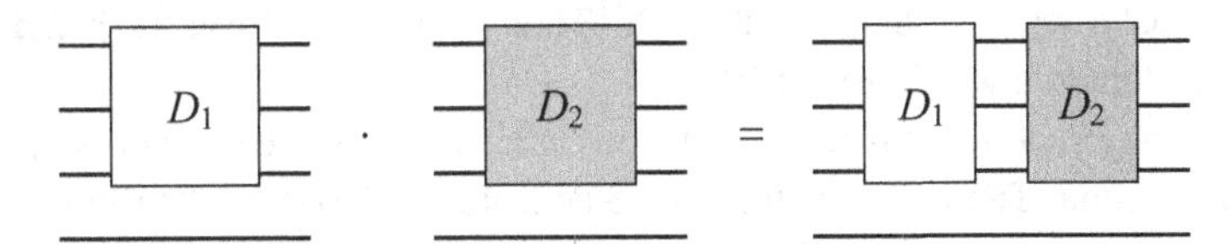

(and the symmetric adding of a strand to the right – in this figure 'on top' – corresponds to the fact that e_n is also a homomorphism of B_n into B_{n+1}).
(ii) As in Exercise 1.1.22, the two operations of adding a strand on the right by e_n and adding a strand on the left by shift_n commute: in terms of braid diagrams, we have the relation $\mathsf{shift}_n(e_n(D)) \simeq e_n(\mathsf{shift}_n(D))$. Thus, if g belongs to B_n, it is not ambiguous to define $\mathsf{shift}(g)$ by $\mathsf{shift}(g) := \mathsf{shift}_n(g)$. By (i), the mapping shift is compatible with the product, that is, it is an endomorphism of B_∞. By construction, shift sends σ_i to σ_{i+1} for every i, and, as the group B_∞ is generated by the braids σ_i, it is the only endomorphism with this property. This endomorphism is not surjective, as σ_1 is not in the image of shift. □

Exercise 2.1.27 (Reversing) What transformation of geometric braids corresponds to a reversing?

Solution In terms of geometric braids, a reversing corresponds to a rotation of angle $\pi/2$ about the vertical axis passing through the median point $(1/2, (n+1)/2, 0)$ – while an inversion (operation $\beta \mapsto \beta^{-1}$) corresponds to a symmetry with respect to the plane $\{1/2\}\times\mathbb{R}^2$. In both cases, the fact that we have an antiautomorphism corresponds to an inversion of the order of the generators; however, the reversing of σ_i is σ_i, whereas the inversion of σ_i is $\overline{\sigma_i}$, as suggested by the following figures:

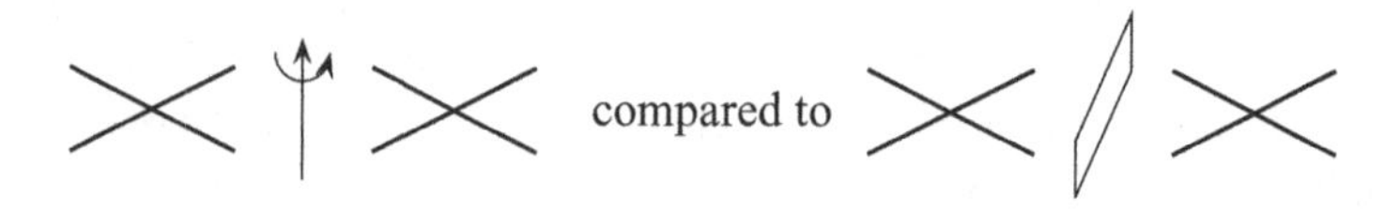

□

Exercise 2.1.28 (Coxeter presentation of $\mathfrak{S}_3$) Show that the quotient G of the group B_3 by the congruence generated by the supplementary relations $\sigma_1^2 = \sigma_2^2 = 1$ is (isomorphic to) the symmetric group $\mathfrak{S}_3$. [Hint: Use the relations to show that the cardinality of G is at most 6.]

Solution Let G be the group $\langle \mathsf{a}, \mathsf{b} \mid \mathsf{a}^2 = \mathsf{b}^2 = 1, \mathsf{aba} = \mathsf{bab}\rangle$. Let $\equiv$ be the congruence generated by the relations of the presentation, including the free group relations $\mathsf{aA} = 1$, etc. First of all, $\mathsf{a}^2 \equiv 1$ implies $\mathsf{A} \equiv \mathsf{a}$ and, analogously, $\mathsf{b}^2 \equiv 1$ implies $\mathsf{B} \equiv \mathsf{b}$. Thus every word over $\{\mathsf{a}, \mathsf{b}, \mathsf{A}, \mathsf{B}\}$ is $\equiv$-equivalent to a word over $\{\mathsf{a}, \mathsf{b}\}$.

Next, any such word is $\equiv$-equivalent to at least one of the six words ε, a, b, ab, ba, or aba. The reason is that, if we concatenate a or b to the right of any of these six words, the resulting word is again $\equiv$-equivalent to one of these six words. For example, $\mathsf{ab} \cdot \mathsf{b} \equiv \mathsf{a}$, and $\mathsf{aba} \cdot \mathsf{b} \equiv \mathsf{aaba} \equiv \mathsf{ba}$. Hence, $\equiv$ has at most six equivalence classes, and G has at most six elements.

Showing that G has exactly six elements, and identifying G requires showing that the six words ε, a, b, ab, ba, and aba are two-by-two non $\equiv$-equivalent. For this we can use a $\equiv$-invariant. Define $I\colon \{\mathsf{a}, \mathsf{b}\} \to \mathfrak{S}_3$ by $I(\mathsf{a}) = (1, 2)$ and $I(\mathsf{b}) = (2, 3)$. The transpositions $(1, 2)$ and $(2, 3)$ satisfy the relations of the presentation: we have $(1, 2)^2 = (2, 3)^2 = \mathsf{id}$ and $(1, 2)(2, 3)(1, 2) = (2, 3)(1, 2)(2, 3)$, thus I induces a homomorphism of G into $\mathfrak{S}_3$. We then verify that the images by I of the words ε, a, b, ab, ba, and aba are distinct permutations of $\{1, 2, 3\}$, so these words are two-by-two non $\equiv$-equivalent. Hence G has exactly six elements. As I is surjective, it is also injective, and I is an isomorphism of G onto $\mathfrak{S}_3$. In other words,

$$\langle \mathsf{a}, \mathsf{b} \mid \mathsf{a}^2 = \mathsf{b}^2 = 1, \mathsf{aba} = \mathsf{bab}\rangle$$

is a presentation of the symmetric group $\mathfrak{S}_3$. □

Exercise 2.2.1 (Neutral element) Show that a monoid has only a single neutral element.

Solution If M is a monoid, and if 1 and $1'$ are neutral elements of M, the hypotheses imply $1 = 1 \cdot 1' = 1'$. □

Exercise 2.2.6 (Monoid) Show that $(S^*, \cdot, \varepsilon)$ is a monoid (Proposition 2.2.5).

Solution For any S-word w and for $1 \leqslant i \leqslant |w|$, denote $w[i]$ the ith letter of w, that is, the ith element of w seen as a sequence of length $|w|$, or again, as a mapping of $\{1, ..., |w|\}$ into S. Let u, v, w be words over S. As the addition of integers is associative, we first have

$$|u(vw)| = |u| + |uv| = |u| + (|v| + |w|) = (|u| + |v|) + |w| = |uv| + |w| = |(uv)w|,$$

hence $u(vw)$ and $(uv)w$ are of the same length.

Next, for $1 \leqslant i \leqslant |u|$, we find

$$(u(vw))[i] = u[i] = (uv)[i] = ((uv)w)[i].$$

Similarly, for $|u| + 1 \leqslant i \leqslant |u| + |v|$,

$$(u(vw))[i] = (vw)[i - |u|] = v[i - |u|] = (uv)[i].$$

Then, for $|u| + |v| + 1 \leqslant i \leqslant |u| + |v| + |w|$,

$$(u(vw))[i] = (vw)[i - |u|] = w[(i - |u|) - |v|] = w[i - |uv|] = (uv)w[i].$$

Hence the words $u(vw)$ and $(uv)w$ always coincide, and consequently concatenation is associative.

Finally, let w be an arbitrary S-word. Then the length of the word $w\varepsilon$ is $|w| + |\varepsilon|$, or $|w|$. Moreover, for $1 \leqslant i \leqslant |w|$, we have, by definition, $(w\varepsilon)[i] = w[i]$, so the words $w\varepsilon$ and w coincide. A similar verification shows that εw and w are also coincident. Consequently, ε is the neutral element for concatenation, and $(S^*, \cdot, \varepsilon)$ is a monoid. □

Exercise 2.2.7 (Generating subsets of S^*) Show that a subset X of a monoid S^* is a generating set S^* if and only if X includes S. In particular, S generates S^*.

Solution An S-word of length ℓ, say $s_1 \cdots s_\ell$, is the product of the ℓ words $s_1, ..., s_\ell$, of length 1. Hence, as soon as X contains S, any nonempty word of S^* is a finite product of elements of X, and consequently, X generates S^*.

Conversely, suppose X generates M, and let s be an element of S. There must exist $w_1, ..., w_m$ in X satisfying $s = w_1 \cdots w_m$, hence, in particular, $1 = |w_1| + \cdots + |w_m|$: the only possibility is that all the words w_i are empty except one, say w_{i_0}, which is equal to s. But if $w_{i_0} = s$, then $s \in X$, and S is contained in X. □

Exercise 2.2.9 (Free monoid) Prove Proposition 2.2.8: 'For any mapping ϕ of S into a monoid M, there exists a unique homomorphism $\widehat{\phi}$ of S^* into M extending ϕ, defined by $\widehat{\phi}(\varepsilon) := 1$ and, for $w = s_1 \cdots s_\ell$ with $s_1, ..., s_\ell$ in S, by $\widehat{\phi}(w) := \phi(s_1) \cdots \phi(s_\ell)$.'

Solution The definition of $\widehat{\phi}$ makes sense since a word in S^* admits only one decomposition as a product of letters in S: if $w = s_1 \cdots s_\ell = t_1 \cdots t_m$ with $s_1, ..., t_m$ in S, then necessarily $\ell = |w| = m$, and $s_i = w(i) = t_i$ for $1 \leqslant i \leqslant \ell$. Next, $\widehat{\phi}$ extends ϕ, and is a homomorphism since, for $u = s_1 \cdots s_\ell$ and for $v = t_1 \cdots t_m$, we find

$$\begin{aligned}\widehat{\phi}(uv) &= \widehat{\phi}(s_1 \cdots s_\ell t_1 \cdots t_m) = \phi(s_1) \cdots \phi(s_\ell)\phi(t_1) \cdots \phi(t_m) \\ &= (\phi(s_1) \cdots \phi(s_\ell))\phi((t_1) \cdots \phi(t_m)) = \widehat{\phi}(u)\widehat{\phi}(v),\end{aligned}$$

and, similarly,

$$\widehat{\phi}(u\varepsilon) = \widehat{\phi}(u) = \widehat{\phi}(u)\widehat{\phi}(\varepsilon) \quad \text{and} \quad \widehat{\phi}(\varepsilon u) = \widehat{\phi}(u) = \widehat{\phi}(\varepsilon)\widehat{\phi}(u).$$

Finally, $\widehat{\phi}$ is unique because, if ϕ^* extends ϕ on S and is a homomorphism, it must obey the relation $\phi^*(w) := \phi(s_1) \cdots \phi(s_\ell)$. □

Exercise 2.2.13 (Quotient monoid) Prove Lemma 2.2.12: 'If M is a monoid and $\equiv$ an equivalence relation on M, then the mapping sending an element to its class for $\equiv$ induces a quotient monoid structure on $M/\equiv$ if and only if $\equiv$ is a congruence on M'.

Solution Denote $[a]$ the class of an element a of M with respect to $\equiv$. Suppose $\equiv$ induces a quotient monoid structure $(M/\equiv, *)$, meaning, by definition, that the mapping $\phi\colon a \mapsto [a]$ defines a homomorphism of M with values in $(M/\equiv, *)$. The product $*$ on $M/\equiv$ must thus be satisfied:

$$[a] * [b] = [ab]. \tag{9.1}$$

For this equality to make sense, $[ab]$ must be the same when a and b are replaced by other elements in their same equivalence classes, so if $a' \equiv a$ and $b' \equiv b$, then also $a'b' \equiv ab$. In other words, $\equiv$ must be a congruence.

Conversely, suppose $\equiv$ is a congruence on M. Then (9.1) provides a binary operation well-defined on $M/\equiv$. Then, for any a, b, c in M, we have

$$([a] * [b]) * [c] = [ab] * [c] = [abc] = [a] * [bc] = [a] * ([b] * [c]),$$

$$[a] * [1] = [a1] = [a] = [1a] = [1] * [a],$$

hence $(M/\equiv, *, [1])$ is a monoid. □

Exercise 2.2.15 (Image) Show that, if M, M' are monoids, the existence of a surjective homomorphism of M onto M' is equivalent to that of a congruence $\equiv$ on M such that M' is isomorphic to the quotient $M/\equiv$.

Solution Suppose ϕ is a surjective homomorphism of M onto M'. Let $\equiv$ be the binary relation on M defined by $a \equiv a' \Leftrightarrow \phi(a) = \phi(a')$. Then $\equiv$ is a congruence on M (show this!), and ϕ induces a well-defined mapping $\overline{\phi}$ of the quotient $M/\equiv$ into M' since two elements of the same class have the same image. By construction $\overline{\phi}$ is a homomorphism, and is bijective (check this!).

Conversely, suppose $\equiv$ is a congruence on M. Then the mapping associating with each element of M its class for $\equiv$ is, by definition, a surjective homomorphism of M onto the quotient monoid $M/\equiv$. Thus there exists a surjective homomorphism of M onto $M/\equiv$ and hence, onto any monoid isomorphic to $M/\equiv$. □

Exercise 2.2.21 Prove Lemma 2.2.20: 'If M is a monoid, then for any subset R of $M \times M$, there exists a smallest congruence including R, namely the relation "there exists an R-derivation from a to a'".'

Solution Denote $a \equiv_R a'$ for 'there exists an R-derivation from a to a''. Then $\equiv_R$ is reflexive, since for any a, the sequence (a) is an R-derivation from a to a. It is symmetric, as if $(a_0, ..., a_m)$ is an R-derivation from a to a', then $(a_m, ..., a_0)$ is an R-derivation from a' to a. Finally, $\equiv_R$ is transitive since, if $(a_0, ..., a_m)$ is an R-derivation of a to a' and if $(b_0, ..., b_p)$ is an R-derivation from a' to a'', then $a_m = a' = b_0$, and $(a_0, ..., a_m, b_1, ..., b_p)$ is an R-derivation from a to a''. Furthermore, $\equiv_R$ is compatible with left- and right-multiplication, as if $(a_0, ..., a_m)$ is an R-derivation from a to a', then for any x, y in M, the sequence $(xa_0y, ..., xa_my)$ is an R-derivation from xay to $xa'y$. Hence $\equiv_R$ is a congruence on M. Finally, if (b, b') belong to R, then (b, b') is an R-derivation from b to b': $\equiv_R$ contains R (seen as a set of pairs).

Conversely, suppose $\sim$ is a congruence on M including R, and suppose $(a_0, ..., a_m)$ is an R-derivation from a to a'. For each i, there exists (b, b') in R and x, y in M satisfying $a_{i-1} = xby$, $a_i = xb'y$, or vice versa. Since $\sim$ includes R and is symmetric, $b \sim b'$, and as $\sim$ is compatible with the product, we also have $xby \sim xb'y$, thus $a_{i-1} \sim a_i$. Since $\sim$ is transitive, we deduce $a \sim a'$. Hence $\sim$ includes $\equiv_R$. □

Exercise 2.2.22 (Generated congruence) Show that, in the case of Example 2.2.19, the congruence generated by the pair (ab, ba) is the relation 'having the same numbers of the letters a and b'.

Solution Denote $\equiv$ the generated congruence, and $|w|_s$ the number of letters s in the word w. Then $|\mathsf{ab}|_\mathsf{a} = |\mathsf{ba}|_\mathsf{a} = 1$, and $|\mathsf{ab}|_\mathsf{b} = |\mathsf{ba}|_\mathsf{b} = 1$. From this, by induction on the length of a derivation, $w \equiv w'$ implies $|w|_\mathsf{a} = |w'|_\mathsf{a}$ and $|w|_\mathsf{b} = |w'|_\mathsf{b}$.

Conversely, we show by induction on $|w|$ the equivalence $w \equiv \mathsf{a}^{|w|_\mathsf{a}}\mathsf{b}^{|w|_\mathsf{b}}$ for any word w over $\{\mathsf{a}, \mathsf{b}\}$. Hence the conjunction of $|w|_\mathsf{a} = |w'|_\mathsf{a}$ and $|w|_\mathsf{b} = |w'|_\mathsf{b}$ implies

$$w \equiv \mathsf{a}^{|w|_\mathsf{a}}\mathsf{b}^{|w|_\mathsf{b}} \equiv \mathsf{a}^{|w'|_\mathsf{a}}\mathsf{b}^{|w'|_\mathsf{b}} \equiv w'.$$

□

Exercise 2.2.26 (Presented monoids)
(i) Let S be arbitrary, and R the set of relations $s = 1$ for s in S. Show that $\langle S \mid R\rangle^+$ is the trivial monoid of one element.
(ii) Let S be a nonempty set, and R the set of relations $s = t$ for s, t in S. Show that $\langle S \mid R\rangle^+$ is isomorphic to $(\mathbb{N}, +)$. [Hint: Show that the congruence on S^* generated by R is the relation 'have the same length'.]
(iii) Show that $\langle \mathsf{a}, \mathsf{b} \mid \mathsf{ab} = \mathsf{ba}\rangle^+$ is the direct product of two copies of the monoid $(\mathbb{N}, +)$. [Hint: Use Exercise 2.2.22 and show that the mapping $w \mapsto (\|w\|_\mathsf{a}, \|w\|_\mathsf{b})$, where $\|w\|_s$ denotes the number of s in w, induces a bijection of $\{\mathsf{a}, \mathsf{b}\}^*/\equiv$ onto $\mathbb{N}^2$, and then that this bijection is a homomorphism when $\mathbb{N}^2$ is equipped with addition coordinate by coordinate.]
(iv) Show that the monoid $\langle \mathsf{a} \mid \mathsf{a}^p = 1\rangle^+$ is isomorphic to the cyclic group $\mathbb{Z}/p\mathbb{Z}$. [Hint: Show that the congruence $\equiv$ on $\{\mathsf{a}\}^*$ generated by $(\mathsf{a}^p, \varepsilon)$ is the relation $|w| = |w'| \pmod{p}$.] Describe in the same manner the monoid $\langle \mathsf{a} \mid \mathsf{a}^p = \mathsf{a}^q\rangle^+$ for fixed $p, q \geqslant 0$.

Solution
(i) Let $\equiv$ be the congruence on S^* generated by $S \times \{\varepsilon\}$, that is, the relations $s = 1$ for s in S. An induction on $|w|$ shows $w \equiv \varepsilon$ for every S-word w. The relation $\equiv$ thus has only one class, so $\langle S \mid S^* \times \{\varepsilon\}\rangle^+ = \{1\}$.
(ii) Let $\equiv$ be the generated congruence, and let a be a fixed letter of S. For every word w, by induction on $|w|$, we have $w \equiv \mathsf{a}^{|w|}$. Hence the relation $w \equiv w'$ is equivalent to $|w| = |w'|$. The mapping associating p in $\mathbb{N}$ with a^p is surjective onto $\langle S \mid R\rangle^+$ since every word is $\equiv$-equivalent to a word a^p, and it is injective as two words with different lengths are not $\equiv$-equivalent.
(iii) Let $\equiv$ be the congruence on $\{\mathsf{a}, \mathsf{b}\}^*$ generated by the unique pair $(\mathsf{ab}, \mathsf{ba})$. We saw in Exercise 2.2.22 that $w \equiv w'$ is satisfied if and only if w and w' have the same number of a's and b's. So let ϕ be the mapping $w \mapsto (|w|_\mathsf{a}, |w|_\mathsf{b})$, where $|w|_s$ denotes the number of s in w. Then ϕ induces a bijection of $\{\mathsf{a}, \mathsf{b}\}^*/\equiv$ onto $\mathbb{N}^2$. Next, $\phi(uv) = (|uv|_\mathsf{a}, |uv|_\mathsf{b}) = (|u|_\mathsf{a}, |u|_\mathsf{b}) + (|v|_\mathsf{a}, |v|_\mathsf{b})$, thus ϕ is a homomorphism and hence an isomorphism with image $(\mathbb{N}, +) \times (\mathbb{N}, +)$.

(iv) Let $\equiv$ be the congruence on $\{a\}^*$ generated by the single pair (a^p, ε). By induction on the length of a derivation, we show that $w \equiv w'$ implies $|w| = |w'| \pmod{p}$. Conversely, any word w is $\equiv$-equivalent to a^q, so q is the unique integer of $\{0, 1, ..., p-1\}$ congruent to $|w| \bmod p$.

Then the mapping ϕ associating q with a^q for $0 \leqslant q < p$ induces an isomorphism of $\langle a \mid a^p = 1\rangle^+$ with the cyclic group $\mathbb{Z}/p\mathbb{Z}$.

Similarly let $\equiv$ be the congruence on $\{a\}^*$ generated by the couple (a^p, a^q), where we suppose $q > p$. By induction on the length of a derivation, we can show that $w \equiv w'$ implies $|w| = |w'| \bmod (q-p)$. Because of their length, the words ε, a, ..., a^{p-1} are isolated, that is, they are the only elements in their respective equivalence class. Indeed, if a word w is of length strictly inferior to p, and hence to q, no relation applies to w. Moreover, any word w of length $\geqslant p$ is $\equiv$-equivalent to a^{p+r}, where r is the unique integer of $\{0, 1, ..., (q-p-1)\}$ congruent to $|w| \bmod (q-p)$. The monoid $\langle S \mid R\rangle^+$ thus has q elements, the classes of ε, a, ..., a^{q-1}, and the multiplication table is given by $[a^r] \cdot [a^s] := [a^{f(r+s)}]$ where

$$f(t) := \begin{cases} t & \text{for } t < p, \\ \text{the unique element of } p, ..., q-1 & \\ \qquad\qquad \text{congruent to } t \bmod (q-p) & \text{otherwise.} \end{cases}$$

□

Exercise 2.2.28 (Quotient) Prove Proposition 2.2.27: 'A monoid M generated by a set S is a quotient of the monoid $\langle S \mid R\rangle^+$ if and only if all the relations of R are satisfied in M.'

Solution If $\equiv$ is the congruence on S^* generated by R, and if $\sim$ is the congruence on S^* such that $u \sim v$ is satisfied if and only if u and v have the same evaluation in M, then the two conditions are equivalent to the inclusion of $\equiv$ in $\sim$ (seen as sets of pairs). □

Exercise 2.2.30 Prove Lemma 2.2.29: 'A homomorphism ϕ defined on a monoid S^* to a monoid M induces a homomorphism of $\langle S \mid R\rangle^+$ into M if and only if $\phi(u) = \phi(v)$ for each relation $u = v$ of R.'

Solution Denote π the canonical projection of S^* onto $\langle S \mid R\rangle^+$, that is, onto $S^*/\equiv_R$. The question is to know if ϕ *factorizes* by π, meaning there exists a homomorphism $\dot{\phi}$ satisfying $\phi = \dot{\phi} \circ \pi$. Denote $[w]$ the class of the word w for the congruence $\equiv$ generated by R. If $\dot{\phi}$ exists, the only definition possible would be $\dot{\phi}([w]) := \phi(w)$. The problem is to know if the value thus defined depends only on $[w]$: this is the case if and only if $\pi(w) = \pi(w')$, that is, $w \equiv w'$ implies $\phi(w) = \phi(w')$. □

Exercise 2.2.34 (Normal form) Prove Proposition 2.2.33: 'Suppose M is a monoid generated by a set S, and L is a subset of S^* containing exactly one element per class of the congruence $\sim$ where M is $S^*/\sim$. For w in S^*, let $\text{NF}(w)$ be the unique element $\sim$-equivalent to w in L. Then M is isomorphic to $(L, *, \text{NF}(\varepsilon))$, with $*$ defined by $u * v := \text{NF}(uv)$. Moreover, if NF is computable, the word problem of M with respect to S is decidable.'

Solution Denote $[w]$ the element of M represented by an S-word w. Then, by hypothesis, the mapping $w \mapsto [w]$ induces a bijection ι of L to M. Then, for any u, v in L, we have $[u] \cdot [v] = [uv] = [\text{NF}(uv)]$, so ι is an isomorphism of $(L, *)$ onto $(M, *)$. Finally, by construction, we have $[\varepsilon] = 1 = [\text{NF}(\varepsilon)]$, hence $\text{NF}(\varepsilon)$ is the unit element of $(L, *)$.

If the mapping NF is computable, it provides a direct solution to the word problem: two words u, v represent the same element of M if and only if $\text{NF}(u) = \text{NF}(v)$. It thus suffices to compute the words $\text{NF}(u)$ and $\text{NF}(v)$, and test whether or not they are the same. □

Exercise 2.2.36 (Normal form) Describe the monoid $\langle \mathsf{a}, \mathsf{b} \mid \mathsf{ab}^2 = \mathsf{ba} \rangle^+$ following the model given in Example 2.2.35.

Solution Let $M := \langle \mathsf{a}, \mathsf{b} \mid \mathsf{ab}^2 = r\mathsf{ba} \rangle^+$. Let $\equiv$ be the congruence generated by $\mathsf{ab}^2 = \mathsf{ba}$, and set $L := \{\mathsf{a}^p\mathsf{b}^q \mid p, q \geqslant 0\}$. By induction on the length, every word is $\equiv$-equivalent to a word in L.

We must show that two distinct words of L are not $\equiv$-equivalent. A first $\equiv$-invariant is $I_1(w) := |w|_\mathsf{a}$. Then define $I_2(w)$ as the sum of $|w|_\mathsf{a}$ (number of letters a in w) and, for each letter b, of 2^n, where n is the number of a to the right of the letter b under consideration. For example, $I_2(\mathsf{a}^2\mathsf{baba}) = 1 + 1 + 4 + 1 + 2 + 1 = 10$. Then I_2 is an $\equiv$-invariant since $I_2(\mathsf{ab}^2) = I_2(\mathsf{ba}) = 3$ and, more generally, $I_2(u\mathsf{ab}^2v) = I_2(u\mathsf{ba}v)$ for any words u, v: the contribution of a is in both cases $+1$, while the contributions of b^2 and b are $2 \times 2^{|v|_\mathsf{b}}$. Hence two distinct words of L are not $\equiv$-equivalents, as $I_1(\mathsf{a}^r\mathsf{b}^s) = r$, and $I_2(\mathsf{a}^r\mathsf{b}^s) = r + s$.

The monoid M is the semidirect product of two copies of $(\mathbb{N}, +)$, where the action of the second on the first is by multiplication by 2: the law of multiplication is given by $\mathsf{a}^r\mathsf{b}^s \cdot \mathsf{a}^{r'}\mathsf{b}^{s'} := \mathsf{a}^{r+r'}\mathsf{b}^{s2^{r'}+s'}$. □

Exercise 2.3.4 (Group) Prove Proposition 2.3.3: 'For any set S, the monoid F_S is a group, and as a group, it is generated by $[S]$. Moreover, the mapping $x \mapsto [x]$ is an injection of $S \cup \overline{S}$ into F_S.'

Solution Let w be a word over $S \cup \overline{S}$, say $w = x_1 \cdots x_\ell$ with $x_1, ..., x_\ell$ in $S \cup \overline{S}$. In F_S, we find

$$[w] \cdot [\overline{w}] = [x_1] \cdots [x_\ell] \cdot [\overline{x_\ell}] \cdots [\overline{x_1}]. \tag{9.2}$$

However, as noted previously, for every letter x of $S \cup \overline{S}$, we have $[x] \cdot [\overline{x}] = 1$ in F_S: the relations of Definition 2.3.1 were chosen for this. Consequently, $[x_\ell]$ and $[\overline{x_\ell}]$ cancel each other, as do $[x_{\ell-1}]$ and $[\overline{x_{\ell-1}}]$, etc., and, finally, we arrive to $[w] \cdot [\overline{w}] = 1$. The computation is similar for $[\overline{w}] \cdot [w]$. Hence, $[\overline{w}]$ is an inverse for $[w]$, and F_S is a group.

By definition, a set X generates a group G if G is the smallest subgroup of G including X. As the latter is the set of elements of G that can be written as products of elements of X and inverses of elements of X, we conclude that X generates G if and only if every element of G can be written as a product of elements of X and inverses of elements of X. Here, every element g of F_S is the class of a word over $S \cup \overline{S}$, say $[x_1 \cdots x_\ell]$ with $x_1, ..., x_\ell$ in $S \cup \overline{S}$. By definition, we have $g = g_1 \cdots g_\ell$ with $g_i = [s]$ if x_i is a letter s of S, and $g_i = [s]^{-1}$ if x_i is a letter $\overline{s}$ of $\overline{S}$. Hence F_S is generated by $[S]$.

Finally, let w be a word satisfying $[w] = [x]$ with x in $S \cup \overline{S}$. By Lemma 2.2.20, w is linked to x by a derivation with respect to the relations of Definition 2.3.1. Denote $|w|_s$ the number of letters s in the word w. An induction on the length of the derivation implies, for any word w thus linked to x,

$$|w|_x = |w|_{\overline{x}} + 1 \quad \text{and} \quad |w|_s = |w|_{\overline{s}} \quad \text{for } s \neq x \text{ in } S \cup \overline{S}.$$

However, these equalities are not satisfied by any letter s distinct from x. The mapping $s \mapsto [s]$, of $S \cup \overline{S}$ into F_S, is thus injective. □

Exercise 2.3.6 (Universal property) Prove Proposition 2.3.3: 'For any mapping ϕ of S to a group G, there exists a unique homomorphism $\widehat{\phi}$ of F_S into G extending ϕ. It is defined by $\widehat{\phi}(1) := 1$, $\widehat{\phi}(\overline{s}) := \phi(s)^{-1}$, and $\widehat{\phi}(w) := \phi(x_1) \cdots \phi(x_\ell)$ for $w = x_1 \cdots x_\ell$ with $x_1, ..., x_\ell \in S \cup \overline{S}$.'

Solution Let ϕ be a mapping of S to a group G. We first extend ϕ to $S \cup \overline{S}$ by setting $\phi(\overline{s}) := \phi(s)^{-1}$ for $\overline{s}$ in $\overline{S}$. By Proposition 2.2.8, there exists a unique homomorphism, here denoted ϕ^*, of the monoid $(S \cup \overline{S})^*$ into G (seen as a monoid) extending ϕ. Now, for any s in S, by construction, we have

$$\begin{aligned} \phi^*(s\overline{s}) &= \phi(s)\phi(\overline{s}) = \phi(s)\phi(s)^{-1} = 1 = \phi^*(\varepsilon), \\ \text{and} \quad \phi^*(\overline{s}s) &= \phi(\overline{s})\phi(s) = \phi(s)^{-1}\phi(s) = 1 = \phi^*(\varepsilon). \end{aligned}$$

Again applying Proposition 2.2.8, we deduce that ϕ^* induces a homomorphism $\widehat{\phi}$ of the monoid F_S into G. This homomorphism is automatically a group homomorphism since, for any g in F_S, we have

$$\widehat{\phi}(g)\widehat{\phi}(g^{-1}) = \widehat{\phi}(gg^{-1}) = \widehat{\phi}(1) = 1,$$

hence, necessarily, $\widehat{\phi}(g^{-1}) = \widehat{\phi}(g)^{-1}$. Finally, $\widehat{\phi}$ is unique, as any homomorphism ψ of a group F_S in G extending ϕ must satisfy $\psi(1) := 1, \psi(\overline{s}) := \psi(s)^{-1}$, and $\psi(w) := \psi(x_1) \cdots \psi(x_\ell)$ for $w = x_1 \cdots x_\ell$ with $x_1, ..., x_\ell \in S \cup \overline{S}$. □

Exercise 2.3.7 (Cyclic groups)
(i) Show that $\mathbb{Z}$ is a free group with a single generator.
(ii) Show that $\mathbb{Z}/n\mathbb{Z}$ is not free.

Solution
(i) By Definition 2.3.1, the monogenic free group is the monoid

$$M := \langle \mathtt{a}, \mathtt{A} \mid \mathtt{aA} = \mathtt{Aa} = 1 \rangle^{+},$$

and by Proposition 2.3.36, the reduced words constitute a (unique) normal form for M. Define then $\phi\colon \mathsf{Red}_{\mathtt{a}} \to \mathbb{Z}$ by

$$\phi(\mathtt{a}^p) := p, \quad \phi(\varepsilon) := 0, \quad \phi(\mathtt{A}^p) := -p.$$

The mapping ϕ is bijective, and is a homomorphism. Nine combinations must be considered as a function of the letters $\mathtt{a}$, ε, or $\mathtt{A}$. Typically, for $\mathtt{a}$ and $\mathtt{A}$, we find

$$\phi(\mathtt{a}^p) * \phi(\mathtt{A}^q) = \begin{cases} \mathtt{a}^{p-q} & \text{for } p > q, \\ \varepsilon & \text{for } p = q, \\ \mathtt{A}^{q-p} & \text{for } p < q, \end{cases}$$

and, to start, $\phi(\mathtt{a}^p) * \phi(\mathtt{A}^q) = \phi(p - q)$ in all three cases. The eight other combinations are similar, hence ϕ is an isomorphism. Consequently, $(\mathbb{Z}, +)$ is a free group with a single generator.
(ii) We have $[p] = [0]$, whereas in the group $\mathbb{Z}$, we do not have $p = 0$: thus there cannot exist a homomorphism of $\mathbb{Z}/p\mathbb{Z}$ into $\mathbb{Z}$, and the universal property of a free group is not respected. □

Exercise 2.3.9 (Congruence) Show that, if G is a group and $\equiv$ is a monoid congruence on G, then $\equiv$ is a group congruence, that is, it is necessarily compatible with the inverse operation.

Solution Suppose $g \equiv g'$. We deduce

$$g^{-1} = g^{-1}g'g'^{-1} \equiv g^{-1}gg'^{-1} = g'^{-1},$$

hence $\equiv$ is compatible with the inverse operation. □

Exercise 2.3.11 (Group congruences) Prove Proposition 2.3.10: ‘Let G be a group.

(i) If $\equiv$ is a congruence on G, the equivalence class of 1 is a distinguished subgroup H of G, and for every g, g' in G, the relation $g \equiv g'$ is equivalent to $g^{-1}g' \in H$.
(ii) Conversely, if H is a distinguished subgroup of G, the relation $\equiv_H$ defined by $g^{-1}g' \in H$ is a congruence on G, hence H is the class of 1.'

Solution The proof is well known.
(i) Let H be the class of 1 for $\equiv$. For $g \equiv 1$ and $h \equiv 1$, we have $gh \equiv 1$ and $h^{-1} \equiv 1$ (see Exercise 2.3.9). Hence H is a subgroup of G. Moreover, for $g \equiv 1$ and arbitrary h, we have $hgh^{-1} \equiv hh^{-1} \equiv 1$, so H is distinguished in G. Then, for any g, g', the relation $g \equiv g'$ is equivalent to $g^{-1}g \equiv g^{-1}g'$, hence to $g^{-1}g \equiv 1$, or to $g^{-1}g' \in H$.
(ii) Let H be a distinguished subgroup of G, and g, g', g'' arbitrary in G. Then $g^{-1}g = 1 \in H$, hence $g \equiv_H g$, and $\equiv_H$ is reflexive. If $g^{-1}g' \in H$, then also $g'^{-1}g = (g^{-1}g')^{-1} \in H$, hence $g' \equiv_H g$, and $\equiv_H$ is symmetric. If $g^{-1}g' \in H$ and $g'^{-1}g'' \in H$, then also $(g^{-1}g')(g'^{-1}g'') = g^{-1}g'' \in H$, hence $g \equiv_H g''$, and $\equiv_H$ is transitive. Next, suppose $g \equiv_H g'$, and let h, h' be arbitrary in G. We find $(hgh')^{-1}(hg'h') = h'^{-1}g^{-1}g'h'$, and the hypothesis that $g^{-1}g'$ belongs to H and that H is distinguished implies that $(hgh')^{-1}(hg'h')$ is in H, hence $hgh' \equiv_H hg'h'$. Thus $\equiv_H$ is a congruence. Finally, $g \equiv_H 1$ is equivalent by definition to $g^{-1} \in H$, so also to $g \in H$.

We leave to the reader the verification, very simple, that the two correspondences 'distinguished subgroups' $\leftrightarrow$ 'congruences' are inverses one of the other. □

Exercise 2.3.12 (Images and quotients) Show that, if G and G' are groups, then there exists a surjective homomorphism of G onto G' if and only if there exists a congruence $\equiv$ on G such that G' is isomorphic to $G/\equiv$.

Solution Suppose ϕ is a surjective homomorphism of G onto G', and define $g \equiv g'$ by $\phi(g) = \phi(g')$. By definition, $\equiv$ is an equivalence relation. Suppose $g \equiv g'$ and let h, h' be arbitrary. Then

$$\phi(hgh') = \phi(h)\phi(g)\phi(h') = \phi(h)\phi(g')\phi(h') = \phi(hg'h'),$$

hence $hgh' \equiv hg'h'$, and $\equiv$ is a congruence. Next, denote $[g]$ the $\equiv$ class of g. The mapping $\overline{\phi} : [g] \mapsto \phi(g)$ is well-defined and injective, as $g \equiv g'$ is equivalent to $\phi(g) = \phi(g')$. It is surjective since every element of G' is in the image, and is a homomorphism because

$$\overline{\phi}([gg']) = \phi(gg') = \phi(g) \cdot \phi(g') = \overline{\phi}([g]) \cdot \overline{\phi}([g']).$$

Thus G' is isomorphic to $G/\equiv$.

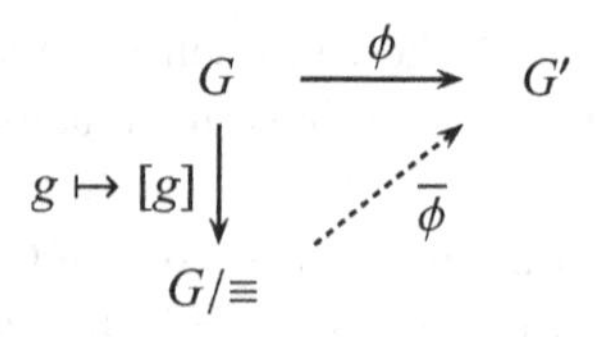

Conversely, suppose G' is isomorphic to $G/\equiv$, and consider $\phi : g \mapsto [g]$. By construction, the mapping ϕ of G into $G/\equiv$ is surjective. Since $\equiv$ is a congruence, ϕ is a homomorphism of G onto $G/\equiv$: indeed, the conjunction of $g \equiv g'$ and $h \equiv h'$ implies $gh \equiv g'h \equiv g'h'$, hence $gh \equiv g'h'$, and consequently, $[g] \cdot [h] = [gh]$. Thus there exists a surjective homomorphism of G onto G'. □

Exercise 2.3.15 Prove Lemma 2.3.14: 'Two signed S-words representing the same element of F_S have the same evaluation in every group including S.'

Solution By definition, for any letter s in S, we have

$$\mathsf{eval}_G(s\overline{s}) = ss^{-1} = 1 = \mathsf{eval}_G(\varepsilon),$$
$$\mathsf{eval}_G(\overline{s}s) = s^{-1}s = 1 = \mathsf{eval}_G(\varepsilon).$$

Applying Proposition 2.2.16 to the alphabet $S \cup \overline{S}$ and to eval_G, we conclude that eval_G induces a well-defined homomorphism on $\langle S \cup \overline{S} \mid \mathsf{Sym}(S)\rangle^+$, that is, on F_S. □

Exercise 2.3.18 Prove Lemma 2.3.17: 'A homomorphism ϕ defined on a monoid $(S \cup \overline{S})^*$ towards a group G induces a homomorphism of $\langle S \mid R\rangle$ into G if and only if $\phi(u) = \phi(v)$ for every relation $u = v$ of R and every free group relation.'

Solution It suffices to use Lemma 2.2.29 (Exercise 2.2.30): if G is a group generated by a family S, the free group relations $\mathsf{Sym}(S)$ are automatically satisfied. □

Exercise 2.3.21 (Presented groups) Revisit the presentations of Exercise 2.2.26 and, in each case, identify the corresponding group $\langle S \mid R\rangle$.

Solution
(i) Let $\equiv$ be the congruence on $(S \cup \overline{S})^*$ generated by $S \times \{\varepsilon\}$, that is, by all the relations $s = 1$ for s in S, and $\mathsf{Sym}(S)$. For any letter s, we have $\overline{s} \equiv \overline{s}s \equiv 1$, by $s \equiv 1$ and $\mathsf{Sym}(S)$. An induction on $|w|$ shows $w \equiv \varepsilon$ for any signed S-word w. Thus the relation $\equiv$ has only a single class, so $\langle S \mid S^* \times \{\varepsilon\}\rangle = \{1\}$.
(ii) Let $\equiv$ be the generated congruence, and let a be a fixed letter of S. For any letter s, we have $\overline{s}s \equiv \mathtt{Aa} \equiv 1$ by $\mathsf{Sym}(S)$ and then $\overline{s}s \equiv \overline{s}\mathtt{a}$ by the relations,

hence $\bar{s}\mathrm{a} \equiv \mathrm{Aa}$, then $\bar{s} \equiv \bar{s}\mathrm{aA} \equiv \mathrm{AaA} \equiv \mathrm{A}$. For any word w, by induction on $|w|$, we have $w \equiv \mathrm{a}^p$ if the number of letters of S in w is more than the number of letters of $\overline{S}$ by a strictly positive p_1, $w \equiv \varepsilon$ if the number of letters of S in w is equal to the number of letters of $\overline{S}$, and $w \equiv \mathrm{A}^q$ if the number of letters of $\overline{S}$ in w is more than the number of letters of S by a strictly positive q. The mapping associating p in $\mathbb{Z}$ with a^p for $p > 0$, ε for $p = 0$, and $\mathrm{A}^{|p|}$ for $p < 0$, is surjective on $\langle S \mid R\rangle$ as every word is $\equiv$-equivalent to a word a^p, ε, or A^q, and is injective as two words of different lengths are not $\equiv$-equivalent.

(iii) Let $\equiv$ be the congruence on $\{\mathrm{a}, \mathrm{b}, \mathrm{A}, \mathrm{B}\}^*$ generated by the unique pair $(\mathrm{ab}, \mathrm{ba})$. Recalling Exercise 2.2.22, we show that $w \equiv w'$ is satisfied if and only if w and w' have the same values of algebraic differences $|w|_\mathrm{a} - |w|_\mathrm{A}$ and $|w|_\mathrm{b} - |w|_\mathrm{B}$. Let ϕ be the mapping $w \mapsto (|w|_\mathrm{a} - |w|_\mathrm{A}, |w|_\mathrm{b} - |w|_\mathrm{B})$. Then ϕ induces a bijection of $\{\mathrm{a}, \mathrm{b}, \mathrm{A}, \mathrm{B}\}^*/\equiv$ onto $\mathbb{Z}^2$, and $\phi(uv)$ is easily shown to be equal to $\phi(u) + \phi(v)$ (everything is additive). Hence, ϕ is an isomorphism with image $(\mathbb{Z}, +) \times (\mathbb{Z}, +)$.

(iv) Let $\equiv$ be the congruence on $\{\mathrm{a}, \mathrm{A}\}^*$ generated by the unique pair $(\mathrm{a}^p, \varepsilon)$ and by $\mathsf{Sym}(\mathrm{a})$. By induction on the length of w, we show that any word w is $\equiv$-equivalent to ε or to a word a^r for $r \in \{1, ..., p{-}1\}$. In other words, it is permissible to ignore the letter A. Thus the rest is identical with the case of the monoid $\langle \mathrm{a} \mid \mathrm{a}^p = 1\rangle^+$, and we again find the cyclic group $\mathbb{Z}/p\mathbb{Z}$.

Similarly, let $\equiv$ be the congruence on $\{\mathrm{a}, \mathrm{A}\}^*$ generated by the unique pair $(\mathrm{a}^p, \mathrm{a}^q)$ and by $\mathsf{Sym}(\mathrm{a})$. We suppose $q > p$. By induction on the length of w, we show that any w is $\equiv$-equivalent to ε or to a word a^r for $r \in \{1, ..., q{-}1\}$, hence it is permissible to ignore the letter A. This time the rest is identical to the case of the monoid $\langle \mathrm{a} \mid \mathrm{a}^p = \mathrm{a}^q\rangle^+$, and we once again find the description of Exercise 2.2.26(iv). □

Exercise 2.3.23 (Presentation) Prove Proposition 2.3.22: 'For any presentation (S, R), the group $\langle S \mid R\rangle$ is isomorphic to $F_S/{\approx_R}$, where $\approx_R$ is the congruence on F_S generated by the relations $[u] = [v]$ for $u = v$ in R.'

Solution As usual, let $\equiv$ be the congruence on the signed S-words generated by $\mathsf{Sym}(S)$ and $\equiv_R$ the one generated by $R \cup \mathsf{Sym}(S)$. We show that $w \equiv_R w'$ is equivalent to $[w] \approx_R [w']$. First suppose $w \equiv_R w'$. Then there exists an $(R \cup \mathsf{Sym}(S))$-derivation linking w to $w'0$, say $(w_0, ..., w_m)$. In such a derivation, the elementary steps bring into play either relations of $\mathsf{Sym}(S)$, or relations of R; in the latter case we write $\rightarrow_R$ ('applying a relation of R'). Grouping the steps by type, we obtain indices $i_0 = 0$, j_0, $i_1 = j_0 + 1$, j_1, ..., $i_p = j_{p-1} + 1$, $j_p = m$ satisfying

$$w = w_{i_0} \equiv w_{j_0} \rightarrow_R w_{i_1} \equiv w_{j_1} \rightarrow_R \cdots \rightarrow_R w_{i_p} \equiv w_{j_p} = w'.$$

Projecting in F_S, we deduce

$$[w] = [w_{i_0}] = [w_{j_0}] \to_{[R]} [w_{i_1}] = [w_{j_1}] \to_{[R]} \cdots \to_{[R]} [w_{i_p}] = [w_{j_p}] = [w'],$$

hence $[w] \approx_R [w']$.

Conversely, suppose $[w] \approx_R [w']$. Then there exists a $[R]$-derivation of $[w]$ to $[w']$ in the group F_S, say $([w_0], ..., [w_p])$ with $[w_0] = [w]$ and $[w_p] = [w']$. The equality $[w_0] = [w]$ implies $w \equiv w_0$, and $[w_p] = [w']$ implies $w_p \equiv w'$. Next, for $0 \leqslant k < p$, we have $[w_k] \to_{[R]} [w_{k+1}]$: by definition, this means there exist signed S-words u_k and v_k satisfying $w_k \equiv u_k \to_R v_k \equiv w_{k+1}$. Then the sequence

$$(w, w_0, w_1, u_1, v_1, w_2, ..., w_{p-1}, u_{p-1}, v_{p-1}, w_p, w')$$

is an $(R \cup \mathsf{Sym}(S))$-derivation of w to w', and we deduce $w \equiv_R w'$.

From this, we can conclude that the quotient monoid $(S \cup \overline{S})^*/{\equiv_R}$, that is, the group $\langle S \mid R \rangle$, is isomorphic to the quotient group $F_S/{\approx_R}$. □

Exercise 2.3.25 (Quotient) Prove Proposition 2.3.24: 'A group G generated by a set S is a quotient of the group $\langle S \mid R \rangle$ if and only if all the relations of R are satisfied in G.'

Solution By Proposition 2.2.27, the group G seen as a monoid is the quotient of the monoid $\langle S \cup \overline{S} \mid R \cup \mathsf{Sym}(S) \rangle^+$ if and only if the relations of R and those of $\mathsf{Sym}(S)$ are satisfied in G. Since G is a group, the relations of $\mathsf{Sym}(S)$ are satisfied by hypothesis; there remains only those of R. □

Exercise 2.3.27 (Presentation) Prove Proposition 2.3.26: 'Suppose G is a group generated by a set S, R is a list of relations satisfied in G, and E is a set of signed S-words such that

(i) every signed S-word is R-equivalent to a word of E,

(ii) the mapping eval_G is injective on E.

Then G admits the presentation $\langle S \mid R \rangle$.'

Solution Let $\equiv$ be the congruence on the signed S-words such that the group $\langle S \mid R \rangle$ is $(S \cup \overline{S})^*/{\equiv}$, and $\sim$ the congruence such that $\langle S \mid R \rangle$ is $(S \cup \overline{S})^*/{\sim}$. Since the relations of R are satisified in G, the relation $w \equiv w$ implies $w \sim w'$ (hence G is a quotient of $\langle S \mid R \rangle$). Conversely, suppose $w \sim w'$. By (i), there exists w_0 and w'_0 in E satisfying $w \equiv w_0$ and $w' \equiv w'_0$. As $\equiv$ is contained in $\sim$, we also have $w \sim w_0$ and $w' \sim w'_0$, thus $w_0 \sim w'_0$. By (ii), we deduce that $w_0 = w'_0$, and hence $w \equiv w_0 = w'_0 \equiv w'$, so $w \equiv w'$. Consequently, the congruences $\equiv$ and $\sim$ coincide, thus G and $\langle S \mid R \rangle$ are isomorphic. □

Exercise 2.3.29 Following the model of Algorithm 2.2.32, give a solution to the word problem for the presented group $\langle \mathsf{a}, \mathsf{b} \mid \mathsf{ab} = \mathsf{ba} \rangle$.

Solution It suffices to replace the number of a in the word w considered by the algebraic difference $|w|_{\mathtt{a}} - |w|_{\mathtt{A}}$ and, similarly, the number of b in w by the algebraic $|w|_{\mathtt{b}} - |w|_{\mathtt{B}}$. Here is a suggestion:

Algorithm 9.2.1 (Word problem for $\langle \mathtt{a}, \mathtt{b} \mid \mathtt{ab} = \mathtt{ba} \rangle$)

Input: a signed word w in $\{\mathtt{a}, \mathtt{b}, \mathtt{A}, \mathtt{B}\}^*$
Output: **yes** if w represents 1 in $\langle \mathtt{a}, \mathtt{b} \mid \mathtt{ab} = \mathtt{ba} \rangle$,
no otherwise

1: $p \leftarrow |w|_{\mathtt{a}} - |w|_{\mathtt{A}}$
2: $q \leftarrow |w|_{\mathtt{b}} - |w|_{\mathtt{B}}$
3: **if** $p = 0$ **and** $q = 0$ **then**
4: **return yes**
5: **else**
6: **return no**
7: **end if**

□

Exercise 2.3.34 Prove Lemma 2.3.33: ‘Every reduction of a word w in $()S \cup \overline{S})^*$ leads the same reduced word.’

Solution We establish by induction on $\ell \geqslant 0$ the property

$$(\mathcal{P}_\ell) \quad \text{If } w \to^* w' \text{ and } w \to^* w'' \text{ with } w' \text{ reduced and } |w| - |w'| = 2\ell, \text{ then } w'' \to^* w'.$$

First suppose $\ell = 0$. Then $w' = w$, so w is reduced, hence $w \to^* w''$ implies $w'' = w$, thus, trivially, $w'' \to^* w'$.

Now suppose $\ell > 0$. Let $w \to w'_1 \to^* w'$ be a reduction of w to w'. If we have $w'' = w$, then $w'' \to^* w'$ is satisfied by hypothesis. If not, let $w \to w''_1 \to^* w''$ be a reduction of w to w''. If the words w'_1 and w''_1 coincide, then $w'_1 \to^* w'$ and $w'_1 \to^* w''$ with w' reduced and $|w'_1| - |w'| = 2\ell - 2$. Then $(\mathcal{P}_{\ell-1})$, satisfied by the induction hypothesis, implies $w'' \to^* w'$.

There remains the case $w'_1 \neq w''_1$. This can only occur if we reduce two different factors to go from w to w'_1 and from w to w''_1. Two cases are a priori possible. One is that the two reduced factors overlap, in which case they are necessarily of the type $s\overline{s}$ and $\overline{s}s$ with the letter $\overline{s}$ in common, and hence there exist two words x and y satisfying $w = xs\overline{s}sy$, leading to $w'_1 = xsy = w''_1$, which contradicts the hypothesis $w'_1 \neq w''_1$.

The other case, the only remaining possibility, is that the two reduced factors do not overlap. Then there exist two letters s, t of $S \cup \overline{S}$ and words x, y, z satisfying $w = xs\overline{s}yt\overline{t}z$, $w'_1 = xyt\overline{t}z$, and $w'_2 = xs\overline{s}yz$, or vice versa. Then we have $w'_1 \to xyz$ and $w''_1 \to xyz$. Since $w'_1 \to^* w_1$ and $w'_1 \to xyz$ with w_1

irreducible and $|w_1'|-|w'| = 2\ell-2$, the property $(\mathcal{P}_{\ell-1})$, satisfied by the induction hypothesis, implies $xyz \to^* w'$. By transitivity, we deduce $w_1'' \to^* w'$. But then, $w_1'' \to^* w'$ and $w_1'' \to w''$ with w' irreducible and $|w_1''| - |w'| = 2\ell - 2$, and another invocation of the induction hypothesis $(\mathcal{P}_{\ell-1})$ gives $w'' \to^* w'$, as desired.

Ignoring ℓ in $(\mathcal{P}_\ell)$, we deduce that, if $w \to^* w'$ and $w \to^* w''$ with w' reduced, then $w'' \to^* w'$. The desired uniqueness thus follows. Indeed, suppose we have at the same time $w \to^* w'$ and $w \to^* w''$ with w' and w'' reduced. By the above, we must have $w'' \to^* w'$. Since w'' is reduced, the only possibility is $w'' = w'$. □

Exercise 2.3.37 Prove Proposition 2.3.36: 'The reduced words provide a normal form for F_S.'

Solution Denote $\equiv$ the congruence on the monoid $(S \cup \overline{S})^*$ generated by the relations of Definition 2.3.1. We must show that each equivalence class for $\equiv$ contains one and only one reduced word.

First, by definition, we have $s\overline{s} \equiv \varepsilon$ and $\overline{s}s \equiv \varepsilon$ for any letter s, hence $w \to w'$. Then, by a simple induction, $w \to^* w'$ implies $w \equiv w$. Consequently, the reductions take place without changing equivalence class. Since every word reduces to a reduced word, every equivalence class contains at least one reduced word. The point is to show that there can be only one.

For this, denote $w \sim w'$ for $\mathsf{red}(w) = \mathsf{red}(w')$, where red is the function introduced in Definition 2.3.35. Then $\sim$ is an equivalence relation on $(S \cup \overline{S})^*$. We now show that it is even a congruence, that is, it is compatible with the left and right products. Suppose $u \sim u'$ and $v \sim v'$. Then $uv \to^* \mathsf{red}(u)\mathsf{red}(v)$, hence, by Lemma 2.3.33, $\mathsf{red}(uv) = \mathsf{red}(\mathsf{red}(u)\mathsf{red}(v))$ and, similarly, $\mathsf{red}(u'v') = \mathsf{red}(\mathsf{red}(u')\mathsf{red}(v'))$, so $\mathsf{red}(uv) = \mathsf{red}(u'v')$, or $uv \sim u'v'$.

Moreover, we have $\mathsf{red}(s\overline{s}) = \mathsf{red}(\overline{s}s) = \varepsilon = \mathsf{red}(\varepsilon)$, or $s\overline{s} \sim \varepsilon$ and $\overline{s}s \sim \varepsilon$ for every letter s of S. Hence $\sim$ is a congruence on $(S \cup \overline{S})^*$ containing all the pairs $(s\overline{s}, \varepsilon)$ and $(\overline{s}s, \varepsilon)$. Since $\equiv$ is the smallest congruence with these properties, we deduce that $\sim$ includes $\equiv$, that is, $w \equiv w'$ implies $w \sim w'$, or $\mathsf{red}(w) = \mathsf{red}(w')$. Consequently, an equivalence class for $\equiv$ can only contain one reduced word. □

Exercise 2.3.40 (Presented group) Show that the group $\langle S \mid R\rangle$ is the quotient of the free group F_S by the distinguished subgroup generated by the elements $\mathsf{red}(u^{-1}v)$ for the relations $u = v$ of R.

Solution Let $\equiv$ be the congruence such that $\langle S \mid R\rangle$ is $F_S/\equiv$, and H the distinguished subgroup generated by the elements $\mathsf{red}(u^{-1}v)$ for the relations $u=v$ of R. By induction on the length ℓ of a derivation from g to g', we show that $g^{-1}g'$ belongs to H^{ℓ}, hence to H, in other words $g \equiv g'$ implies $g^{-1}g' \in H$.

Conversely, $g^{-1}g' \in H$ implies the existence of a finite sequence of elements of the form $h_i^{-1}\mathsf{red}(u_i^{-1}v_i)h_i$ with $u_i = v_i$ a relation of R of which $g^{-1}g'$ is the product: for each relation $u_i = v_i$ of R, we have

$$\mathsf{red}(u_i) \equiv u_i \equiv v_i \equiv \mathsf{red}(v_i),$$

consequently, $\mathsf{red}(u_i^{-1}v_i) \equiv 1$, hence $h_i^{-1}\mathsf{red}(u_i^{-1}v_i)h_i \equiv 1$, and thus, by product, $g^{-1}g' \equiv 1$, or $g \equiv g'$.

Remark as well that by Proposition 2.3.10 (Exercise 2.3.11), the subgroup H coincides with the canonical subgroup associated with the congruence $\equiv$. □

9.3 Exercises of Chapter 3

Exercise 3.1.9 (Inverse) Show that for every $n \leqslant \infty$ the unit element 1 is the only invertible element of B_n^+.

Solution Let $a \neq 1$ in B_n^+. For any element b, we have $|ab| = |a| + |b| \geqslant |a| > 0$, hence $ab \neq 1$, which shows that a does not have an inverse in B_n^+. □

Exercise 3.1.11 (Word problem) Write the above solution in the form of a pseudocode algorithm.

Solution The principle is to exhaustively construct the equivalence class $[w]^+$ of w by saturation of the application of the braid relations, independently of the word w', and to finally test if w' belongs to $[w]^+$. Given that the number of braid words of $\mathcal{BW}_n^+$ with length ℓ is finite and equal to $(n-1)^{\ell}$, the number of steps leading to saturation is bounded by $(n-1)^{\ell}$, so a simple **for** loop suffices, rather than a **while** loop.

For u, v positive braid words, denote $u \rightarrow v$ if there exists a braid relation applied to u whose result is v (there is not in general uniqueness: several applications of braid relation to u are possible). A proposed implementation is:

Algorithm 9.3.1 (Word problem for B_∞^+)

Input: two words w, w' in $\mathcal{BW}_\infty^+$

Output: **yes** if w and w' represent the same element of B_∞^+,
no otherwise

1: **Select** n such that $w \in \mathcal{BW}_n^+$
2: $\ell \leftarrow |w|$
3: $m \leftarrow (n-1)^\ell$
4: $C \leftarrow \{w\}$
5: **for** p from 1 to m **do**
6: $\quad C \leftarrow C \cup \{v \in \mathcal{BW}_n^+ \mid u \in C$ **and** $u \to v\}$
7: **end for**
▹ At this stage, C is the complete equivalence class of w
8: **if** $w' \in C$ **then**
9: $\quad$ **return yes**
10: **else**
11: $\quad$ **return no**
12: **end if** □

Exercise 3.1.14 (Shift)
(i) Show that the mapping $\sigma_i \mapsto \sigma_{i+1}$ induces an endomorphism dec^+ of B_∞^+ into itself.
(ii) Show that dec^+ is injective, but not surjective.

Solution
(i) If (u, v) is a pair of braid words of the form $(\sigma_i\sigma_{i+1}\sigma_i, \sigma_{i+1}\sigma_i\sigma_{i+1})$ or $(\sigma_i\sigma_j, \sigma_j\sigma_i)$ with $|i - j| \geqslant 2$, then so is $(\mathsf{dec}^+(u), \mathsf{dec}^+(v))$. Hence, by induction on the number of braid relations in play, $u \equiv^+ v$ implies $\mathsf{dec}^+(u) \equiv^+ \mathsf{dec}^+(v)$. By Proposition 2.2.8, it follows that dec^+ induces an endomorphism of B_∞^+.
(ii) By construction, σ_1 is not in the image of dec^+.

On the other hand, let $\mathcal{BW}_{\geqslant 2}^+$ be the set of positive braid words from which σ_1 is absent and, for w in $\mathcal{BW}_{\geqslant 2}^+$, let $f(w)$ be the image of w by $\sigma_i \mapsto \sigma_{i-1}$. As above, mutatis mutandis, $w \equiv^+ w'$ implies $f(w) \equiv^+ f(w')$. Suppose then $\mathsf{dec}^+(u) \equiv^+ \mathsf{dec}^+(v)$. By construction, $\mathsf{dec}^+(u)$ and $\mathsf{dec}^+(v)$ belong to $\mathcal{BW}_{\geqslant 2}^+$, and $u = f(\mathsf{dec}^+(u)) \equiv^+ f(\mathsf{dec}^+(v)) = v$. □

Exercise 3.1.16 (Enveloping group)
(i) Show that, for any monoid presentation (S, R), the identity of S induces a homomorphism ι of the monoid $\langle S \mid R\rangle^+$ into the group $\langle S \mid R\rangle$.

(ii) Hence show that any homomorphism of $\langle S \mid R\rangle^+$ towards a group can be factored by ι.

Solution
(i) As all the relations of R are satisfied in $\langle S \mid R\rangle$, it follows from Proposition 2.2.27 that the identity of S induces a homomorphism ι of the monoid $\langle S \mid R\rangle^+$ into the group $\langle S \mid R\rangle$.
(ii) Let ϕ be a homomorphism of the monoid $\langle S \mid R\rangle^+$ into a group G. Denote $\equiv^+$ and $\equiv$ the congruences in play. To see that ϕ factorizes by $\langle S \mid R\rangle$, it suffices to show that $[g] = [g']$, that is, $g \equiv g'$, implies $\phi(g) = \phi(g')$. However, if $g \equiv g'$, there exists an $R \cup \mathsf{Sym}(S)$-derivation from g to g'. In the case of R with a single relation, we have $\phi(g) = \phi(g')$ by hypothesis, and, in the case of $\mathsf{Sym}(S)$ and for $g = hs\bar{s}h'$ and $g' = hh'$ (and other analogous cases), we have, since ϕ has values in a group,

$$\phi(g) = \phi(hs\bar{s}h') = \phi(h)\phi(s)\phi(s)^{-1}\phi(h') = \phi(hh') = \phi(g').$$

By induction on the length of the derivation from g to g', we deduce $g \equiv g'$ implies $\phi(g) = \phi(g')$.

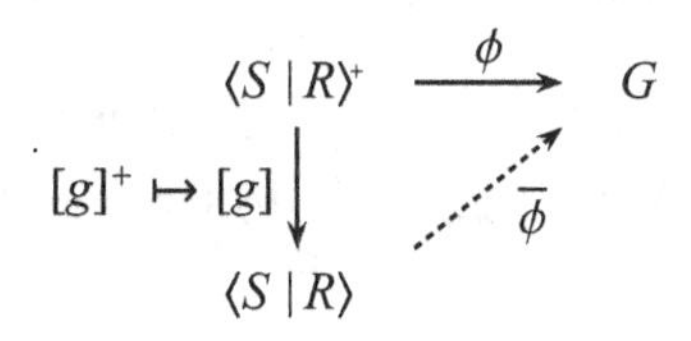

□

Exercise 3.1.18 (Non-embedding) Let M be the monoid $\langle \mathsf{a}, \mathsf{b}, \mathsf{c} \mid \mathsf{ab} = \mathsf{ac}\rangle^+$ and G the group $\langle \mathsf{a}, \mathsf{b}, \mathsf{c} \mid \mathsf{ab} = \mathsf{ac}\rangle$. Show that M cannot be embedded into G. [Hint: Show that in the group, we have $\mathsf{b} = \mathsf{c}$, whereas in the monoid, $\mathsf{b} \neq \mathsf{c}$.]

Solution In the group G, we have $\mathsf{ab} = \mathsf{ac}$, thus $\mathsf{b} = \mathsf{c}$, whereas in the monoid M, we do not have the equality $\mathsf{b} = \mathsf{c}$, and the morphism induced by the identity on $\{\mathsf{a}, \mathsf{b}, \mathsf{c}\}$ is not injective. Indeed, denoting $\equiv^+$ and $\equiv$ the monoid and group congruences involved, we obtain $\mathsf{b} \equiv \mathsf{a}^{-1}\mathsf{ab} \equiv \mathsf{a}^{-1}\mathsf{ac} \equiv \mathsf{c}$, whereas $\mathsf{b} \equiv^+ \mathsf{c}$ is false as there is no derivation applicable here to words of length one. □

Exercise 3.1.20 (Cancellative) Show that a free monoid is cancellative.

Solution Let S^* be the free monoid generated by S. It suffices, by induction on the length of word to be left-simplified, to show that, for any letter of S and any words u, v of S^*, the relation $su = sv$ implies $u = v$. However, u is the suffix of length $|su| - 1$ of the word su, and v is the suffix of length $|sv| - 1$ of the word sv: the hypothesis $su = sv$ thus implies $u = v$. The argument is symmetric on the right. □

Solution Let S^* be the free monoid generated by S. It suffices to show that, for any letter of S and any words u, v of S^*, the relation $su = sv$ implies $u = v$.

Recall that for S-words x and y to be equal, we need $|x| = |y|$ and $x[i] = y[i]$ for $1 \leqslant i \leqslant |x|$ (see Exercise 2.2.6). Hence if $su = sv$, then $|u| = |v| = |su| - 1$. Moreover, we must have $s = s$ (trivially true), and $u[i] = v[i]$ for $1 \leqslant i \leqslant |u|$. Thus $u = v$. The argument is symmetric on the right. □

Exercise 3.1.27 (Automorphism) Show that, for any a in B_n^+, we have the equality $a\Delta_n = \Delta_n\phi_n(a)$.

Solution Lemma 3.1.26 gives the result for a of length 1. For the general case, we proceed by induction on the length of a. Suppose $a = b\sigma_i$. Then

$$\begin{aligned} a\Delta_n = b\sigma_i\Delta_n &= b\Delta_n\sigma_{n-i} && \text{by Lemma 3.1.26} \\ &= \Delta_n\phi_n(b)\sigma_{n-i} && \text{by the induction hypothesis} \\ &= \Delta_n\phi_n(a). \end{aligned}$$

□

Exercise 3.1.29 (Words $\underline{\partial}_4(\sigma_i)$) Determine the words $\underline{\partial}_n(\sigma_i)$ for $n \geqslant 5$ and $i < n$.

Solution Apart from the first values, the words $\underline{\partial}_n(\sigma_i)$ are not unique. We calculate here the values produced by the inductive proof of Lemma 3.1.28. To begin, we find $\underline{\Delta}_3 := \sigma_1\sigma_2\sigma_1$, directly implying $\underline{\partial}_3(\sigma_1) := \sigma_2\sigma_1$ and $\underline{\partial}_3(\sigma_2) := \sigma_1\sigma_2$ (unique values).

Suppose $n := 4$, and first consider $i \leqslant 2$.The proof provides the solution $\underline{\partial}_4(\sigma_i) := \underline{\partial}_3(\sigma_i)\underline{\sigma}_{4,1}$, or

$\underline{\partial}_4(\sigma_1) := \sigma_2\sigma_1 \cdot \sigma_3\sigma_2\sigma_1 = \sigma_2\sigma_1\sigma_3\sigma_2\sigma_1$,

$\underline{\partial}_4(\sigma_2) := \sigma_1\sigma_2 \cdot \sigma_3\sigma_2\sigma_1 = \sigma_1\sigma_2\sigma_3\sigma_2\sigma_1$.

Next suppose $i := 3$. Let $\mathsf{shift}(\underline{\Delta}_3)$ be the word obtained from $\underline{\Delta}_3$ by shifting all the indices of the generators σ_i by $+1$, giving $\sigma_2\sigma_3\sigma_2$. The proof now provides the solution $\underline{\partial}_4(\sigma_3) := \underline{\sigma}_{3,1}\mathsf{shift}(\underline{\Delta}_3)$, or

$\underline{\partial}_4(\sigma_3) := \sigma_2\sigma_1 \cdot \sigma_2\sigma_3\sigma_2 = \sigma_2\sigma_1\sigma_3\sigma_2\sigma_1$.

The case $n := 5$ is similar. For $i \leqslant 3$, we find

$$\underline{\partial}_5(\sigma_1) := \underline{\partial}_4(\sigma_1)\,\underline{\sigma}_{5,1} = \sigma_2\sigma_1\sigma_3\sigma_2\sigma_1\sigma_4\sigma_3\sigma_2\sigma_1,$$
$$\underline{\partial}_5(\sigma_2) := \underline{\partial}_4(\sigma_2)\,\underline{\sigma}_{5,1} = \sigma_1\sigma_2\sigma_3\sigma_2\sigma_1\sigma_4\sigma_3\sigma_2\sigma_1,$$
$$\underline{\partial}_5(\sigma_3) := \underline{\partial}_4(\sigma_3)\,\underline{\sigma}_{5,1} = \sigma_2\sigma_1\sigma_3\sigma_2\sigma_1\sigma_4\sigma_3\sigma_2\sigma_1,$$

and, for $i := 4$,

$$\underline{\partial}_5(\sigma_4) := \underline{\sigma}_{4,1}\,\mathsf{shift}(\underline{\Delta}_4) = \sigma_3\sigma_2\sigma_1\sigma_2\sigma_3\sigma_4\sigma_2\sigma_3\sigma_2.$$ □

Exercise 3.3.2 (Shift) Show the compatibility of the endomorphism dec^+ of the monoid B_∞^+ and the endomorphism shift of the group B_∞.

Solution By Proposition 3.3.1, the embedding ι of B_∞^+ into B_∞ is injective, and all is immediate: for a in B_∞^+, we have $\iota(\mathsf{dec}^+(a)) = \mathsf{shift}(a)$. Thus there is no risk in identifying dec^+ with the restriction of shift to B_∞^+. □

Exercise 3.3.7 (Centre)
(i) Show that Δ_n^2 is central in the monoid B_n^+ (i.e. commutes with every element).
(ii) Show that an element g of B_n is central if and only if it can be written $\Delta_n^{2m}a$ with a central in B_n^+.

Solution
(i) For $1 \leqslant i \leqslant n-1$, Lemma 3.1.26 implies

$$\sigma_i\underline{\Delta}_n^2 \equiv^+ \underline{\Delta}_n\sigma_{n-i}\underline{\Delta}_n \equiv^+ \underline{\Delta}_n^2\sigma_{n-(n-i)} = \underline{\Delta}_n^2\sigma_i,$$

so $\sigma_i\Delta_n^2 = \Delta_n^2\sigma_i$ in B_n^+: hence Δ_n^2 commutes with $\sigma_1, \ldots, \sigma_{n-1}$. Thus Δ_n^2 commutes with every element of B_n^+, as any such element is a finite product of elements of $\{\sigma_1, \ldots, \sigma_{n-1}\}$.
(ii) We have just seen that Δ_n^2 is central in B_n^+. It is thus central in B_n, since B_n is the group of fractions of B_n^+. Hence the condition is sufficient. Conversely, suppose g is central in B_n. By Proposition 3.3.5, there exist m in $\mathbb{Z}$ and a in B_n^+ satisfying $g = \Delta_n^{2m}a$: the proof of Proposition 3.3.5 shows that it is permissible to choose an even exponent of Δ_n since, if $\Delta_n^m a$ is an expression of g, then so is $\Delta_n^{m-1}(\Delta_n a)$.

Let b be arbitrary in B_n^+. Then $\Delta_n^2 ab = b\Delta_n^2 a$, so $\Delta_n^2 ab = \Delta_n^2 ba$, and $ab = ba$, hence a is central in B_n^+. □

Exercise 3.3.9 (Automorphism) Show that $\sigma_i \mapsto \sigma_{n-i}$ can be extended to an automorphism of the monoid B_n^+, then to an automorphism of the group B_n, in fact the inner automorphism associated with conjugation by Δ_n.

Solution Denote ϕ_n^+ the involution of $\mathcal{BW}_n^+$ replacing every generator σ_i by σ_{n-i}. By a direct invocation of Lemma 3.1.13, ϕ_n^+ induces an involutive automorphism of B_n^+.

Let $g \in B_n$. There exist a, b in B_n^+ satisfying $g = ab^{-1}$. Define $\phi_n(g)$ by $\phi_n(g) := \phi_n^+(a)\phi_n^+(b)^{-1}$. The definition makes sense. Indeed, the fractionary decomposition is not unique, but $g = ab^{-1} = a'b'^{-1}$ is equivalent to the existence of c, c' in B_n^+ satisfying $ac = a'c'$ and $bc = b'c'$ (see the proof of Ore's theorem). Suppose $g = ab^{-1} = a'b'^{-1}$. There exist c, c' satisfying $ac = a'c'$ and $bc = b'c'$. Hence $\phi_n^+(a)\phi_n^+(c) = \phi_n^+(a')\phi_n^+(c')$ and $\phi_n^+(b)\phi_n^+(c) = \phi_n^+(b')\phi_n^+(c')$, then $\phi_n^+(a)\phi_n^+(b)^{-1} = \phi_n^+(a')\phi_n^+(b')^{-1}$.

Next, ϕ_n is involutive, hence bijective. Let $g, h \in B_n^+$. There exist $a, b, c, d \in B_n^+$ satisfying $g = ab^{-1}$ and $h = cd^{-1}$. Then $gh = aef^{-1}d^{-1}$ where $be = cf$. Thus

$$\phi_n(gh) = \phi_n^+(a)\phi_n^+(e)\phi_n^+(f)^{-1}\phi_n^+(b)^{-1},$$

$$\phi_n(g)\phi_n(h) = \phi_n^+(a)\phi_n^+(b)^{-1}\phi_n^+(c)\phi_n^+(b)^{-1}.$$

As $be = cf$ implies $\phi_n^+(b)\phi_n^+(e) = \phi_n^+(c)\phi_n^+(f)$, bringing together the above equalities implies $\phi_n(gh) = \phi_n(g)\phi_n(h)$, and ϕ_n is an automorphism of B_n.

Finally, by Lemma 3.1.26, we have $g\Delta_n = \Delta_n\phi_n(g)$ when g is a single generator σ_i, and consequently, when g is arbitrary in B_n as ϕ_n is a homomorphism. We deduce $\phi_n(g) = \Delta_n^{-1}g\Delta_n$.

Remark: Instead of extending ϕ_n^+ by coming back to the fact that B_n is the group of fractions of B_n^+, we could reason directly at the level of the B_n: by Lemma 3.1.26, the inner automorphism associated with Δ_n sends σ_i to σ_{n-i}, and it necessarily coincides with ϕ_n. □

Exercise 3.3.12 (No such grid) Show that there does not exist a grid with source $(\mathtt{a}, \mathtt{bc})$ with respect to the presentation

$$\langle \mathtt{a}, \mathtt{b}, \mathtt{c} \mid \mathtt{aba} = \mathtt{bab}, \mathtt{bcb} = \mathtt{cbc}, \mathtt{cac} = \mathtt{aca} \rangle^+.$$

Solution Note that, as with the presentation of B_4^+, there is exactly one relation of the type $s... = t...$ for each pair of letters. Thus, for any pair of words (u, v), either there exists a unique grid with source (u, v), or such a grid does not exist.

To find an eventual grid with source $(\mathtt{a}, \mathtt{bc})$, we consider the reversing of $\mathtt{Abc}$:

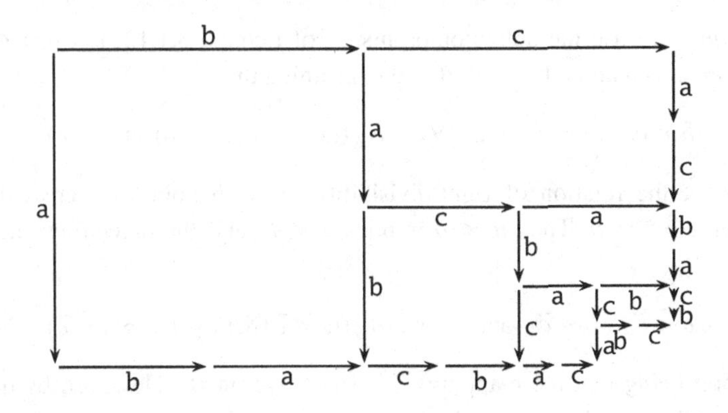

... and we end up in an infinite loop: the pattern Abc reappears. More precisely, setting u := bacbac and v := acbacb, we obtain Abc $\curvearrowright$ $u(\text{Abc})\overline{v}$, and then, for any k,

$$\text{Abc} \curvearrowright u^k(\text{Abc})\overline{v}^k :$$

clearly the process never ends. Thus there is no grid with source (a, bc).

Hence seemingly minor modifications in the presentation can lead to wildly different behaviour of the reversing relation: recall that for the presentation of B_4^+, the reversing always terminates in finite time, for example Abc $\curvearrowright$ bacbCBA. □

9.4 Exercises of Chapter 4

Exercise 4.1.7 (Gcd) Show that every set (finite or infinite) of positive braids admits a left-gcd.

Solution Let S be a nonempty family of braids. Let $\mathcal{X}$ be the set of left-divisors of all the elements of S. As 1 is in $\mathcal{X}$, the set $\mathcal{X}$ is not empty. Then let a be an element of maximal length in $\mathcal{X}$, and b arbitrary in $\mathcal{X}$. The elements a and b have a right-lcm, say $ab' = ba' = c$. By construction, c belongs to $\mathcal{X}$: for any s in S, we have $a \preccurlyeq s$ and $b \preccurlyeq s$, hence $c \preccurlyeq s$. By maximality of the length of a, we thus have $|c| \leqslant |a|$, implying $c \preccurlyeq a$, that is, $c = a$. In other words, $b \preccurlyeq a$ for every b in $\mathcal{X}$. By definition, this means that a is a left-gcd of the whole of the family S. □

Exercise 4.1.9 (Left-lcm, right-gcd) Prove the result: 'For every $n \leqslant \infty$, any two elements of B_n^+ admit a unique left-lcm, and a unique right-gcd.'

Solution We use the antiautomorphism $\widetilde{\ }$ of Lemma 3.1.12. Let $a, b \in B_\infty^+$. The elements $\widetilde{a}$ and $\widetilde{b}$ have a left-gcd c, meaning that

$$\widetilde{a} \preccurlyeq c, \quad \widetilde{b} \preccurlyeq c \quad \text{and} \quad \forall x \in B_\infty^+ \, ((x \preccurlyeq a) \,\&\, (x \preccurlyeq b) \Rightarrow x \preccurlyeq c).$$

Denote $\widetilde{\preccurlyeq}$ the relation of right-divisibility: $a \mathrel{\widetilde{\preccurlyeq}} b$ holds if there exists x satisfying $xa = b$. Then $a \preccurlyeq b$ implies $\widetilde{a} \mathrel{\widetilde{\preccurlyeq}} \widetilde{b}$, and the preceding relations imply

$$a \mathrel{\widetilde{\preccurlyeq}} \widetilde{c}, \quad b \preccurlyeq \widetilde{c} \quad \text{and} \quad \forall x \in B_\infty^+ \, ((x \mathrel{\widetilde{\preccurlyeq}} \widetilde{a}) \,\&\, (x \mathrel{\widetilde{\preccurlyeq}} \widetilde{b}) \Rightarrow x \mathrel{\widetilde{\preccurlyeq}} \widetilde{c}),$$

the point being that the mapping $\widetilde{\ }$ is surjective on B_n^+. However, by definition, $\widetilde{c}$ is the right-lcm of a and b.

The argument for the left-lcm is similar. □

Exercise 4.1.10 (Computation of the gcd)
(i) Let $a, b \in B_\infty^+$. Suppose $ad = bc$ is the right-lcm of a and b, and $ed = fc$ the left-lcm of c and d. Show that there exists g satisfying $a = ge$ and $b = gf$, and that g is the left-gcd of a and b.
(ii) Deduce an algorithm based on the reversing procedure of Definition 3.3.13 to compute the gcd in B_∞^+. [Hint: Use the antiautomorphism $\widetilde{\ }$ for the computation of the left-lcm.]

Solution
(i) We have $ad = bc$, thus $ad = bc$ is a common left-multiple of c and d. Also, $ed = fc$ is the left-lcm of c and d. By the definition of the right-lcm, ad is a left-multiple cd: there exists g satisfying $ad = ged$. By right-cancellativity, we deduce $a = ge$. Moreover, we have $bc = ad = ged = gfc$, hence $b = gf$.

By construction, g is a common left-divisor of a and b. Suppose $x \preccurlyeq a$ and $x \preccurlyeq b$, so $a = xy$ and $b = xz$. We deduce $xyd = ad = bc = xzc$. The element yd, which is also zc, is a common left-multiple of c and d, hence $yd = zc$ is a left-multiple of $ed = fc$. Consequently, x is a left-divisor of g, and g is the left-gcd of a and b.

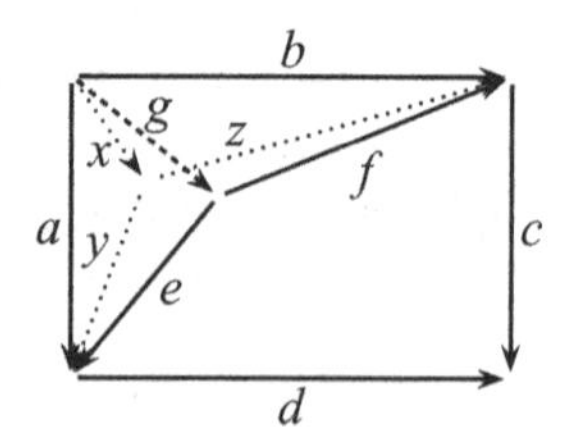

(ii) We use here the procedure of left reversing symmetric to right reversing.

Then $\overline{u}v$ right reverses to $v'\overline{u'}$ if and only if $\widetilde{v}\,\overline{\widetilde{u}}$ left reverses to $\overline{\widetilde{u'}}\widetilde{v'}$, and the two procedures are entirely symmetric the one to the other.

Then, if a and b are two arbitrary elements of B_∞^+, then by construction, a first right reversing of $\overline{a}b$ gives $d\overline{c}$. Then a left reversing of $d\overline{c}$ results in $\overline{e}d$. Finally, a left reversing of $a\overline{e}$ gives g. □

Exercise 4.1.11 (Centre)
(i) An element $a \in B_n^+$ is said to be *quasi-central* if it satisfies

$$\forall b \in B_n^+\, \exists c \in B_n^+\, (ba = ac).$$

Show that every element Δ_n^m is quasi-central.
(ii) Conversely, show that, if a is quasi-central in B_n^+ and $\sigma_i \preccurlyeq a$, then also $\sigma_j \preccurlyeq a$ for $|i - j| = 1$.
(iii) Deduce that, if a is quasi-central, then it is a right-multiple Δ_n, and then show that a is a power of Δ_n.
(iv) Deduce that, for $n \geqslant 3$, the centre of B_n^+ is the submonoid generated by Δ_n^2, then, by using Exercise 3.3.6, that the centre of B_n is the subgroup generated by Δ_n^2.

Solution
(i) We have already observed that, for any braid b in B_n^+ and every integer m, $b\Delta_n^m = \Delta_n^m \phi_n^m(b)$, hence Δ_n^m is quasi-central.
(ii) Suppose a is quasi-central and $\sigma_i \preccurlyeq a$. Let $j := i \pm 1$. By definition, there exists c satisfying $\sigma_j\sigma_i a = ac$. Since ac is a right-multiple of σ_i and $\sigma_j\sigma_i$, it is a right-multiple of their right-lcm, which is $\sigma_j\sigma_i\sigma_j$. Thus $\sigma_i\sigma_j\sigma_j \preccurlyeq \sigma_j\sigma_i a$, hence $\sigma_j \preccurlyeq a$.
(iii) Suppose a is quasi-central with $a \neq 1$. There exists at least one integer i satisfying $\sigma_i \preccurlyeq a$. By (ii), we step by step deduce $\sigma_j \preccurlyeq a$ for every j in $\{1, \ldots, n-1\}$. Thus a must be a right multiple of the right lcm of $\sigma_1, \ldots, \sigma_{n-1}$, which is Δ_n. Write $a = \Delta_n a'$. Then a' is quasi-central. Indeed, let b be arbitrary. There exists c satisfying $\phi_n(b)\Delta_n a' = \Delta_n a' c$, or $\Delta_n b a' = \Delta_n a' c$, so $ba' = a'c$. An induction on ℓ then shows that, if a is quasi-central of length ℓ, then there exists m such that a is Δ_n^m.
(iv) A central element is quasi-central. It only remains to see which powers of Δ_n are central in B_n^+. For $n \geqslant 3$, we have $\sigma_1\Delta_n = \Delta_n\sigma_{n-1} \neq \Delta_n\sigma_1$, so Δ_n is not central. However, we have already seen that Δ_n^2 is central, and the centre of B_n^+ is thus $\{\Delta_n^{2m} \mid m \geqslant 0\}$. Next, by Exercise 3.3.6, an element g of B_n is central if and only if it can be written $\Delta_n^{2m}a$ with a central in B_n^+. Hence the centre of the group B_n is $\{\Delta_n^{2m} \mid m \in \mathbb{Z}\}$. □

Exercise 4.2.8 (Iterated lcm) Give a direct proof of the general result: 'If a, b, c are elements of a left-cancellable monoid M, and if $ab' = ba' = \mathsf{lcm}(a, b)$ and $a'c' = ca'' = \mathsf{lcm}(a', c)$, then $\mathsf{lcm}(a, bc)$ exists and $\mathsf{lcm}(a, bc) = ab'c' = bca''$.'

Solution First, we see $ab'c' = ba'c' = bca''$, so $a(b'c') = (bc)a''$.

Suppose $ay = (bc)x$. Since ba' is the right-lcm of a and b, there exists z satisfying $ay = bcx = ba'z$, or $a'z = cx$ by simplifying by b on the left. Since $a'c'$ is the right-lcm of a' and c, there exists z' satisfying $a'z = a'c'z'$, thus $ay = ba'z = ba'c'z'$, and $ab'c' \preccurlyeq ay$. Hence $ab'c'$ is the right-lcm of a and bc.

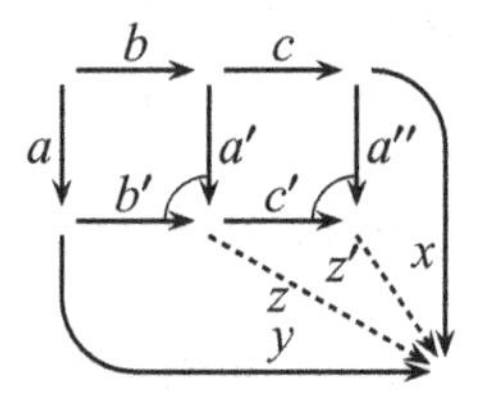

□

Exercise 4.2.13 (Normal forms) Determine the normal forms of σ_1^k, Δ_n^k, $\sigma_1\sigma_2$, and $\sigma_1^2\sigma_2^2$.

Solution The normal form of σ_1^k is $(\sigma_1, \ldots, \sigma_1)$, with k terms, as it is a decomposition of σ_1^k and is normal.

Similarly, the normal form of Δ_n^k, is $(\Delta_n, \ldots, \Delta_n)$, with k terms: it is a decomposition of Δ_n^k and is normal.

The normal form of $\sigma_1\sigma_2$ is $(\sigma_1\sigma_2)$, as $\sigma_1\sigma_2$ is simple.

Finally, the normal form of $\sigma_1^2\sigma_2^2$ is $(\sigma_1, \sigma_1\sigma_2, \sigma_2)$, of length 3, as $(\sigma_1, \sigma_1\sigma_2, \sigma_2)$ is a decomposition of $\sigma_1^2\sigma_2^2$: the braids σ_1, $\sigma_1\sigma_2$ and σ_2 are simple, $H(\sigma_1^2\sigma_2^2) = \sigma_1$, and $H(\sigma_1\sigma_2^2) = \sigma_1\sigma_2$. □

Exercise 4.2.14 (Algorithm) Write pseudocode for the computation of the normal form on B_n^+ based on the direct inductive definition.

Solution Denote CONCAT(S, T) the concatenation of two finite sequences S and T, and REVERSING(w) the result of a (right) reversing of the word w. In this way, if $u \preccurlyeq v$, say $v \equiv^+ uu'$, then REVERSING$(\overline{u}v) \equiv^+ u'$: the reversing determines the left quotient. Here is a proposal.

Note that the loop is a **while**; however, the condition $a \neq \varepsilon$ guarantees that the number of iterations is bounded by $|w|$.

Algorithm 9.4.1 (Normal form on B_n^+)
Input: a word w of $\mathcal{BW}_n^+$
Output: the normal form of $[w]^+$

1: $S \leftarrow ()$
2: **while** $w \neq \varepsilon$ **do**
3: $\quad a \leftarrow H(w)$ $(= \text{GCD}(a, \Delta_n))$
4: $\quad S \leftarrow \text{CONCAT}(S, (a))$
5: $\quad w \leftarrow \text{REVERSING}(\overline{a}w)$
6: **end while**
7: **return** S

□

Exercise 4.2.18 (Normality) Show that (s_1, s_2) is normal if and only if for any left-divisor σ_i of s_2, the braid $s_1\sigma_i$ is not simple.

Solution By Lemma 4.2.17, (s_1, s_2) is normal if and only if $\mathsf{gcd}(\partial_n s_1, s_2,)$ is trivial, meaning, for every left-divisor σ_i of s_2, we have $\sigma_i \not\preccurlyeq \partial_n s_1$, or $s_1\sigma_i \not\preccurlyeq s_1\partial_n s_1$, hence $s_1\sigma_i \not\preccurlyeq \Delta_n$. As s_1 is simple, $s_1\sigma_i \not\preccurlyeq \Delta_n$ is equivalent to the non-simplicity of $s_1\sigma_i$. □

Exercise 4.3.4 (Multiplication by Δ_n^e) Show that, if the Δ_n-normal form of a braid g is $(\Delta_n^m \mid s_1, ..., s_d)$, then for any integer e, that of $\Delta_n^e g$ is $(\Delta_n^{m+e} \mid s_1, ..., s_d)$.

Solution The sequence $(\Delta_n^{p+e} \mid s_1, ..., s_d)$ satisfies the conditions to be a Δ-normal sequence as soon as $(\Delta_n^p \mid s_1, ..., s_d)$ does, and it provides a decomposition of $\Delta_n^e g$ as soon as $(\Delta_n^p \mid s_1, ..., s_d)$ provides a decomposition of g. By unicity of the Δ_n-normal form, it follows that $(\Delta_n^{p+e} \mid s_1, ..., s_d)$ is the Δ_n-normal form of $\Delta_e g$. □

Exercise 4.3.8 (Algorithm) Write the pseudocode for an algorithm determining the Δ_n-normal form of a braid in B_n by introducing the two procedures for left-multiplication and left-division by a simple braid.

Solution For simple braids s, t, denote $\mathsf{DecNormal}(t, s)$ the unique normal pair (s', t') satisfying $ts = s't'$, determined by $s' := H(ts)$ and $ts = s't'$. Here is the proposed algorithm:

Algorithm 9.4.2 (Δ_n-normal form in B_n)

Input: a signed word w in $\mathcal{BW}_n$

Output: the Δ_n-normal form $(m \mid S)$ of the braid $[w]$

```
1:  (m | S) ← (0| )
2:  for p decreasing from |w| to 1 do
3:      if    Positive(w(p)) then
4:          S ← LeftMultΔ(m, S, w(p))
5:      else
6:          S ← LeftDivΔ(m, S, w(p))
7:      end if
8:  end for
9:  while LastTerm(S) = 1 do                ▹ Suppress the last letter of w
10:     SUPPRESS LastTerm(S)
11: end while
12: return S
```

```
1:  function LEFTMULTΔ((m, S): Δn-normal sequence, t: simple braid)
2:      t' ← φn^m(t)
3:      S' ← ()
4:      for k increasing from 1 to |S| do
5:          (s, t) ← DecNormal(t, S(k))
6:          S' ← CONCAT((s), S')
7:      end for
8:      S' ← CONCAT(S', (t))
9:      if S'(1) ≠ Δn then
10:         return (m | S')
11:     else
12:         return (m+1 | sequence(S'))
13:     end if
14: end function
```

```
1: function LeftDivΔ((m, S): Δn-normal sequence, t: simple braid)
2:     r := Δn t^-1
3:     r' := φn^m(s)
4:     S' := ()
5:     for k increasing from 1 to |S| do
6:         (s, t) := DecNormal(t, S(k))
7:         S' := concat((s), S')
8:     end for
9:     S' := concat(S', (t))
10:    if S'(1) ≠ Δn then
11:        return (m | S')
12:    else
13:        return (m+1 | sequence(S'))
14:    end if
15: end function
```

□

9.5 Exercises of Chapter 5

Exercise 5.1.3 (Permutation) Show that isotopic homeomorphisms of $\mathbb{D}_n$ necessarily induce the same permutation of the marked points.

Solution For any homeomorphism ϕ of $\mathbb{D}_n$, let $\mathsf{perm}(\phi)$ be the associated permutation of the marked points, determined by $\phi(P_k) = P_{\mathsf{perm}(\phi)(k)}$ for every k. By definition, ϕ depends continuously on the isotopy class of ϕ, thus, by composition, so does the mapping $\phi \mapsto \mathsf{perm}(\phi)$, with values in the symmetric group $\mathfrak{S}_n$. The latter is a discrete space (with $n!$ elements), hence the value of $\mathsf{perm}(\phi)$ is constant on each isotopy class. □

Exercise 5.2.5 (Trichotomy)
(i) Admitting the acyclicity property, show that, for any braid g, the properties 'g is σ_i-positive', 'g is σ_i-neutral', and 'g is σ_i-negative' are mutually exclusive.
(ii) Admitting the acyclicity property, show that, for any braid g, the properties 'g is σ-positive', 'g is trivial', and 'g is σ-negative' are mutually exclusive.
(iii) Show that a braid can be σ_i-positive for at most one value of i.

Solution
(i) Let g be an arbitrary braid, and suppose g simultaneously σ_i-positive and σ_i-negative. By definition, there exists a σ_i-positive w representative of g, and a σ_i-negative representative w' of g. The words w and $\overline{w'}$ are thus σ_i-positive, as is the product $w\overline{w'}$. Thus, by definition, the braid $[w\overline{w'}]$ is σ-positive. However, this braid is gg^{-1}, so is trivial. By the acyclicity property, a σ-positive braid cannot be trivial. This contradicts the hypothesis.

The argument is the same for 'g is σ_i-positive', 'g is σ_i-neutral': suppose $g = [w] = [w']$ with w σ_i-positive and w' σ_i-neutral, the word $w\overline{w'}$ is σ_i-positive and represents 1, leading to the same contradiction.

Suppose now $g = [w] = [w']$ with w σ_i-negative and w' σ_i-neutral, the word $\overline{w}w'$ is σ_i-positive and represents $g^{-1}g = 1$, again a contradiction.
(ii) We reason in the same manner. Suppose g is simultaneously σ-positive and σ-negative. By definition, there exists a σ_i-positive representative w of g, and a σ_j-negative w' representative of g. If $i \leqslant j$, the word w' is either σ_i-neutral, or σ_i-negative, and $w\overline{w'}$ is σ_i-positive. However, it represents $gg^{-1} = 1$, exposing the same contradiction as in (i). If $i > j$, the word w is σ_j-negative, and $\overline{w}w'$ is σ_j-positive. However, it represents $gg^{-1} = 1$, leading always to the same contradiction as in (i).

The cases bringing into play 'g is trivial' are direct consequences of the acyclicity property.
(iii) Let w be a σ_i-positive word and w' a σ_j-positive word, with $i < j$. Suppose $g = [w] = [w']$. Then w' is σ_i-neutral. Thus $w\overline{w'}$ is a σ_i-positive word, and $[w\overline{w'}]$ is a σ_i-positive braid, hence σ-positive. However we have $[w\overline{w'}] = gg^{-1} = 1$, contradicting the acyclicity property. □

Exercise 5.2.8 (Image) Show that $\widehat{\sigma_1}$ sends any reduced word terminating in x_1 to a reduced word terminating in x_1^{-1}.

Solution Consider a reduced word terminating in x_1, say vx_1. Then

$$\widehat{\sigma_1}(vx_1) = \mathsf{red}(\widehat{\sigma_1}(v)x_1x_2x_1^{-1}).$$

Suppose $\widehat{\sigma_1}(vx_1)$ does not end in x_1^{-1}. This means that, in the above reduction, the final letter x_1^{-1} disappears by reduction with a letter x_1 which, as it could not be the one preceding x_2, must necessarily come from $\widehat{\sigma_1}(v)$. By the definition of $\widehat{\sigma_1}$, a letter x_1 in $\widehat{\sigma_1}(v)$ comes either from a letter $x_1^{\pm 1}$ in v, or from a letter x_2.

Suppose the letter x_1 in play comes from a letter x_1^e, $e = \pm 1$, in v, so $v = v_1x_1^ev_2$. We thus find

$$\widehat{\sigma_1}(vx_1) = \mathsf{red}(\widehat{\sigma_1}(v_1)\,x_1\,\underline{x_2^e x_1^{-1}\widehat{\sigma_1}(v_2)x_1x_2\,x_1^{-1}}),$$

where the underlined factor reduces to the empty word. We thus have $\widehat{\sigma_1}(v_2) = x_1 x_2^{-1-e} x_1^{-1}$, which, by definition of $\widehat{\sigma_1}$, requires $v_2 = x_1^{-2}$ for $e = 1$, and $v_2 = 1$ for $e = -1$. Then the word vx_1 is either $v_1 x_1 x_1^{-2} x_1$, or $v_1 x_1^{-1} x_1$, and neither is reduced. This case is thus impossible.

Suppose now that the letter x_1 in play comes from a letter x_2 in v, so $v = v_1 x_2 v_2$. Then

$$\widehat{\sigma_1}(vx_1) = \mathsf{red}(\widehat{\sigma_1}(v_1)\, x_1\, \underline{\widehat{\sigma_1}(v_2) x_1 x_2\, x_1}),$$

where the underlined factor reduces to the empty word. Hence $\widehat{\sigma_1}(v_2) = x_2^{-1} x_1^{-1}$, implying $v_2 = x_2^{-1} x_1^{-1}$. Then $vx_1 = v_1 x_2 x_2^{-1} x_1^{-1} x_1$, which is not reduced. Hence this case is also impossible, thus the hypothesis that $\widehat{\sigma_1}(vx_1)$ does not finish with x_1^{-1} is contradictory. □

Exercise 5.3.7 (Fundamental group)
(i) Show that the fundamental group is an invariant of homeomorphism: two homeomorphic spaces have isomorphic fundamental groups.
(ii) Show that, if $\mathcal{X}$ is a subspace of $\mathbb{R}^n$ such that, for any P in $\mathcal{X}$, the segment $[0, P]$ is included in $\mathcal{X}$ ('starred' space), then the fundamental group of $\mathcal{X}$ is trivial.

Solution We show that every loop in $\mathcal{X}$ can be continuously deformed into the trivial loop. Indeed, let γ be a loop in $\mathcal{X}$, and γ_s the image of γ by the homothety of centre O and ratio s for $s \in [0, 1]$. The hypothesis that $\mathcal{X}$ is starred guarantees that γ_s is included in $\mathcal{X}$. Hence, the family $(\gamma_s)_{s\in[0,1]}$ is a homotopy joining γ to the trivial loop. □

9.6 Exercises of Chapter 6

Exercise 6.1.7 (Reduction) Verify algebraically that every permitted handle is equivalent to its reduct.

Solution By definition, a permitted σ_i-handle is a word of the form $v = \sigma_i^e u \sigma_i^{-e}$ with $e = \pm 1$, u σ_i-neutral and the letters σ_{i+1} and σ_{i+1}^{-1} not both present in u. Making apparent the (eventual) letters $\sigma_{i+1}^{\pm 1}$ in u, we can write

$$v = \sigma_i^e\, u_0\, \sigma_{i+1}^d\, u_1\, \cdots\, u_{r-1}\, \sigma_{i+1}^d\, u_r\, \sigma_i^{-e},$$

with $d = \pm 1$ and $u_0, ..., u_r$ σ_{i+1}-neutral. The reduct of v is then

$$v' = u_0\, \sigma_{i+1}\sigma_i\sigma_{i+1}^{-1}\, u_1\, \cdots\, u_{r-1}\, \sigma_{i+1}\sigma_i\sigma_{i+1}^{-1}\, u_r\, \sigma_i^{-1}\sigma_i.$$

Suppose $e = -1$ and $d = 1$, so

$$v = \sigma_i^{-1}\, u_0\, \sigma_{i+1}\, u_1 \,\cdots\, u_{r-1}\, \sigma_{i+1}\, u_r\, \sigma_i.$$

The idea is to migrate the letter σ_i^{-1} from left to right. As the only generators appearing in u_0 are $\sigma_j^{\pm 1}$ with $j \geqslant i{+}2$, we have $\sigma_i^{-1} u_0 \equiv u_0 \sigma_i^{-1}$, and thus

$$v \equiv u_0\, \sigma_i^{-1}\, \sigma_{i+1}\, u_1 \,\cdots\, u_{r-1}\, \sigma_{i+1}\, u_r\, \sigma_i.$$

Then $\sigma_i^{-1}\, \sigma_{i+1} \equiv \sigma_{i+1}\sigma_i\sigma_{i+1}^{-1}\, \sigma_i^{-1}$, hence

$$v \equiv u_0\, \sigma_{i+1}\sigma_i\sigma_{i+1}^{-1}\, \sigma_i^{-1}\, u_1 \,\cdots\, u_{r-1}\, \sigma_{i+1}\, u_r\, \sigma_i.$$

Repeating the same with $\sigma_i^{-1} u_1$, we have $\sigma_i^{-1} u_1 \equiv u_1 \sigma_i^{-1}$, so

$$v \equiv u_0\, \sigma_{i+1}\sigma_i\sigma_{i+1}^{-1}\, u_1\, \sigma_i^{-1} \,\cdots\, u_{r-1}\, \sigma_{i+1}\, u_r\, \sigma_i.$$

We continue to migrate the letter σ_i^{-1} from left to right by commutation with the words u_k and by $\sigma_i^{-1} \cdot \sigma_{i+1} \equiv \sigma_{i+1}\sigma_i\sigma_{i+1}^{-1} \cdot \sigma_i^{-1}$, giving

$$v \equiv u_0\, \sigma_{i+1}\sigma_i\sigma_{i+1}^{-1}\, u_1 \,\cdots\, u_{r-1}\, \sigma_{i+1}\sigma_i\sigma_{i+1}^{-1}\, u_r\, \sigma_i^{-1}\, \sigma_i.$$

A free reduction $\sigma_i^{-1}\, \sigma_i \equiv \varepsilon$ finally gives

$$v \equiv u_0\, \sigma_{i+1}\sigma_i\sigma_{i+1}^{-1}\, u_1 \,\cdots\, u_{r-1}\, \sigma_{i+1}\sigma_i\sigma_{i+1}^{-1}\, u_r = v'.$$

The case $e = 1$ and $d = -1$ is treated in the same manner, using $\sigma_i \cdot \sigma_{i+1}^{-1} \equiv \sigma_{i+1}^{-1}\sigma_i^{-1}\sigma_{i+1} \cdot \sigma_i$ in place of $\sigma_i^{-1} \cdot \sigma_{i+1} \equiv \sigma_{i+1}\sigma_i\sigma_{i+1}^{-1} \cdot \sigma_i^{-1}$.

The other two cases correspond to migrating the final letter $\sigma_i^{\pm 1}$ from right to left, appealing to, respectively, for $e = d = -1$, $\sigma_{i+1}^{-1} \cdot \sigma_i \equiv \sigma_i \cdot \sigma_{i+1}\, \sigma_i^{-1}\, \sigma_{i+1}^{-1}$, and for $e = d = 1$, $\sigma_{i+1} \cdot \sigma_i^{-1} \equiv \sigma_i^{-1} \cdot \sigma_{i+1}^{-1}\, \sigma_i\, \sigma_{i+1}$. □

Exercise 6.3.3 (Injectivity) Suppose F is a function on B_n such that $F(g) \neq F(1)$ implies $F(\mathsf{dec}(g)) \neq F(1)$ and $F(g^{-1}) \neq F(1)$. Show that, if $F(g) \neq F(1)$ for every σ_1-positive braid g, then $F(g) \neq F(1)$ for any non-trivial braid.

Solution Let g be a non-trivial braid. Then g is either σ-positive or σ-negative. Suppose g is σ-positive. There exists i such that g is σ_i-positive. For $i = 1$, we have $F(g) \neq F(1)$ by hypothesis. For $i \geqslant 2$, there exists a σ_1-positive g' satisfying $g = \mathsf{dec}^{i-1}(g')$. By hypothesis, $F(g') \neq F(1)$, which implies
$F(g) = F(\mathsf{dec}^{i-1}(g') \neq F(1)$.

Now suppose g is σ-negative. There exists i such that g is σ_i-negative. For $i = 1$, the braid g^{-1} is σ_1-positive. We have $F(g^{-1}) \neq F(1)$ by hypothesis, which implies $F(g) \neq F(1)$. For $i \geqslant 2$, there exists a σ_1-positive g' satisfying

$g = \mathrm{dec}^{i-1}(g'^{-1}) = (\mathrm{dec}^{i-1}(g'))^{-1}$. By hypothesis, $F(g') \neq F(1)$, hence $F(g) = F((\mathrm{dec}^{i-1}(g'))^{-1})$, consequently $F(g) \neq F(1)$. □

Exercise 6.3.9 (Bi-orderability) Suppose $n \geqslant 3$. Show that no total order on B_n can be compatible with both the left and right products.

Solution Suppose $<$ is a total order compatible with the left and right products. Suppose $\sigma_1 < \sigma_2$. By compatibility with the left product, we deduce $\Delta_3^{-1}\sigma_1 < \Delta_3^{-1}\sigma_2$, then, by compatibility with the right product, $\Delta_3^{-1}\sigma_1\Delta_3 < \Delta_3^{-1}\sigma_2\Delta_3$, meaning $\sigma_2 < \sigma_1$, in contradiction with the hypothesis. Symmetrically, $\sigma_2 < \sigma_1$ would imply $\sigma_1 < \sigma_2$. Hence the existence of an order compatible with the left and right products is untenable. □

Exercise 6.3.10 (Right orderability) Show that the relation $a^{-1} < b^{-1}$ is a total order on braids compatible with the right product.

Solution Denote $a \prec b$ the relation $a^{-1} < b^{-1}$. Then $\prec$ is a total order on B_∞. Suppose $a \prec b$, and let c be arbitrary. By hypothesis, ab^{-1}, or $(a^{-1})^{-1}b^{-1}$, is a σ-positive braid. Consequently, $acc^{-1}b^{-1}$, or $((ac)^{-1})^{-1}(bc)^{-1}$, is σ-positive. From this, we deduce $(ac)^{-1} < (bc)^{-1}$, hence $ac \prec bc$: the relation $\prec$ is compatible with the right product. □

Exercise 6.3.11 (Non-Conradian order) Let $a := \sigma_2^{-1}\sigma_1$ and $b := \sigma_2^{-2}\sigma_1$. Show that $a > 1$ and nonetheless $ba^p < b$ for every p ('non-Conradian order').

Solution The words $a = \sigma_2^{-1}\sigma_1$ and $b = \sigma_2^{-2}\sigma_1$ are σ_1-positive, witnessing $a > 1$ and $b > 1$. We show that for every $p \geqslant 0$, the word $a^{-1}ba^p$ has a σ_1-negative representative, witnessing $a^{-1}ba^p < 1$, and, equivalently, $ba^p < a$. Hence there is an integer p with $a < ba^p$.

For $p = 0$, we find

$$a^{-1}b = \sigma_1^{-1}\sigma_2\sigma_2^{-2}\sigma_1 = \sigma_1^{-1}\sigma_2^{-1}\sigma_1 = \sigma_2\sigma_1^{-1}\sigma_2^{-1},$$

explicitly σ_1-negative. For $p = 1$, we find

$$a^{-1}ba = \sigma_1^{-1}\sigma_2\sigma_2^{-2}\sigma_1\sigma_2^{-1}\sigma_1 = \sigma_1^{-1}\sigma_2^{-1}\sigma_1\sigma_2^{-1}\sigma_1 = \sigma_2\sigma_1^{-1}\sigma_2^{-2}\sigma_1 = \sigma_2^{2}\sigma_1^{-2}\sigma_2^{-1},$$

again explicitly σ_1-negative. Finally, we establish

$$a^{-1}ba^p = \sigma_2^{2}(\sigma_1^{-1}\sigma_2)^{p-1}\sigma_1^{-2}\sigma_2^{-1}$$

by induction on $p \geqslant 1$. The formula is satisfied for $p = 1$. For $p \geqslant 2$, using the induction hypothesis and the equality $\sigma_1^{-1}\sigma_2^{-2}\sigma_1 = \sigma_2\sigma_1^{-2}\sigma_2^{-1}$, we find

$$\begin{aligned} a^{-1}ba^p &= (\sigma_2^2(\sigma_1^{-1}\sigma_2)^{p-2}\sigma_1^{-2}\sigma_2^{-1})(\sigma_2^{-1}\sigma_1) \\ &= \sigma_2^2(\sigma_1^{-1}\sigma_2)^{p-2}\sigma_1^{-1}\sigma_2\sigma_1^{-2}\sigma_2^{-1} \\ &= \sigma_2^2(\sigma_1^{-1}\sigma_2)^{p-1}\sigma_1^{-2}\sigma_2^{-1}, \end{aligned}$$

again a σ_1-negative expression. □

Exercise 6.3.12 (Flipped order)
(i) Show that the order $<$ is compatible with the shift on B_∞.
(ii) For a, b in B_n, declare $a \mathrel{\tilde{<}_n} b$ when $\phi_n(a) < \phi_n(b)$, where ϕ_n is the automorphism exchanging σ_i and σ_{n-i} for every i. Show that $\tilde{<}_n$ is a total order on B_n, compatible with the left product, and that the order $\tilde{<}_n$ on B_n is the restriction of the order $\tilde{<}_{n+1}$ on B_{n+1}. What is the smallest element of B_n^+ according to $\tilde{<}$?

Solution
(i) Recall that shift is an injective endomorphism of B_∞ sending σ_i to σ_{i+1} for every i (cf. Exercises 3.1.14 and 3.3.2). The property 'c is σ_i-positive' is equivalent to '$\mathrm{dec}(c)$ σ_{i+1}-positive', so 'c is σ-positive' is equivalent to '$\mathrm{dec}(c)$ is σ-positive'. Hence $a < b$ is equivalent to $\mathrm{dec}(a) < \mathrm{dec}(b)$.
(ii) The relation $\tilde{<}$ is a total order on B_n. Suppose $a \tilde{<} b$ and let c be arbitrary: we have $\phi_n(a) < \phi_n(b)$, thus $\phi_n(c)\phi_n(a) < \phi_n(c)\phi_n(b)$, which is $\phi_n(ca) < \phi_n(cb)$, alias $ca \tilde{<} cb$: the order $\tilde{<}$ is compatible with the left product.

Suppose $a, b \in B_n$ and $a \mathrel{\tilde{<}_n} b$. By definition, $\phi_n(a) < \phi_n(b)$. By (i) we deduce $\mathrm{dec}(\phi_n(a)) < \mathrm{dec}(\phi_n(b))$. However, by construction, we have $\mathrm{dec}(\phi_n(a)) = \phi_{n+1}(a)$ and $\mathrm{dec}(\phi_n(b)) = \phi_{n+1}(b)$, thus $\phi_{n+1}(a) < \phi_{n+1}(b)$, or $a \mathrel{\tilde{<}_{n+1}} b$. Conversely, suppose $a \mathrel{\tilde{<}_{n+1}} b$. If we did not have $a \mathrel{\tilde{<}_n} b$, then we would necessarily have $a = b$ or $b \mathrel{\tilde{<}_n} a$, which excludes $a \mathrel{\tilde{<}_{n+1}} b$. Thus $a \mathrel{\tilde{<}_n} b \Leftrightarrow a \mathrel{\tilde{<}_{n+1}} b$, and the order $\tilde{<}_n$ is the restriction of $\tilde{<}_{n+1}$ to B_n. As a result there exists a total order $\tilde{<}$, well-defined on B_∞, whose restriction to B_n is $\tilde{<}_n$. By construction, we have

$$1 \tilde{<} \sigma_1 \tilde{<} \sigma_1^2 \tilde{<} \sigma_1^3 \tilde{<} \cdots \tilde{<} \sigma_2 \tilde{<} \sigma_3 \tilde{<} \cdots.$$

Let $g \in B_\infty^+ \setminus B_2^+$: every expression of g contains a generator σ_i with $i \geqslant 2$, witnessing the relation $\sigma_1 < g$: the smallest non-trivial element of B_∞^+ is σ_1. □

9.7 Exercises of Chapter 7

Exercise 7.1.12 (Dynnikov formulas) Starting from $(0, 1, 0, 1, ..., 0, 1)$, verify that the formulas of Proposition 7.1.11 give the values $(1, 0, 0, 2, 0, 1, ..., 0, 1)$ for σ_1, read from Figure 7.3. What are the values for σ_1^{-1}?

Solution Let S be the sequence of coordinates of σ_1 in B_n. Applying the formulas of Proposition 7.1.11, S is equal to $F^+(0, 1, 0, 1)$, followed by $(0, 1)$ $n-3$ times. The parameter 't_1' is $t_1 := 0 - 1^- - 0 + 1^+ := 1$. We thus find

$$
\begin{aligned}
F_1^+(0, 1, 0, 1) &:= 0 + 1^+ + 1^+ - 1^+ := 1,\\
F_2^+(0, 1, 0, 1) &:= 1 - 1^+ := 0,\\
F_3^+(0, 1, 0, 1) &:= 0 + 1^- + 1^- + 1^- := 0,\\
F_4^+(0, 1, 0, 1) &:= 1 + 1^+ := 2,
\end{aligned}
$$

hence $S = (1, 0, 0, 2, 0, 1, ..., 0, 1)$.

Similarly, let S' be the sequence of coordinates of σ_1^{-1} in B_n. This time, S' is equal to $F^-(0, 1, 0, 1)$, followed by $n-3$ times $(0, 1)$. The parameter 't_2' is $t_2 := 0 + 1^- - 0 - 1^+ := -1$.We thus find

$$
\begin{aligned}
F_1^-(0, 1, 0, 1) &:= 0 - 1^+ - 1^+ - 1^+ := -1,\\
F_2^-(0, 1, 0, 1) &:= 1 + (-1)^- := 0,\\
F_3^-(0, 1, 0, 1) &:= 0 - 1^- + 0^- - (-1)^- := 0,\\
F_4^-(0, 1, 0, 1) &:= 1 - (-1)^- := 2,
\end{aligned}
$$

hence $S' = (-1, 0, 0, 2, 0, 1, ..., 0, 1)$.

More generally, we could verify that, for k positive, we have

$$
\begin{aligned}
\rho_D(\sigma_1^k) &:= (1, -k+1, 0, k+1, 0, 1, ..., 0, 1),\\
\rho_D(\sigma_1^{-k}) &:= (-1, -k+1, 0, k+1, 0, 1, ..., 0, 1).
\end{aligned}
$$

□

Exercise 7.2.3 (σ-negative) Show that the first non-zero odd coordinate of a σ-negative braid is a strictly negative integer.

Solution Let w be a σ_1-negative word, $w = w_0\sigma_1^{-1}w_1\sigma_1^{-1} \cdots \sigma_1^{-1}w_p$ where $\sigma_1^{\pm 1}$ is absent from the words w_k. We follow the first two coordinates of $(0, 1, ..., 0, 1){\bullet}w$ as the letters are progressively added, starting initially with $(0, 1)$. For as long as σ_1^{-1} is not encountered, we remain with $(0, 1)$. As soon as the first letter σ_1^{-1} is met, the first coordinate becomes $F_1^-(0, 1, x_2, y_2)$, thus, by definition, $0-1^+-z^+$ for a certain integer z (in fact $-x_2$), hence a strictly negative integer. Next, crossing a word w_k does not change the first coordinate, while crossing a letter σ_1^{-1} can only diminish it, as, by definition, we only subtract a non-negative integer from it. Hence the result is established for a σ_1-negative braid.

Now let w be a σ_i^{-1}-negative braid word with $i \geqslant 2$. Then w is $\mathsf{dec}^{i-1}(w')$ where w' is σ_1-negative. By construction, the sequence of coordinates of w is that of w' preceded by $(0, 1, ..., 0, 1)$ with $2i - 2$ terms. The first $i - 1$ odd coordinates of w are thus zero, whereas the ith is the first coordinate of w', strictly negative as shown above. □

9.8 Exercises of Chapter 8

Exercise 8.1.1 (Generating family)
(i) Show that $\{\sigma_1, ..., \sigma_{n-1}\}$ is a minimal generating family of B_n, in the sense that no proper subfamily is generating.
(ii) Show that, for every $n \geqslant 3$, the family $\{\sigma_1, \cdots \sigma_{n-1}\}$ generates B_n.

Solution
(i) Let G be the subgroup of B_n generated by $\{\sigma_1, ..., \sigma_{i-1}\} \cup \{\sigma_{i+1}, ..., \sigma_n\}$. We want to show that σ_i is not an element of G. Let g be arbitrary in G. By induction on the length of a representative of g, we show that $\mathsf{perm}(g)$ is of the form ff', where f belongs to $\mathfrak{S}_{\{1,...,i\}}$ and f' belongs to $\mathfrak{S}_{\{i+1,...,n\}}$: in other words, $\mathsf{perm}(g)$ leaves the sets $\{1, ..., i\}$ and $\{i+1, ..., n\}$ globally invariant. Hence σ_i does not belong to G, as $\mathsf{perm}(\sigma_i)$, which is the transposition s_i, does not leave $\{1, ..., i\}$ and $\{i+1, ..., n\}$ invariant as it sends i to $i+1$.
(ii) By Lemma 3.1.23, for $2 \leqslant i \leqslant n$, we have $\sigma_2 \cdot \sigma_1 \cdots \sigma_n = \sigma_1 \cdots \sigma_n \cdot \sigma_1$, hence the conjugate of σ_1 by $\sigma_1 \cdots \sigma_{n-1}$ is σ_2:

$$\sigma_2 = (\sigma_1 \cdots \sigma_n)\, \sigma_1\, (\sigma_1 \cdots \sigma_n)^{-1},$$

and, similarly, for $2 \leqslant i \leqslant n$,

$$\sigma_i = (\sigma_1 \cdots \sigma_n)^{i-1}\, \sigma_1\, (\sigma_1 \cdots \sigma_n)^{-i+1}.$$

The subgroup of B_n generated by σ_1 and $\sigma_1 \cdots \sigma_{n-1}$ thus contains $\sigma_1, \sigma_2, ..., \sigma_{n-1}$, hence is the whole of B_n. □

Exercise 8.1.4 Prove the result: 'In terms of the generators $a_{i,j}$, the group B_n is presented by the relations $a_{i,j}a_{i',j'} = a_{i',j'}a_{i,j}$ if the intervals $[i, j]$ and $[i', j']$ are disjoint or nested, and $a_{i,j}a_{j,k} = a_{j,k}a_{i,k} = a_{i,k}a_{i,j}$ for $1 \leqslant i < j < k \leqslant n$.'

Solution It is easy to verify from the definition of the generators $a_{i,j}$ in terms of the $\sigma_i^{\pm 1}$ that the indicated relations are satisfied. If the intervals $[i, j]$ and

$[i', j']$ are disjoint, each of the $\sigma_p^{\pm 1}$ figuring in $a_{i,j}$ commutes with each of the $\sigma_q^{\pm 1}$ figuring in $a_{i',j'}$, hence $a_{i,j}$ and $a_{i',j'}$ commute. If the interval $[i', j']$ is contained in $[i', j']$, each of the $\sigma_q^{\pm 1}$ figuring in $a_{i',j'}$ commutes with $a_{i,j}$: with the notation of Lemma 3.1.23, we have $a_{i,j} = \sigma_{i,j}\sigma_{i,j-1}^{-1}$, and, for $i < k < j$, by Lemma 3.1.23, we have

$$\sigma_k \cdot a_{i,j} = \sigma_k \cdot \sigma_{i,j}\sigma_{i,j-1}^{-1} = \sigma_{i,j} \cdot \sigma_{k-1} \cdot \sigma_{i,j-1}^{-1} = \sigma_{i,j}\sigma_{i,j-1}^{-1} \cdot \sigma_k = a_{i,j} \cdot \sigma_k.$$

Thus $a_{i,j}$ commutes with $a_{i',j'}$. Now let $i < j < k$. Each of the generators $\sigma_p^{\pm 1}$ figuring in $\sigma_{i,j-1}^{-1}$ commutes with each of the generators $\sigma_q^{\pm 1}$ figuring in $a_{j,k}$. We can thus, by introducing a trivial term $\sigma_{i,j}^{-1}\sigma_{i,j}$, write

$$\begin{aligned} a_{i,j}a_{j,k} &= \sigma_{i,j}\sigma_{i,j-1}^{-1}a_{j,k} \\ &= \sigma_{i,j}a_{j,k}\sigma_{i,j-1}^{-1} \\ &= \sigma_{i,j}a_{j,k}\sigma_{i,j}^{-1}\sigma_{i,j}\sigma_{i,j-1}^{-1} \\ &= (\sigma_{i,j}a_{j,k}\sigma_{i,j}^{-1}) \cdot (\sigma_{i,j}\sigma_{i,j-1}^{-1}) = a_{i,k}a_{i,j} \end{aligned}$$

by regrouping the terms. The verification for $a_{i,j}a_{j,k} = a_{j,k}a_{i,k}$ is analogous.

Moreover, the relations above imply the Artin relations. Indeed, a special case of disjoint intervals is $[i, i{+}1]$ and $[j, j{+}1]$ with $j \geqslant i + 2$, in which case we obtain

$$\sigma_i\sigma_j = a_{i,i+1}a_{j,j+1} = a_{j,j+1}a_{i,i+1} = \sigma_j\sigma_i.$$

Also, $a_{i,i+1}a_{i+1,i+2} = a_{i+1,i+2}a_{i,i+1}$, which translates to $\sigma_i\sigma_{i+1} = \sigma_{i+1}\sigma_{i+1}\sigma_i^{-1}\sigma_i^{-1}$, implying the relation $\sigma_i\sigma_{i+1}\sigma_i = \sigma_{i+1}\sigma_{i+1}\sigma_i$. □

Exercise 8.1.7 (Conjugation by Δ_n^*) Show that the conjugation by Δ_n^* is the automorphism of B_n^{+*} sending $a_{i,j}$ to $a_{i+1,j+1}$ for $j < n$, and $a_{i,n}$ to $a_{1,i+1}$. What is its order? What is the associated geometric transformation to the correspondence of Figure 8.2?

Solution Let ϕ_n^* be the conjugation by Δ_n^* defined by $\phi_n^*(g) := \Delta_n^* g \Delta_n^{*-1}$. By definition, we have $\Delta_n^* = \sigma_1 \cdots \sigma_{n-1}$, and, as seen in Exercise 8.1.1, we have $\phi_n^*(\sigma_i) = \sigma_{i+1}$ for $1 \leqslant i < n$. By substituting in the definition the generators $a_{i,j}$, we see that, for $j < n$, $\phi_n^*(a_{i,j}) = a_{i+1,j+1}$. Moreover, we have

$$\phi_n^*(\sigma_{n-1}) := \Delta_n^*\sigma_{n-1}\Delta_n^{*-1} = \sigma_1 \cdots \sigma_{n-1}\sigma_{n-1}\sigma_{n-1}^{-1} \cdots \sigma_1^{-1} = a_{1,n},$$

or $\phi_n^*(a_{n-1,n}) = a_{1,n}$.

We deduce, for $1 \leqslant i < n$,

$$\begin{aligned} \phi_n^*(a_{i,n}) &= \phi_n^*(\sigma_i \cdots \sigma_{n-2}\sigma_{n-1}\sigma_{n-2}^{-1} \cdots \sigma_i^{-1}) \\ &= \sigma_{i+1} \cdots \sigma_{n-1}a_{1,n}\sigma_{n-1}^{-1} \cdots \sigma_{i+1}^{-1}. \end{aligned} \qquad (*)$$

We proceed with a reverse induction on $i < n$. For $i := n-2$, the equality $(*)$ is

$$\phi_n^*(a_{n-1,n}) = \sigma_{n-1}a_{1,n}\sigma_{n-1}^{-1} = a_{n-1,n}a_{1,n}a_{n-1,n}^{-1}:$$

however the relations between the $a_{i,j}$ (Lemma 8.1.3, cf. Exercise 8.1.4) imply $a_{n-1,n}a_{1,n} = a_{1,n-1}a_{n-1,n}$, hence $\phi_n^*(a_{n-2,n}) = a_{1,n-1}$. For $i := n-3$, the equality $(*)$ is

$$\phi_n^*(a_{n-1,n}) = \sigma_{n-2}a_{1,n-1}\sigma_{n-2}^{-1} = a_{n-2,n-1}a_{1,n-1}a_{n-2,n-1}^{-1}:$$

by the relations between the $a_{i,j}$, we have $a_{n-2,n-1}a_{1,n-1} = a_{1,n-2}a_{n-2,n-1}$, thus $\phi_n^*(a_{n-3,n}) = a_{1,n-2}$, and so on...

It follows that, for every i, we have

$$\begin{aligned}(\phi_n^*)^n(\sigma_i) &= (\phi_n^*)^{i+1}(\sigma_{n-1}) = (\phi_n^*)^i(a_{1,n}) \\ &= (\phi_n^*)^{i-1}(a_{1,2}) = (\phi_n^*)^{i-1}(\sigma_1) = \sigma_i :\end{aligned}$$

the automorphism ϕ_n^* is of order n. It corresponds to a clockwise rotation of an angle $2\pi/n$ of the representation of Figure 8.2. In contrast, the automorphism ϕ_n which is the conjugation by Δ_n is of order 2, and corresponds to a symmetry. □

Exercise 8.1.9 Prove the result: 'For $1 \leqslant i < n$, let $\Sigma_{i,n}$ be the $n \times n$ matrix equal to the identity matrix, except for the 2×2 squares at the intersection of the rows and columns i and $i+1$, which are $\begin{pmatrix} 1-t & t \\ 1 & 0 \end{pmatrix}$. Then $\Sigma_{i,n}$ is invertible in $\mathrm{Mat}_n(\mathbb{Z}[t, t^{-1}])$, and the mapping ρ_{B}, sending σ_i to $\Sigma_{i,n}$ for every i, induces a linear representation of B_n.'

Solution The inverse of $\Sigma_{2,2}$ is $\begin{pmatrix} 0 & 1 \\ t^{-1} & 1-t^{-1} \end{pmatrix}$. The invariance of ρ_{B} with respect to the commutation relations is immediate. Given the action of shifts, it suffices to compare $\rho_{\mathsf{B}}(\sigma_1\sigma_2\sigma_1)$ and $\rho_{\mathsf{B}}(\sigma_2\sigma_1\sigma_2)$. However both are equal to

$$\begin{pmatrix} 1-t & t-t^2 & t^2 \\ 1-t & t & 0 \\ 1 & 0 & 0 \end{pmatrix}.$$

□

Glossary

1	unit braid (trivial braid).
$\mathsf{Aut}(G)$	automorphism group of G.
$\mathcal{BD}_n$, $\mathcal{BD}_n^{\mathsf{aff}}$	families of braid diagrams.
B_n^{+*}	dual braid monoid.
B_n^+, B_∞^+	braid monoids.
B_n	space of n-strand braids.
$B_n(\mathbb{A})$	braid group on the torus $\mathbb{A}$.
$B_n(\mathbb{S})$	braid group on the sphere $\mathbb{S}$.
$B_n(\Sigma)$	space (group) of braids on Σ.
$\mathcal{BW}_n$	family of n-strand braid words.
$D_1 \cdot D_2$	product of braid diagrams.
$\mathbb{D}_n$	disk with n marked points.
$\mathbb{D}_n^-$	disk with n holes.
$\mathsf{deg}(a)$	degree of a positive braid.
F_S	free group based on S.
$\mathcal{GB}_n$	family of n-strand geometric braids.
$\mathcal{GB}_n^{\mathsf{aff}}$	piecewise linear braids.
$\mathcal{GB}_n^{\mathsf{aff,reg}}$	regular piecewise linear geometric braids.
$\mathcal{GB}_n^{\mathsf{aff,semireg}}$	semi-regular piecewise linear geometric braids.
$\mathcal{GB}_n^\infty$	C^∞ braids.
$\mathsf{Homeo}^{\mathsf{st}}(\mathcal{X})$	group of stratified homeomorphisms of $\mathcal{X}$.
$\mathsf{inv}(f)$	inversion number of a permutation f.
$L{\bullet}\beta$	action of a braid on a lamination.
L_*	base lamination.
$\mathcal{MCG}_+(\Sigma, \{P_1, ..., P_n\})$	mapping class group of the surface Σ with marked points $P_1, ..., P_n$.
$\overline{w}$	word symmetric to w.

$\Omega_1(\mathcal{X}, P_*)$	set of loops in $\mathcal{X}$ with basepoint P_*.
Red_S	set of reduced words on S.
$\mathsf{Sym}(S)$	relations of the free group over S.
$\mathbb{S}$	2-dimensional sphere.
$\mathsf{SSS}(g)$	super summit set of g.
$\overline{S}$	disjoint copy of the alphabet S.
S^*	set of words on an alphabet S.
T_*	base triangulation.
$H(a)$	head of a.
$V(\beta, \rho)$	tubular ρ-radius neighbourhood of a braid β.
$a \preccurlyeq b$	a left divisor of b.
$D^{[i]}$	parametrization of the ith strand of D.
$\beta^{[i]}$	parametrization of the ith strand of β.
$[w]$	equivalence class of w.
$[w]^+$	equivalence class of the word w with respect to $\equiv^+$.
$d(\beta, \beta')$	distance between two geometric braids.
$e(\beta)$	minimal strand-spacing of a geometric braid β.
ι	canonical morphism of B_n^+ into B_n.
$\equiv$	equivalence of braid words.
$\equiv^+$	congruence defining the monoid B_∞^+.
$\mathsf{eval}_G(w)$	evaluation of a (signed) word w in a group G.
$\mathsf{eval}_M(w)$	evaluation of the word w in a monoid M.
ε	empty word.
$\gamma_1, \ldots, \gamma_n$	loops of the standard basis of $\pi_1(\mathbb{D}_n^-)$.
$\beta_1 \cdot \beta_2$	product of geometric braids.
$\approx^{\mathsf{h}}, \approx, \approx^{\mathsf{nr}}$	homotopy, isotopy, isotopy in the unrestrained sense.
$\simeq$	isotopy of braid diagrams.
$\lambda_{i,j}$	linking number of two strands.
ω_n	flip (permutation).
$\partial\Sigma$	boundary of Σ.
perm	permutation of a braid.
ϕ_n	'flip' automorphism of B_n.
$\phi_n(a)$	image of a by the flip ϕ_n^+.
$\pi_1(\mathcal{X})$	fundamental group of $\mathcal{X}$.
$\pi_1(\mathcal{X}, P_*)$	fundamental groupoid of $\mathcal{X}$ with basepoint P_*.
ρ_{A}	Artin representation.

σ_f	permutation braid of f.
$\sigma_i, \overline{\sigma_i}$	code of a crossing at the position $i/i{+}1$.
$\widehat{\sigma_i}$	image of σ_i by the Artin representation.
s_i	transposition exchanging i and $i{+}1$.
$\underline{\sigma}_f$	braid word of the permutation f.
$\underline{\sigma}_{i,j}$	word $\sigma_i \sigma_{i+1} \cdots \sigma_{j-1}$.
$w(i)$	ith letter of the word w.
$w \to w', w \to^* w'$	reduction of a signed word.
$w_1 \cdot w_2$	product of braid words.

Bibliography

Bigelow, S. 2001. Braid groups are linear. *J. Amer. Math. Soc.*, **14**, 471–486.

Birman, J., K.H. Ko, and S.J. Lee. 1998. A new approach to the word problem in the braid groups. *Adv. Math.*, **139-2**, 322–353.

Dehornoy, P. 1993. *Complexité et décidabilité*, Série Mathématiques & Applications, vol. 12, Springer, Paris.

Dehornoy, P. 2004. Braid-based cryptography. In *Group Theory, Statistics, and Cryptography*, Contemp. Math., vol. 360, Amer. Math. Soc. Providence, RI, pp. 5–33.

Dehornoy, P. 2017. Multifraction reduction I: The 3-Ore case and Artin-Tits groups of type FC. *J. Comb. Algebra*, **1**, 185–228.

Dehornoy, P. 2017. Multifraction reduction II: Conjectures for Artin-Tits groups. *J. Comb. Algebra*, **1**, 229–287.

Dehornoy, P. 2019. A cancellativity criterion for presented monoids. *Semigroup Forum*, **99**, 368–390.

Dehornoy, P., I. Dynnikov, D. Rolfsen, and B. Wiest. 2008. *Ordering Braids*, Mathematical Surveys and Monographs, vol. 148, Amer. Math. Soc., Providence, RI.

Dynnikov, I. 2002. On a Yang-Baxter mapping and the Dehornoy ordering. *Uspekhi Mat. Nauk*, **57-3**, 151–152. English translation in *Russian Math. Surveys*, **57-3**, 592–594.

Epstein, D., J. Cannon, D. Holt, S. Levy, M. Paterson, and W. Thurston. 1992. *Word Processing in Groups*, Jones & Bartlett Publ., Boston.

Fromentin, J. 2011. Every braid admits a short sigma-definite expression. *J. Eur. Math. Soc.*, **13**, 1591–1631.

Fromentin, J. and L. Paris. 2012. A simple algorithm for finding short sigma-definite representatives. *J. Algebra*, **350**, 405–415.

Garside, F.A. 1965. The theory of knots and associated problems, PhD thesis, Oxford.

Garside, F.A. 1969. The braid group and other groups. *Quart. J. Math. Oxford*, **20-78**, 235–254.

Gebhardt, V. and J. González-Meneses. 1969. The cyclic sliding operation in Garside groups. *Math. Z.*, **265**, 85–114.

Guaschi, J. and D. Juan-Pineda. 2015. A survey of surface braid groups and the lower algebraic K-theory of their group rings. In *Handbook of Group Actions, Volume II* Adv. Lect. Math., vol. 32, International Press of Boston Inc., Somerville, MA, pp. 23–75.

Kassel, C. and V. Turaev. 2008. *Braid Groups*, Grad. Texts in Math., Springer, New York.

Ko, K.H., S.J. Lee, J.H. Cheon, J.W. Han, J.S. Kang, and C. Park. 2000. New public-key cryptosystem using braid groups. In *Advances in Cryptology – CRYPTO 2000*, Springer Lect. Notes in Comput. Sci., vol 1880, Springer, Berlin, pp. 166–184.

Krammer, D. 2000. The braid group B_4 is linear. *Invent. Math.*, **142**, 451–486.

Krammerm, D. 2002. Braid groups are linear. *Ann. Math.*, **151-1**, 131–156.

Laver, R. 1996. Braid group actions on left distributive structures, and well orderings in the braid groups. *J. Pure Appl. Algebra*, **108**, 81–98.

Moody, J.A. 1991. The Burau representation of the braid group B_n is unfaithful for large *n*. *Bull. Amer. Math. Soc.*, **25**, 379–384.

Paris, L. 2002. Artin monoids inject in their groups. *Comm. Math. Helv.*, **77**, 609–637.

Schleimer, S. 2008. Polynomial-time word problems. *Comm. Math. Helv.*, **83**, 741–765.

Index

Printed in the United States
by Baker & Taylor Publisher Services